T0006702

THE
SCIENCE
BOOK

BIG IDEAS

THE ART BOOK
THE ASTRONOMY BOOK
THE BIBLE BOOK
THE BUSINESS BOOK
THE CLASSICAL MUSIC BOOK
THE CRIME BOOK
THE ECONOMICS BOOK
THE FEMINISM BOOK
THE HISTORY BOOK
THE LITERATURE BOOK
THE MOVIE BOOK
THE MYTHOLOGY BOOK
THE PHILOSOPHY BOOK
THE POLITICS BOOK
THE PSYCHOLOGY BOOK
THE RELIGIONS BOOK
THE SCIENCE BOOK
THE SHAKESPEARE BOOK
THE SHERLOCK HOLMES BOOK
THE SOCIOLOGY BOOK

SIMPLY EXPLAINED

THE
SCIENCE
BOOK
DK

DK LONDON

PROJECT ART EDITOR
Katie Cavanagh

SENIOR EDITOR
Georgina Palffy

US EDITOR
Jane Perlmutter

US SENIOR EDITOR
Margaret Parrish

MANAGING ART EDITOR
Lee Griffiths

MANAGING EDITOR
Stephanie Farrow

PUBLISHING DIRECTOR
Jonathan Metcalf

ART DIRECTOR
Phil Ormerod

PUBLISHER
Andrew Macintyre

JACKET DESIGNER
Laura Brim

JACKET EDITOR
Maud Whatley

JACKET DESIGN
DEVELOPMENT MANAGER
Sophia MTT

PREPRODUCTION PRODUCER
Adam Stoneham

PRODUCER
Nancy-Jane Maun

ILLUSTRATIONS
James Graham, Peter Liddiard

produced for DK by

TALL TREE LTD.

EDITORS
Rob Colson
Camilla Hallinan
David John

DESIGN AND ART DIRECTION
Ben Ruocco

DK DELHI

PROJECT EDITOR
Priyaneet Singh

ASSISTANT ART EDITOR
Vidit Vashisht

DTP DESIGNER
Jaypal Chauhan

MANAGING EDITOR
Kingshuk Ghoshal

MANAGING ART EDITOR
Govind Mittal

PREPRODUCTION MANAGER
Balwant Singh

original styling by

STUDIO 8

This American Edition, 2019
First American Edition, 2014
Published in the United States
by DK Publishing
1450 Broadway
Suite 801, New York, NY 10018

21 22 23 10 9 8 7 6 5
077–312858–Feb/2019

A catalog record for this book is available from the Library of Congress.
ISBN: 978-1-4654-8122-1

Printed and bound in China

CONTRIBUTORS

ADAM HART-DAVIS, CONSULTANT EDITOR

Adam Hart-Davis trained as a chemist at the universities of Oxford, York, and Alberta (Canada). He spent five years editing science books and has been making television and radio programs about science, technology, mathematics, and history, as producer and host, for 30 years. He has written 30 books on science, technology, and history.

JOHN FARNDON

John Farndon is a science writer whose books have been short-listed for the Royal Society junior science book prize four times and for the Society of Authors Education Award. His books include *The Great Scientists* and *The Oceans Atlas*. He was a contributor to DK's *Science* and *Science Year by Year*.

DAN GREEN

Dan Green is an author and science writer. He has an MA in Natural Sciences from Cambridge University and has written more than 40 titles. He received two separate nominations for the Royal Society Young People's Book Prize 2013, and his Basher Science series has sold more than 2 million copies.

DEREK HARVEY

Derek Harvey is a naturalist with a particular interest in evolutionary biology and a writer for titles that include DK's *Science* and *The Natural History Book*. He studied zoology at the University of Liverpool, taught a generation of biologists, and has led expeditions to Costa Rica and Madagascar.

PENNY JOHNSON

Penny Johnson started out as an aeronautical engineer, working on military aircraft for 10 years before becoming a science teacher then a publisher producing science courses for schools. Penny has been a full-time educational writer for more than 10 years.

DOUGLAS PALMER

Douglas Palmer, a science writer based in Cambridge, Britain, has published more than 20 books in the last 14 years—most recently an app (NHM Evolution) for the Natural History Museum, London, and DK's *WOW Dinosaur* book for children. He is also a lecturer for the University of Cambridge Institute of Continuing Education.

STEVE PARKER

Steve Parker is a writer and editor of more than 300 information books specializing in science, particularly biology and allied life sciences. He holds a BSc in zoology, is a Senior Scientific Fellow of the Zoological Society of London, and has authored titles for a range of ages and publishers. Steve has received numerous awards, most recently the 2013 UK School Library Association Information Book Award for *Science Crazy*.

GILES SPARROW

Giles Sparrow studied astronomy at University College London and Science Communication at Imperial College, London, and is a bestselling science and astronomy author. His books include *Cosmos*, *Spaceflight*, *The Universe in 100 Key Discoveries*, and *Physics in Minutes*, as well as contributions to DK books such as *Universe* and *Space*.

CONTENTS

THE BEGINNING OF SCIENCE
600 BCE–1400 CE

SCIENTIFIC REVOLUTION
1400–1700

EXPANDING HORIZONS
1700–1800

A CENTURY OF PROGRESS
1800–1900

A PARADIGM SHIFT
1900–1945

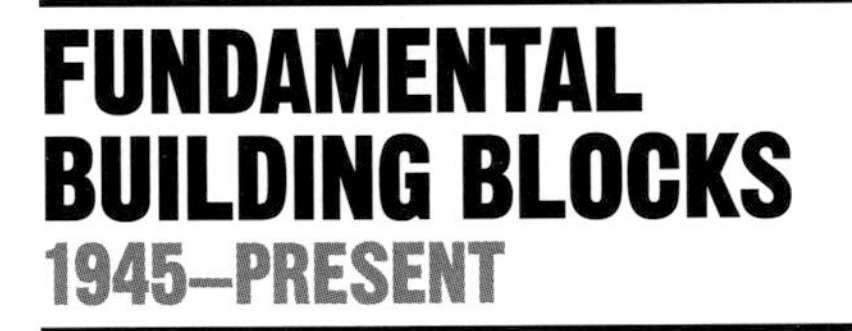

FUNDAMENTAL BUILDING BLOCKS
1945–PRESENT

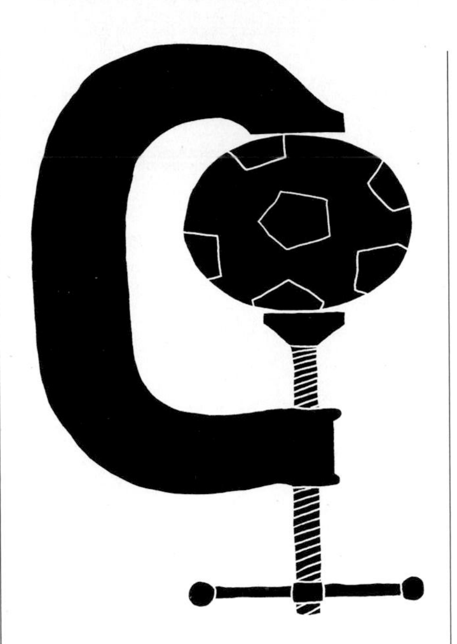

INTRODU

CTION

Science is an ongoing search for truth—a perpetual struggle to discover how the universe works that goes back to the earliest civilizations. Driven by human curiosity, it has relied on reasoning, observation, and experiment. The best known of the ancient Greek philosophers, Aristotle, wrote widely on scientific subjects and laid foundations for much of the work that has followed. He was a good observer of nature, but he relied entirely on thought and argument, and did no experiments. As a result, he got a number of things wrong. He asserted that big objects fall faster than little ones, for example, and that if one object had twice the weight of another, it would fall twice as fast. Although this is mistaken, no one doubted it until the Italian astronomer Galileo Galilei disproved the idea in 1590. While it may seem obvious today that a good scientist must rely on empirical evidence, this was not always apparent.

The scientific method

A logical system for the scientific process was first put forward by the English philosopher Francis Bacon in the early 17th century. Building on the work of the Arab scientist Alhazen 600 years earlier, and soon to be reinforced by the French philosopher René Descartes, Bacon's scientific method requires scientists to make observations, form a theory to explain what is going on, and then conduct an experiment to see whether the theory works. If it seems to be true, then the results may be sent out for peer review, in which people working in the same or a similar field are invited to pick holes in the argument, and so falsify the theory, or to repeat the experiment to make sure that the results are correct.

Making a testable hypothesis or a prediction is always useful. English astronomer Edmond Halley, observing the comet of 1682, realized that it was similar to comets reported in 1531 and 1607, and suggested that all three were the same object, in orbit around the Sun. He predicted that it would return in 1758, and he was right, though only just—it was spotted on December 25. Today, the comet is known as Halley's Comet. Since astronomers are rarely able to perform experiments, evidence can come only from observation.

> All truths are easy to understand once they are discovered; the point is to discover them.
> **Galileo Galilei**

Experiments may test a theory, or be purely speculative. When the New Zealand-born physicist Ernest Rutherford watched his students fire alpha particles at gold leaf in a search for small deflections, he suggested putting the detector beside the source, and to their astonishment some of the alpha particles bounced back off the paper-thin foil. Rutherford said it was as though an artillery shell had bounced back off tissue paper—and this led him to a new idea about the structure of the atom.

An experiment is all the more compelling if the scientist, while proposing a new mechanism or theory, can make a prediction about the outcome. If the experiment produces the predicted result, the scientist then has supporting evidence for the theory. Even so, science can never prove that a theory is correct; as the

20th-century philosopher of science Karl Popper pointed out, it can only disprove things. Every experiment that gives predicted answers is supporting evidence, but one experiment that fails may bring an entire theory crashing down.

Over the centuries, long-held concepts such as a geocentric universe, the four bodily humors, the fire-element phlogiston, and a mysterious medium called ether have all been disproved and replaced with new theories. These in turn are only theories, and may yet be disproved, although in many cases this is unlikely given the evidence in their support.

Progression of ideas

Science rarely proceeds in simple, logical steps. Discoveries may be made simultaneously by scientists working independently, but almost every advance depends in some measure on previous work and theories. One reason for building the vast apparatus known as the Large Hadron Collider, or LHC, was to search for the Higgs particle, whose existence was predicted 40 years earlier, in 1964. That prediction rested on decades of theoretical work on the structure of the atom, going back to Rutherford and the work of Danish physicist Niels Bohr in the 1920s, which depended on the discovery of the electron in 1897, which in turn depended on the discovery of cathode rays in 1869. Those could not have been found without the vacuum pump and, in 1799, the invention of the battery—and so the chain goes back through decades and centuries. The great English physicist Isaac Newton famously said, "If I have seen further, it is by standing on the shoulders of giants." He meant primarily Galileo, but he had probably also seen a copy of Alhazen's *Optics*.

The first scientists

The first philosophers with a scientific outlook were active in the ancient Greek world during the 6th and 5th centuries BCE. Thales of Miletus predicted an eclipse of the Sun in 585 BCE; Pythagoras set up a mathematical school in what is now southern Italy 50 years later, and Xenophanes, after finding seashells on a mountain, reasoned that the whole Earth must at one time have been covered by sea.

In Sicily in the 4th century BCE, Empedocles asserted that earth, air, fire, and water are the "fourfold roots of everything." He also took his followers up to the volcanic crater of Mt. Etna and jumped in, apparently to show he was immortal—and as a result we remember him to this day.

Stargazers

Meanwhile, in India, China, and the Mediterranean, people tried to make sense of the movements of the heavenly bodies. They made star maps—partly as navigational aids—and named stars and groups of stars. They also noted that a few traced irregular paths when viewed against the "fixed stars." The Greeks called these wandering stars "planets." The Chinese spotted Halley's comet in 240 BCE and, in 1054, a supernova that is now known as the Crab Nebula. »

If you would be a real seeker after truth, it is necessary that at least once in your life you doubt, as far as possible, all things.
René Descartes

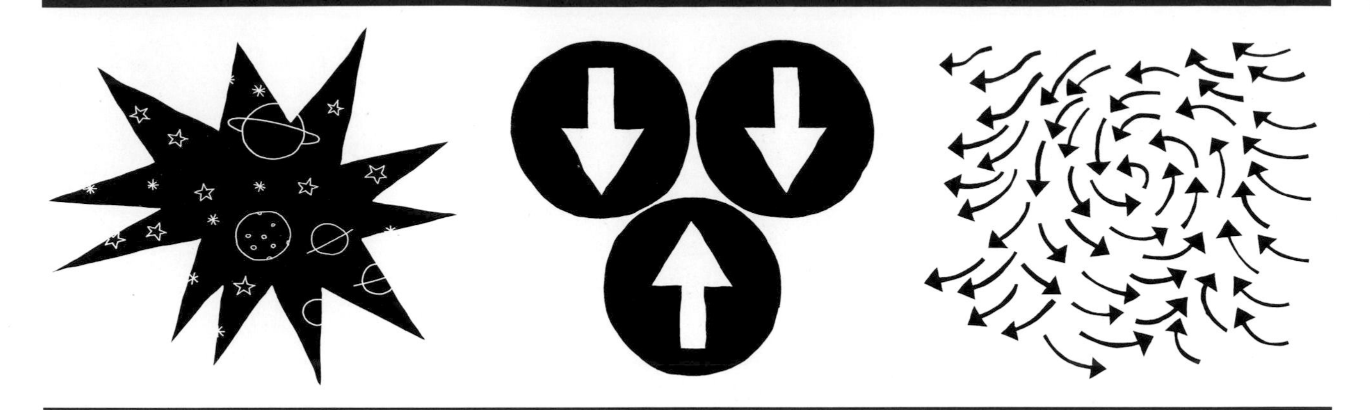

House of Wisdom

In the late 8th century CE, the Abbasid caliphate set up the House of Wisdom, a magnificent library, in its new capital, Baghdad. This inspired rapid advances in Islamic science and technology. Many ingenious mechanical devices were invented, along with the astrolabe, a navigational device that used the positions of the stars. Alchemy flourished, and techniques such as distillation appeared. Scholars at the library collected all the most important books from Greece and from India, and translated them into Arabic, which is how the West later rediscovered the works of the ancients, and learned of the "Arabic" numerals, including zero, that were imported from India.

Birth of modern science

As the monopoly of the Church over scientific truth began to weaken in the Western world, the year 1543 saw the publication of two groundbreaking books. Belgian anatomist Andreas Vesalius produced *De Humani Corporis Fabrica*, which described his dissections of human corpses with exquisite illustrations. In the same year, Polish physician Nicolaus Copernicus published *De Revolutionibus Orbium Coelestium*, which stated firmly that the Sun is the center of the universe, overturning the Earth-centered model figured out by Ptolemy of Alexandria a millennium earlier.

In 1600, English physician William Gilbert published *De Magnete* in which he explained that compass needles point north because Earth itself is a magnet. He even argued that Earth's core is made of iron. In 1623, another English physician, William Harvey, described for the first time how the heart acts as a pump and drives blood around the body, thereby quashing forever earlier theories that dated back 1,400 years to the Greco-Roman physician Galen. In the 1660s, Anglo-Irish chemist Robert Boyle produced a string of books, including *The Sceptical Chymist*, in which he defined a chemical element. This marked the birth of chemistry as a science, as distinct from the mystical alchemy from which it arose.

Robert Hooke, who worked for a time as Boyle's assistant, produced the first scientific best seller, *Micrographia*, in 1665. His superb fold-out illustrations of subjects such as a flea and the eye of a fly opened up a microscopic world no one had seen before. Then in 1687 came what many view as the most important science book of all time, Isaac Newton's *Philosophiæ Naturalis Principia Mathematica*, commonly known as the *Principia*. His laws of motion and principle of universal gravity form the basis for classical physics.

Elements, atoms, evolution

In the 18th century, French chemist Antoine Lavoisier discovered the role of oxygen in combustion, discrediting the old theory of phlogiston. Soon a host of new gases and their properties were being investigated. Thinking about the gases in the atmosphere led British meteorologist John Dalton to

I seem to have been only like a boy playing on the seashore, and diverting myself in now and then finding a smoother pebble…whilst the great ocean of truth lay all undiscovered before me.

Isaac Newton

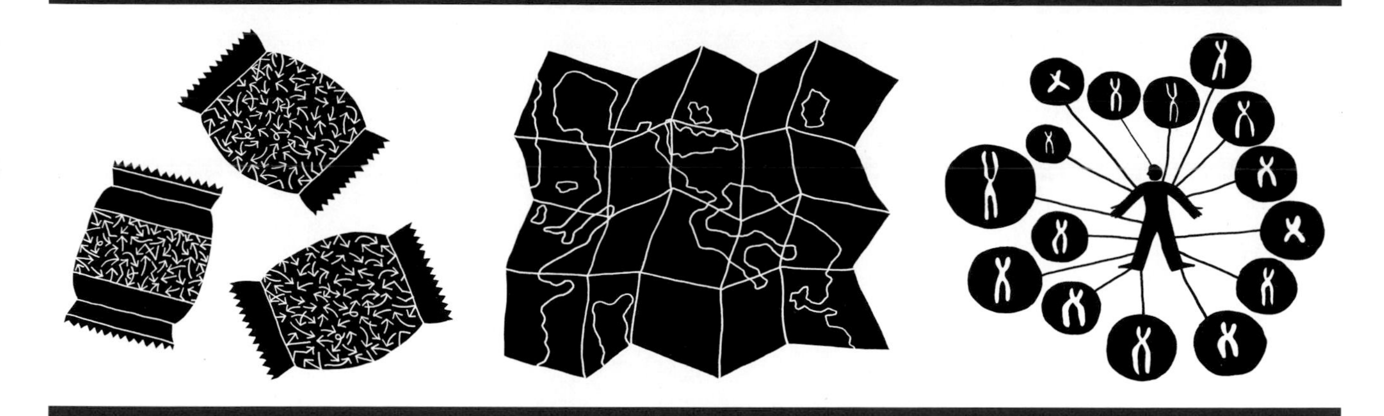

suggest that each element consisted of unique atoms, and propose the idea of atomic weights. Then German chemist August Kekulé developed the basis of molecular structure, while Russian inventor Dmitri Mendeleev laid out the first generally accepted periodic table of the elements.

The invention of the electric battery by Alessandro Volta in Italy in 1799 opened up new fields of science, into which marched Danish physicist Hans Christian Ørsted and British contemporary Michael Faraday, discovering new elements and electromagnetism, which led to the invention of the electric motor. Meanwhile, the ideas of classical physics were applied to the atmosphere, the stars, the speed of light, and the nature of heat, which developed into the science of thermodynamics.

Geologists studying rock strata began to reconstruct Earth's past. Paleontology became fashionable as the remains of extinct creatures began to turn up. Mary Anning, an untutored British girl, became a world-famous assembler of fossil remains. With the dinosaurs came ideas of evolution, most famously from British naturalist Charles Darwin, and new theories on the origins and ecology of life.

Uncertainty and infinity

At the turn of the 20th century, a young German named Albert Einstein proposed his theory of relativity, shaking classical physics and ending the idea of an absolute time and space. New models of the atom were proposed; light was shown to act as both a particle and a wave; and another German, Werner Heisenberg, demonstrated that the universe was uncertain.

What has been most impressive about the last century, however, is how technical advances have enabled science to advance faster than ever before, leap-frogging ideas with increasing precision. Ever more powerful particle colliders revealed new fundamental units of matter. Stronger telescopes showed that the universe is expanding, and started with a Big Bang. The idea of black holes began to take root. Dark matter and dark energy, whatever they were, seemed to fill the universe, and astronomers began to discover new worlds—planets in orbit around distant stars, some of which may even harbor life. British mathematician Alan Turing thought of the universal computing machine, and within 50 years we had personal computers, the worldwide web, and smartphones.

> "Reality is merely an illusion, albeit a very persistent one."
> **Albert Einstein**

Secrets of life

In biology, chromosomes were shown to be the basis of inheritance and the chemical structure of DNA was decoded. Just 40 years later this led to the human genome project, which seemed a daunting task in prospect, and yet, aided by computing, got faster and faster as it progressed. DNA sequencing is now an almost routine laboratory operation, gene therapy has moved from a hope into reality, and the first mammal has been cloned.

As today's scientists build on these and other achievements, the relentless search for the truth continues. It seems likely that there will always be more questions than answers, but future discoveries will surely continue to amaze. ■

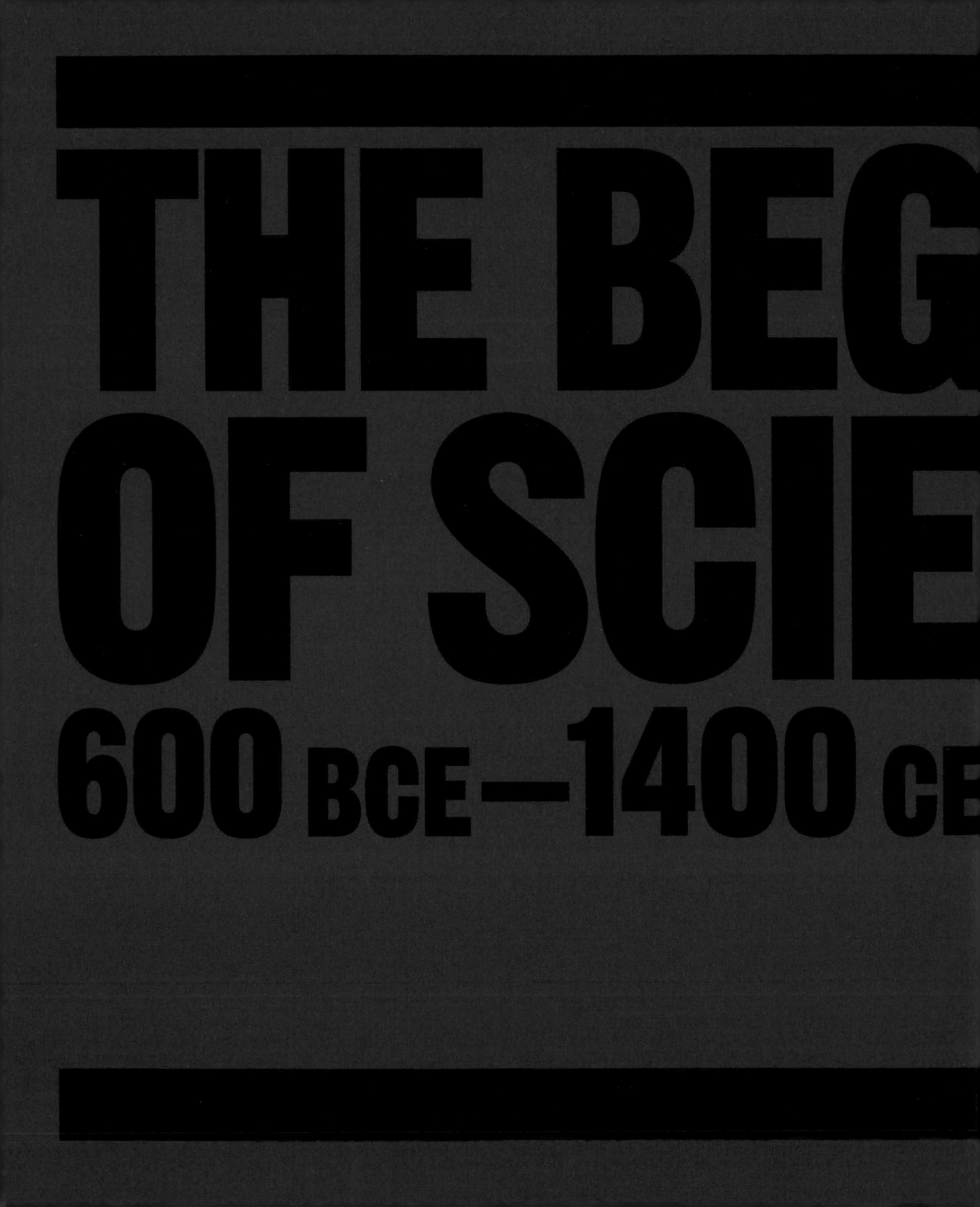
THE BEG
OF SCIE
600 BCE—1400 CE

NNING
NCE

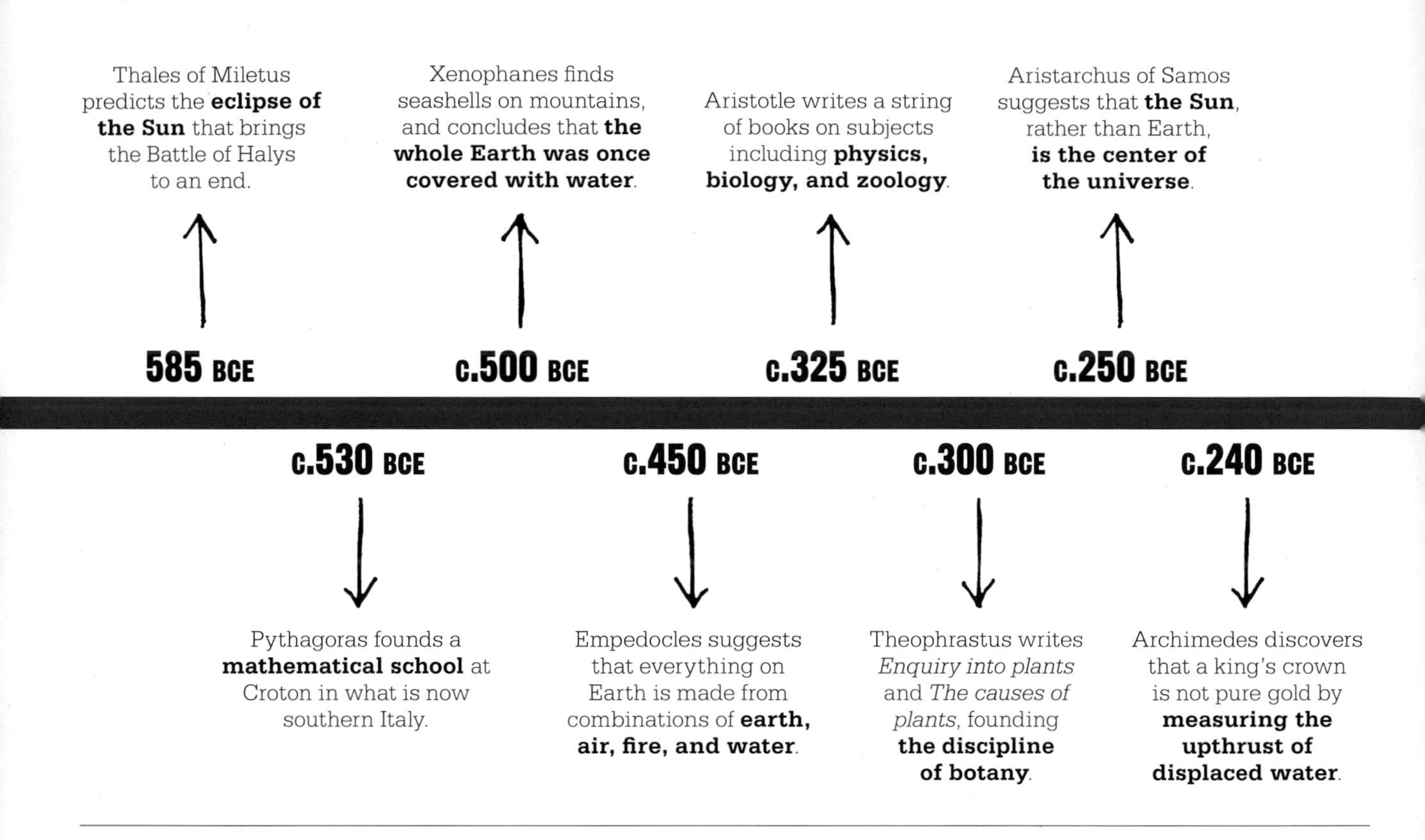

The scientific study of the world has its roots in Mesopotamia. Following the invention of agriculture and writing, people had the time to devote to study and the means to pass the results of those studies on to the next generation. Early science was inspired by the wonder of the night sky. From the fourth millennium BCE, Sumerian priests studied the stars, recording their results on clay tablets. They did not leave records of their methods, but a tablet dating from 1800 BCE shows knowledge of the properties of right-angled triangles.

Ancient Greece

The ancient Greeks did not see science as a separate subject from philosophy, but the first figure whose work is recognizably scientific is probably Thales of Miletus, of whom Plato said that he spent so much time dreaming and looking at the stars that he once fell into a well. Possibly using data from earlier Babylonians, in 585 BCE, Thales predicted a solar eclipse, demonstrating the power of a scientific approach.

Ancient Greece was not a single country, but rather a loose collection of city states. Miletus (now in Turkey) was the birthplace of several noted philosophers. Many other early Greek philosophers studied in Athens. Here, Aristotle was an astute observer, but he did not conduct experiments; he believed that, if he could bring together enough intelligent men, the truth would emerge. The engineer Archimedes, who lived at Syracuse on the island of Sicily, explored the properties of fluids. A new center of learning developed at Alexandria, founded at the mouth of the Nile by Alexander the Great in 331 BCE. Here Eratosthenes measured the size of Earth, Ctesibius made accurate clocks, and Hero invented the steam engine. Meanwhile, the librarians in Alexandria collected the best books they could find to build the best library in the world, which was burned down when Romans and Christians took over the city.

Science in Asia

Science flourished independently in China. The Chinese invented gunpowder—and with it fireworks, rockets, and guns—and made bellows for working metal. They invented the first seismograph and the first compass. In 1054 CE,

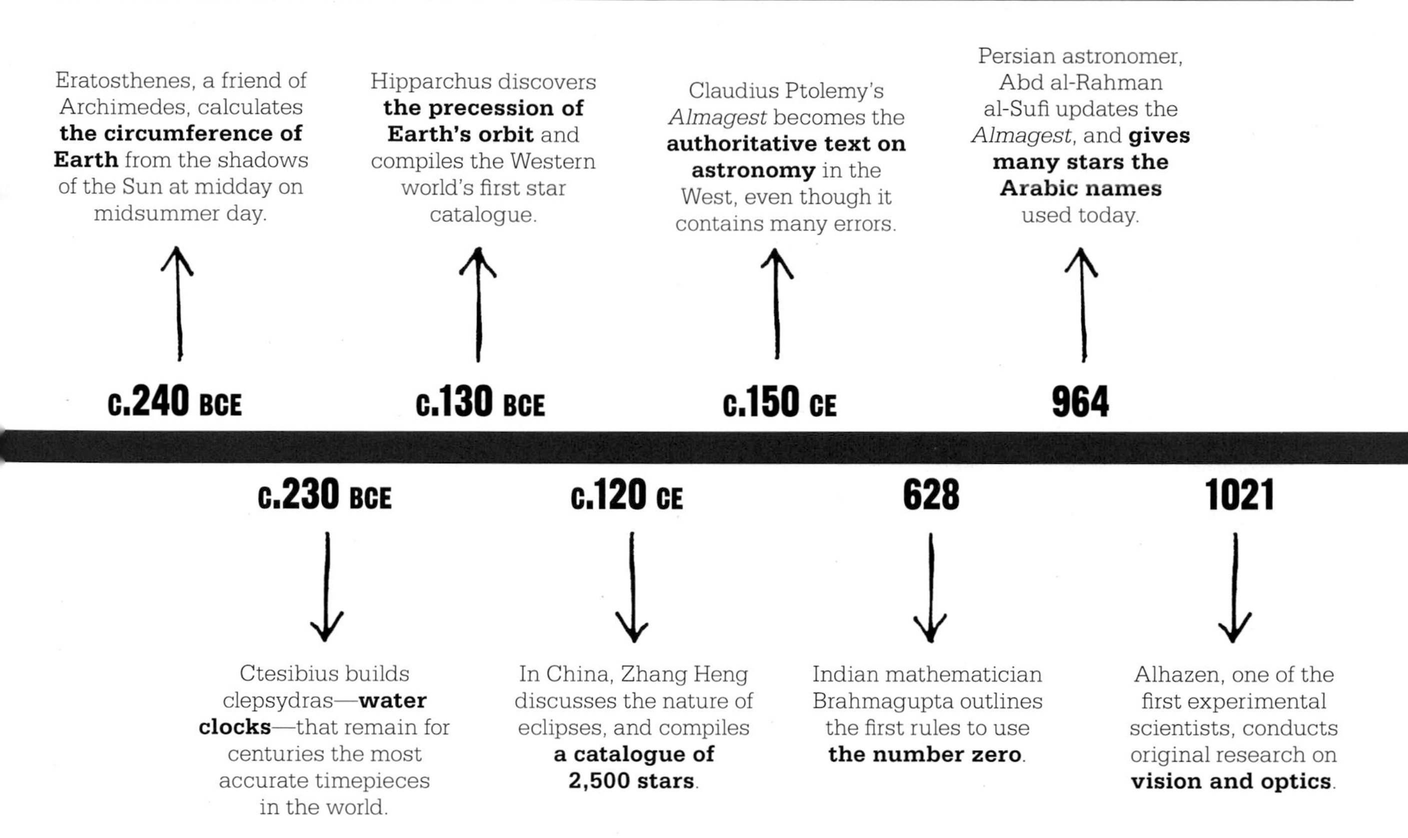

Chinese astronomers observed a supernova, which was identified as the Crab Nebula in 1731.

Some of the most advanced technology in the first millennium CE, including the spinning wheel, was developed in India, and Chinese missions were sent to study Indian farming techniques. Indian mathematicians developed what we now call the "Arabic" number system, including negative numbers and zero, and gave definitions of the trigonometric functions sine and cosine.

The Golden Age of Islam

In the middle of the 8th century, the Islamic Abbasid Caliphate moved the capital of its empire from Damascus to Baghdad. Guided by the Quranic slogan "The ink of a scholar is more holy than the blood of a martyr," Caliph Harun al-Rashid founded the House of Wisdom in his new capital, intending it to be a library and center for research. Scholars collected books from the old Greek city states and India and translated them into Arabic. This is how many of the ancient texts would eventually reach the West, where they were largely unknown in the Middle Ages. By the middle of the 9th century, the library in Baghdad had grown to become a fine successor to the library at Alexandria.

Among those who were inspired by the House of Wisdom were several astronomers, notably al-Sufi, who built on the work of Hipparchus and Ptolemy. Astronomy was of practical use to Arab nomads for navigation, since they steered their camels across the desert at night. Alhazen, born in Basra and educated in Baghdad, was one of the first experimental scientists, and his book on optics has been likened in importance to the work of Isaac Newton. Arab alchemists devised distillation and other new techniques, and coined words such as alkali, aldehyde, and alcohol. Physician al-Razi introduced soap, distinguished for the first time between smallpox and measles, and wrote in one of his many books "The doctor's aim is to do good, even to our enemies." Al-Khwarizmi and other mathematicians invented algebra and algorithms; and engineer al-Jazari invented the crank-connecting rod system, which is still used in bicycles and cars. It would take several centuries for European scientists to catch up with these developments. ■

ECLIPSES OF THE SUN CAN BE PREDICTED

THALES OF MILETUS (624–546 BCE)

IN CONTEXT

BRANCH
Astronomy

BEFORE
c.2000 BCE European monuments such as Stonehenge may have been used to calculate eclipses.

c.1800 BCE In ancient Babylon, astronomers produce the first recorded mathematical description of the movement of heavenly bodies.

2nd millennium BCE Babylonian astronomers develop methods for predicting eclipses, but these are based on observations of the Moon, not mathematical cycles.

AFTER
c.140 BCE Greek astronomer Hipparchus develops a system to predict eclipses using the Saros cycle of movements of the Sun and Moon.

Born in a Greek colony in Asia Minor, Thales of Miletus is often viewed as the founder of Western philosophy, but he was also a key figure in the early development of science. He was recognized in his lifetime for his thinking on mathematics, physics, and astronomy.

Perhaps Thales's most famous achievement is also his most controversial. According to the Greek historian Herodotus, writing more than a century after the event, Thales is said to have predicted a solar eclipse, now dated to May 28, 585 BCE, which famously brought a battle between the warring Lydians and Medes to a halt.

...day became night, and this change of the day Thales the Milesian had foretold...
Herodotus

Contested history

Thales's achievement was not to be repeated for several centuries, and historians of science have long argued about how, and even if, he achieved it. Some argue that Herodotus's account is inaccurate and vague, but Thales's feat seems to have been widely known and was taken as fact by later writers, who knew to treat Herodotus's word with caution. Assuming it is true, it is likely that Thales had discovered an 18-year cycle in the movements of the Sun and Moon, known as the Saros cycle, which was used by later Greek astronomers to predict eclipses.

Whatever method Thales used, his prediction had a dramatic effect on the battle at the river Halys, in modern-day Turkey. The eclipse ended not only the battle, but also a 15-year war between the Medes and the Lydians. ■

See also: Zhang Heng 26–27 ■ Nicolaus Copernicus 34–39 ■ Johannes Kepler 40–41 ■ Jeremiah Horrocks 52

NOW HEAR THE FOURFOLD ROOTS OF EVERYTHING

EMPEDOCLES (490–430 BCE)

IN CONTEXT

BRANCH
Chemistry

BEFORE
c.585 BCE Thales suggests the whole world is made of water.

c.535 BCE Anaximenes thinks that everything is made from air, from which water and then stones are made.

AFTER
c.400 BCE The Greek thinker Democritus proposes that the world is ultimately made of tiny indivisible particles—atoms.

1661 In his work *Sceptical Chymist*, Robert Boyle provides a definition of elements.

1808 John Dalton's atomic theory states that each element has atoms of different masses.

1869 Dmitri Mendeleev proposes a periodic table, arranging the elements in groups according to their shared properties.

The nature of matter concerned many ancient Greek thinkers. Having seen liquid water, solid ice, and gaseous mist, Thales of Miletus believed that everything must be made of water. Aristotle suggested that "nourishment of all things is moist and even the hot is created from the wet and lives by it." Writing two generations after Thales, Anaximenes suggested that the world is made of air, reasoning that when air condenses it produces mist, and then rain, and eventually stones.

Born at Agrigentum on the island of Sicily, the physician and poet Empedocles devised a more complex theory: that everything is made of four roots—he did not use the word elements—namely earth, air, fire, and water. Combining these roots would produce qualities such as heat and wetness to make earth, stone, and all plants and animals. Originally, the four roots formed a perfect sphere, held together by love, the centripetal force. But gradually strife, the centrifugal force, began to pull them apart. For Empedocles, love and strife are the two forces that shape the universe. In this world, strife tends to predominate, which is why life is so difficult.

This relatively simple theory dominated European thought—which referred to the "four humors"—with little refinement until the development of modern chemistry in the 17th century. ■

Empedocles saw the four roots of matter as two pairs of opposites: fire/water and air/earth, which combine to produce everything we see.

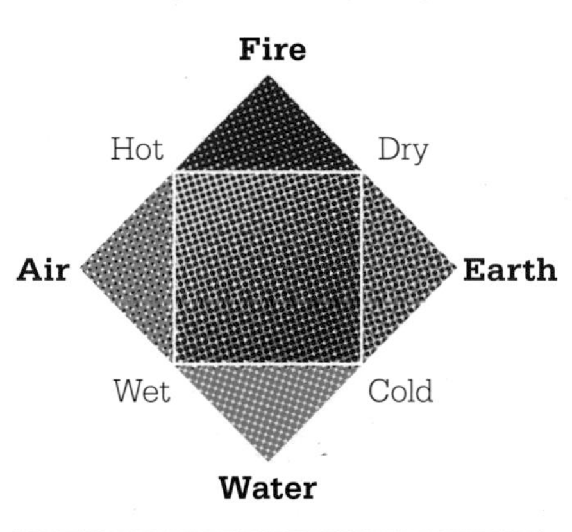

See also: Robert Boyle 46–49 ▪ John Dalton 112–13 ▪ Dmitri Mendeleev 174–79

MEASURING THE CIRCUMFERENCE OF EARTH

ERATOSTHENES (276–194 BCE)

IN CONTEXT

BRANCH
Geography

BEFORE
6th century BCE Greek mathematician Pythagoras suggests Earth may be spherical, not flat.

3rd century BCE Aristarchus of Samos is the first to place the Sun at the center of the known universe and uses a trigonometric method to estimate the relative sizes of the Sun and the Moon and their distances from Earth.

Late 3rd century BCE Eratosthenes introduces the concepts of parallels and meridians to his maps (equivalent to modern longitude and latitude).

AFTER
18th century The true circumference and shape of Earth is found through enormous efforts by French and Spanish scientists.

The Greek astronomer and mathematician Eratosthenes is best remembered as the first person to measure the size of Earth, but he is also regarded as the founder of geography—not only coining the word, but also establishing many of the basic principles used to measure locations on our planet. Born at Cyrene (in modern-day Libya), Eratosthenes traveled widely in the Greek world, studying in Athens and Alexandria, and eventually becoming the librarian of Alexandria's Great Library.

It was in Alexandria that Eratosthenes heard a report that at the town of Swenet, south of Alexandria, the Sun passed directly overhead on the summer solstice (the longest day of the year, when the Sun rises highest in the sky). Assuming the Sun was so distant that its rays were almost parallel to each other when they hit Earth, he used a vertical rod, or "gnomon," to project the Sun's shadow at the same moment in Alexandria. Here, he determined, the Sun was 7.2° south of the zenith—which is 1/50th of the circumference of a circle. Therefore, he reasoned, the separation of the two cities along a north–south meridian must be 1/50th of Earth's circumference. This allowed him to figure out the size of our planet at 230,000 stadia, or 24,662 miles (39,690 km)—an error of less than 2 percent. ■

Sunlight reached Swenet at right angles, but cast a shadow at Alexandria. The angle of the shadow cast by the gnomon allowed Eratosthenes to calculate Earth's circumference.

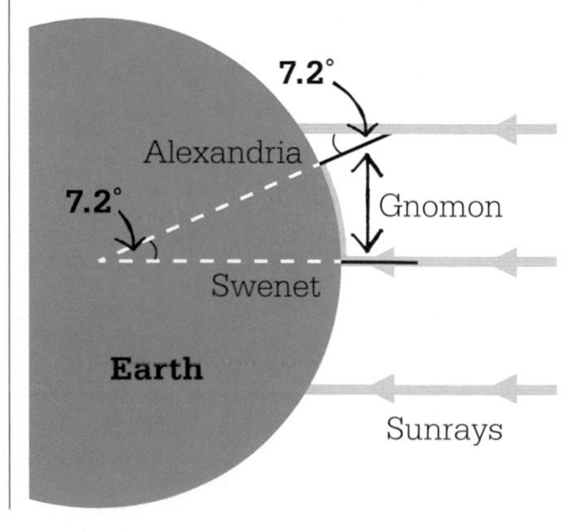

See also: Nicolaus Copernicus 34–39 ▪ Johannes Kepler 40–41

THE HUMAN IS RELATED TO THE LOWER BEINGS

AL-TUSI (1201–1274)

IN CONTEXT

BRANCH
Biology

BEFORE
c.550 BCE Anaximander of Miletus proposes that animal life began in the water, and evolved from there.

c.340 BCE Plato's theory of forms argues that species are unchangeable.

c.300 BCE Epicurus says that many other species have been created in the past, but only the most successful survive to have offspring.

AFTER
1377 Ibn Khaldun writes in *Muqaddimah* that humans developed from monkeys.

1809 Jean-Baptiste Lamarck proposes a theory of evolution of species.

1858 Alfred Russel Wallace and Charles Darwin suggest a theory of evolution by means of natural selection.

A Persian scholar born in Baghdad in 1201, during the Golden Age of Islam, Nazir al-Din al-Tusi was a poet, philosopher, mathematician, and astronomer, and one of the first to propose a system of evolution. He suggested that the universe had once comprised identical elements that had gradually drifted apart, with some becoming minerals and others, changing more quickly, developing into plants and animals.

In *Akhlaq-i-Nasri*, al-Tusi's work on ethics, he set out a hierarchy of life forms, in which animals were higher than plants and humans were higher than other animals. He regarded the conscious will of animals as a step toward the consciousness of humans. Animals are able to move consciously to search for food, and can learn new things. In this ability to learn, al-Tusi saw an ability to reason: "The trained horse or hunting falcon is at a higher point of development in the animal world," he said, adding, "The first steps of human perfection begin from here."

> The organisms that can gain the new features faster are more variable. As a result, they gain advantages over other creatures.
> **al-Tusi**

Al-Tusi believed that organisms changed over time, seeing in that change a progression toward perfection. He thought of humans as being on a "middle step of the evolutionary stairway," potentially able by means of their will to reach a higher developmental level. He was the first to suggest that not only do organisms change over time, but that the whole range of life has evolved from a time when there was no life at all. ■

See also: Carl Linnaeus 74–75 ▪ Jean-Baptiste Lamarck 118 ▪ Charles Darwin 142–49 ▪ Barbara McClintock 271

A FLOATING OBJECT DISPLACES ITS OWN VOLUME IN LIQUID

ARCHIMEDES (287–212 BCE)

IN CONTEXT

BRANCH
Physics

BEFORE
3rd millennium BCE Metalworkers discover that melting metals and mixing them together produces an alloy that is stronger than either of the original metals.

600 BCE In ancient Greece, coins are made from an alloy of gold and silver called electrum.

AFTER
1687 In his *Principia Mathematica*, Isaac Newton outlines his theory of gravity, explaining how there is a force that pulls everything toward the center of Earth—and vice versa.

1738 Swiss mathematician Daniel Bernoulli develops his kinetic theory of fluids, explaining how fluids exert pressure on objects by the random movement of molecules in the fluid.

The Roman author Vitruvius, writing in the 1st century BCE, recounts the possibly apocryphal story of an incident that happened two centuries earlier. Hieron II, the King of Sicily, had ordered a new gold crown. When the crown was delivered, Hieron suspected that the crown maker had substituted silver for some of the gold, melting the silver with the remaining gold so that the color looked the same as pure gold. The king asked his chief scientist, Archimedes, to investigate.

Archimedes puzzled over the problem. The new crown was precious, and must not be damaged

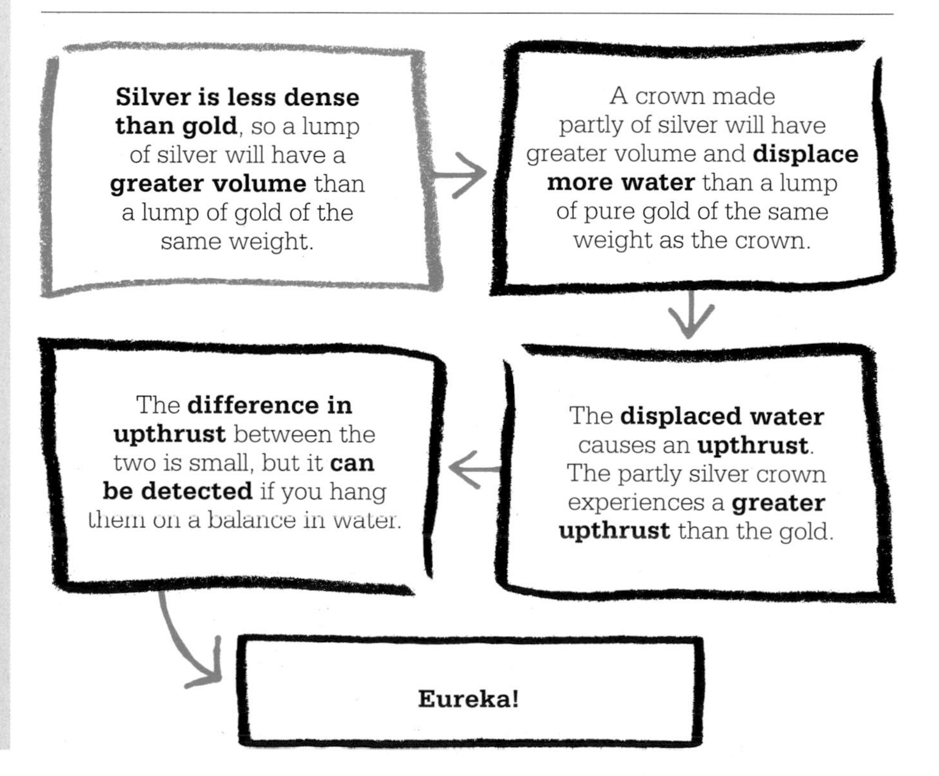

See also: Nicolaus Copernicus 34–39 ▪ Isaac Newton 62–69

in any way. He went to the public baths in Syracuse to ponder the problem. The bath was full to the brim, and when he climbed in, he noticed two things: the water level rose, making some water slop over the side, and he felt weightless. He shouted "Eureka!" (I have found the answer!) and ran home stark naked.

Measuring volume

Archimedes had realized that if he lowered the crown into a bucket filled to the brim with water, it would displace some water—exactly the same amount as its own volume—and he could measure how much water spilled out. This would tell him the volume of the crown. Silver is less dense than gold, so a silver crown of the same weight would be bigger than a gold crown, and would displace more water. Therefore, an adulterated crown would displace more water than a pure gold crown—and more than a lump of gold of the same weight. In practice, the effect would have been small and difficult to measure. But Archimedes had also realized that any object immersed in a liquid experiences an upthrust (upward force) equal to the weight of the liquid it has displaced.

Archimedes probably solved the puzzle by hanging the crown and an equal weight of pure gold on opposite ends of a stick, which he then suspended by its center so that the two weights balanced. Then he lowered the whole thing into a bath of water. If the crown was pure gold, it and the lump of gold would experience an equal upthrust, and the stick would stay horizontal. If the crown contained some silver, however, the volume of the crown would be greater than the volume of the lump of gold—the crown would displace more water, and the stick would tilt sharply.

Archimedes' idea became known as Archimedes' principle, which states that the upthrust on an object in a fluid is equal to the weight of the fluid the object displaces. This principle explains how objects made of dense material can still float on water. A steel ship that weighs one ton will sink until it has displaced one ton of water, but then will sink no further. Its deep, hollow hull has a greater volume and displaces more water than a lump of steel of the same weight, and is therefore buoyed up by a greater upthrust.

Vitruvius tells us that Hieron's crown was indeed found to contain some silver, and that the crown maker was duly punished. ■

> “A solid heavier than a fluid will, if placed in it, descend to the bottom of the fluid, and the solid will, when weighed in the fluid, be lighter than its true weight by the weight of the fluid displaced.
> **Archimedes**”

Archimedes

Archimedes was possibly the greatest mathematician in the ancient world. Born around 287 BCE, he was killed by a soldier when his home town Syracuse was taken by the Romans in 212 BCE. He had devised several fearsome weapons to keep at bay the Roman warships that attacked Syracuse—a catapult, a crane to lift the bows of a ship out of the water, and a death array of mirrors to focus the Sun's rays and set fire to a ship. He probably invented the Archimedes screw, still used today for irrigation, during a stay in Egypt.

Archimedes also calculated an approximation for pi (the ratio of a circle's circumference to its diameter), and wrote down the laws of levers and pulleys. The achievement Archimedes was most proud of was a mathematical proof that the smallest cylinder that any given sphere can fit into has exactly 1.5 times the sphere's volume. A sphere and a cylinder are carved into Archimedes' tombstone.

Key work

c.250 BCE *On Floating Bodies*

THE SUN IS LIKE FIRE, THE MOON IS LIKE WATER

ZHANG HENG (78–139 CE)

IN CONTEXT

BRANCH
Physics

BEFORE
140 BCE Hipparchus figures out how to predict eclipses.

150 CE Ptolemy improves on Hipparchus's work, and produces practical tables for calculating the future positions of the celestial bodies.

AFTER
11th century Shen Kuo writes the *Dream Pool Essays*, in which he uses the waxing and waning of the Moon to demonstrate that all heavenly bodies (though not Earth) are spherical.

1543 Nicolaus Copernicus publishes *On the Revolutions of the Celestial Spheres*, in which he describes a heliocentric system.

1609 Johannes Kepler explains the movements of the planets as free-floating bodies describing ellipses.

During the day **Earth** is **bright**, with **shadows**, because of **sunlight**.

The **Moon** is sometimes **bright**, with **shadows**.

The Moon must be **bright** because of **sunlight**.

Therefore the Sun is like fire, the Moon like water.

In about 140 BCE, the Greek astronomer Hipparchus, probably the finest astronomer of the ancient world, compiled a catalogue of some 850 stars. He also explained how to predict the movements of the Sun and Moon and the dates of eclipses. In his work *Almagest* of about 150 CE, Ptolemy of Alexandria listed 1,000 stars and 48 constellations. Most of this work was effectively an updated version of what Hipparchus had written, but in a more practical form. In the West, the *Almagest* became the standard astronomy text throughout the Middle Ages. Its tables included all the information needed to calculate the future positions of the Sun and Moon, the planets and the major stars, and also eclipses of the Sun and Moon.

In 120 CE, the Chinese polymath Zhang Heng produced a work entitled *Ling Xian*, or *The Spiritual Constitution of the Universe*. In it, he wrote that "the sky is like a hen's egg, and is as round as a crossbow pellet, and Earth is like the yolk of the egg, lying alone at the center. The sky is large and the Earth small." This was, following Hipparchus and Ptolemy, a universe

See also: Nicolaus Copernicus 34–39 ▪ Johannes Kepler 40–41 ▪ Isaac Newton 62–69

The Moon and the planets are Yin; they have shape but no light.
Jing Fang

with Earth at its center. Zhang catalogued 2,500 "brightly shining" stars and 124 constellations, and added that "of the very small stars there are 11,520."

Eclipses of the Moon and planets

Zhang was fascinated by eclipses. He wrote, "The Sun is like fire and the Moon like water. The fire gives out light and the water reflects it. Thus the Moon's brightness is produced from the radiance of the Sun, and the Moon's darkness is due to the light of the Sun being obstructed. The side that faces the Sun is fully lit, and the side that is away from it is dark." Zhang also described a lunar eclipse, where the Sun's light cannot reach the Moon because Earth is in the way. He recognized that the planets were also "like water," reflecting light, and so were also subject to eclipses: "When [a similar effect] happens with a planet, we call it an occultation; when the Moon passes across the Sun's path then there is a solar eclipse."

In the 11th century, another Chinese astronomer, Shen Kuo, expanded on Zhang's work in one significant respect. He showed that observations of the waxing and waning of the Moon proved that the celestial bodies were spherical. ■

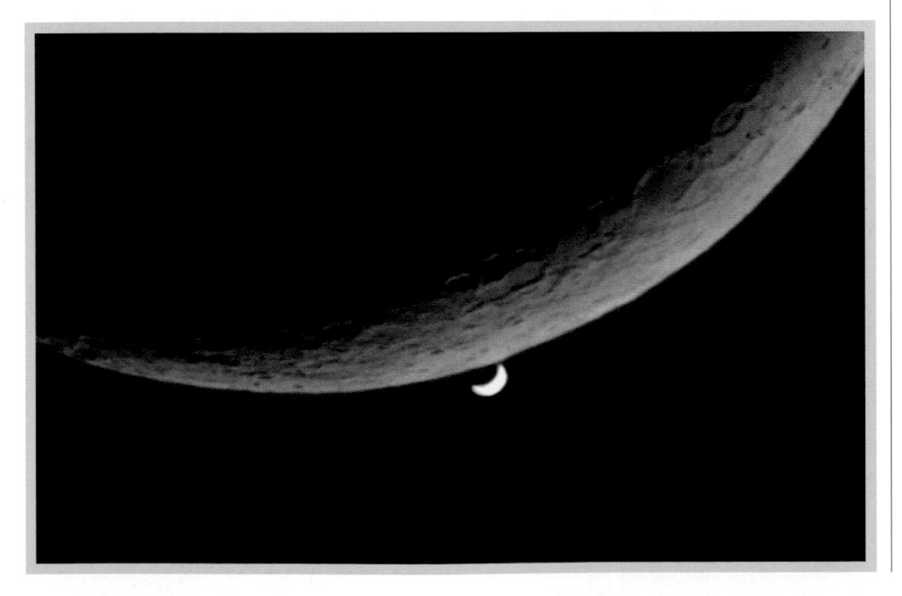

The crescent outline of Venus is about to be occulted by the Moon. Zhang's observations led him to conclude that, like the Moon, the planets did not produce their own light.

Zhang Heng

Zhang Heng was born in 78 CE in the town of Xi'e, in what is now Henan Province, in Han Dynasty China. At 17, he left home to study literature and train to be a writer. By his late 20s, Zhang had become a skilled mathematician and was called to the court of Emperor An-ti, who, in 115 CE, made him Chief Astrologer.

Zhang lived at a time of rapid advances in science. In addition to his astronomical work, he devised a water-powered armillary sphere (a model of the celestial objects) and invented the world's first seismometer, which was ridiculed until, in 138 CE, it successfully recorded an earthquake 250 miles (400 km) away. He also invented the first odometer to measure distances traveled in vehicles, and a nonmagnetic, south-pointing compass in the form of a chariot. Zhang was a distinguished poet, whose works give us vivid insights into the cultural life of his day.

Key works

c.120 CE *The Spiritual Constitution of the Universe*
c.120 CE *The Map of the Ling Xian*

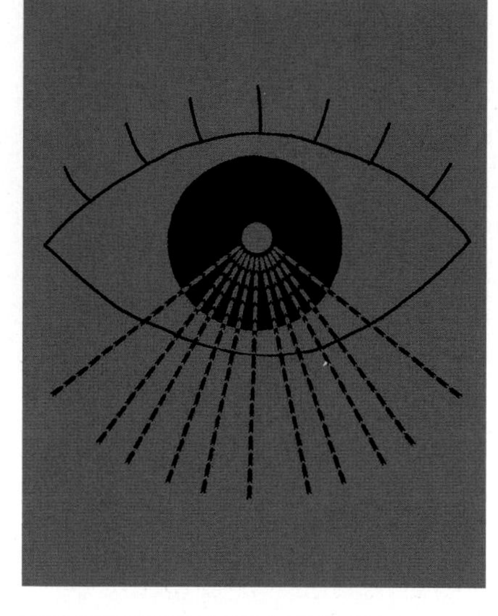

LIGHT TRAVELS IN STRAIGHT LINES INTO OUR EYES

ALHAZEN (c.965–1040)

IN CONTEXT

BRANCH
Physics

BEFORE
350 BCE Aristotle argues that vision derives from physical forms entering the eye from an object.

300 BCE Euclid argues that the eye sends out beams that are bounced back to the eye.

980s Ibn Sahl investigates refraction of light and deduces the laws of refraction.

AFTER
1240 English bishop Robert Grosseteste uses geometry in his experiments with optics and accurately describes the nature of color.

1604 Johannes Kepler's theory of the retinal image is based directly on Alhazen's work.

1620s Alhazen's ideas influence Francis Bacon, who advocates a scientific method based on experiment.

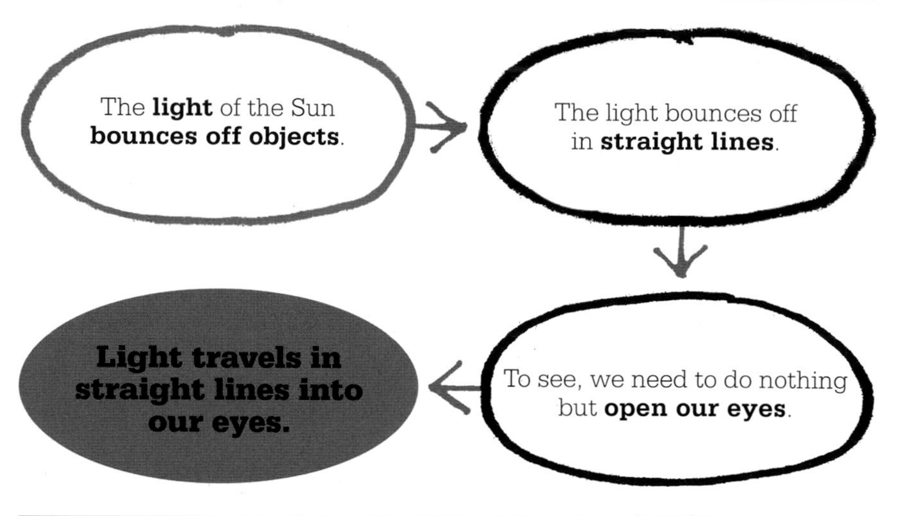

The Arab astronomer and mathematician Alhazen, who lived in Baghdad, in present-day Iraq, during the Golden Age of Islamic civilization, was arguably the world's first experimental scientist. While earlier Greek and Persian thinkers had explained the natural world in various ways, they had arrived at their conclusions through abstract reasoning, not through physical experiments. Alhazen, working in a thriving Islamic culture of curiosity and inquiry, was the first to use what we now call the scientific method: setting up hypotheses and methodically testing them with experiments. As he observed: "The seeker after truth is not one who studies the writings of the ancients and...puts his trust in them, but rather the one who suspects his faith in them and questions what he gathers from them, the one who submits to argument and demonstration."

Understanding vision

Alhazen is remembered today as a founder of the science of optics. His most important works were studies of the structure of the eye and the process of vision. The

See also: Johannes Kepler 40–41 ▪ Francis Bacon 45 ▪ Christiaan Huygens 50–51 ▪ Isaac Newton 62–69

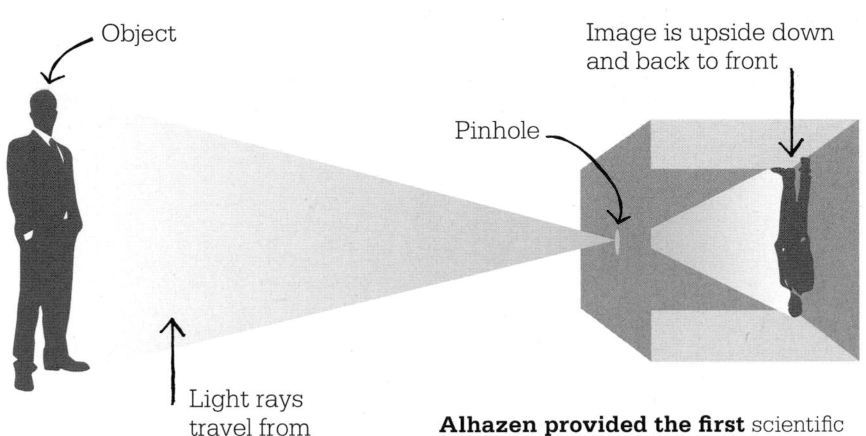

Alhazen provided the first scientific description of a camera obscura, an optical device that projects an upside-down image on a screen.

Greek scholars Euclid and, later, Ptolemy believed that vision derived from "rays" that beamed out of the eye and bounced back from whatever a person was looking at. Alhazen showed, through the observation of shadows and reflection, that light bounces off objects and travels in straight lines into our eyes. Vision was a passive, rather than an active, phenomenon, at least until it reached the retina. He noted that, "from each point of every colored body, illuminated by any light, issue light and color along every straight line that can be drawn from that point." In order to see things, we have only to open our eyes to let in the light. There is no need for the eye to send out rays, even if it could.

Alhazen also found, through his experiments with bulls' eyes, that light enters a small hole (the pupil) and is focused by a lens onto a sensitive surface (the retina) at the back of the eye. However, even though he recognized the eye as a lens, he did not explain how the eye or the brain forms an image.

Experiments with light

Alhazen's monumental, seven-volume *Book of Optics* set out his theory of light and his theory of vision. It remained the main authority on the subject until Newton's *Principia* was published 650 years later. The book explores the interaction of light with lenses, and describes the phenomenon of refraction (change in the direction) of light—700 years before Dutch scientist Willebrord van Roijen Snell's law of refraction. It also examines the refraction of light by the atmosphere, and describes shadows, rainbows, and eclipses. *Optics* greatly influenced later Western scientists, including Francis Bacon, one of the scientists responsible for reviving Alhazen's scientific method during the Renaissance in Europe. ■

The duty of the man who investigates the writings of scientists, if learning the truth is his goal, is to make himself an enemy of all that he reads.

Alhazen

Alhazen

Abu Ali al-Hassan ibn al-Haytham (known in the West as Alhazen) was born in Basra, in present-day Iraq, and educated in Baghdad. As a young man he was given a government job in Basra, but soon became bored. One story has it that, on hearing about the problems resulting from the annual flooding of the Nile in Egypt, he wrote to Caliph al-Hakim offering to build a dam to regulate the deluge, and was received with honor in Cairo. However, when he traveled south of the city, and saw the sheer size of the river—which is almost 1 mile (1.6 km) wide at Aswan—he realized the task was impossible with the technology then available. To avoid the caliph's retribution he feigned insanity and remained under house arrest for 12 years. In that time he did his most important work.

Key works

1011–21 *Book of Optics*
c.1030 *A Discourse on Light*
c.1030 *On the Light of the Moon*

SCIENTI
REVOLU
1400–1700

FIC
TION

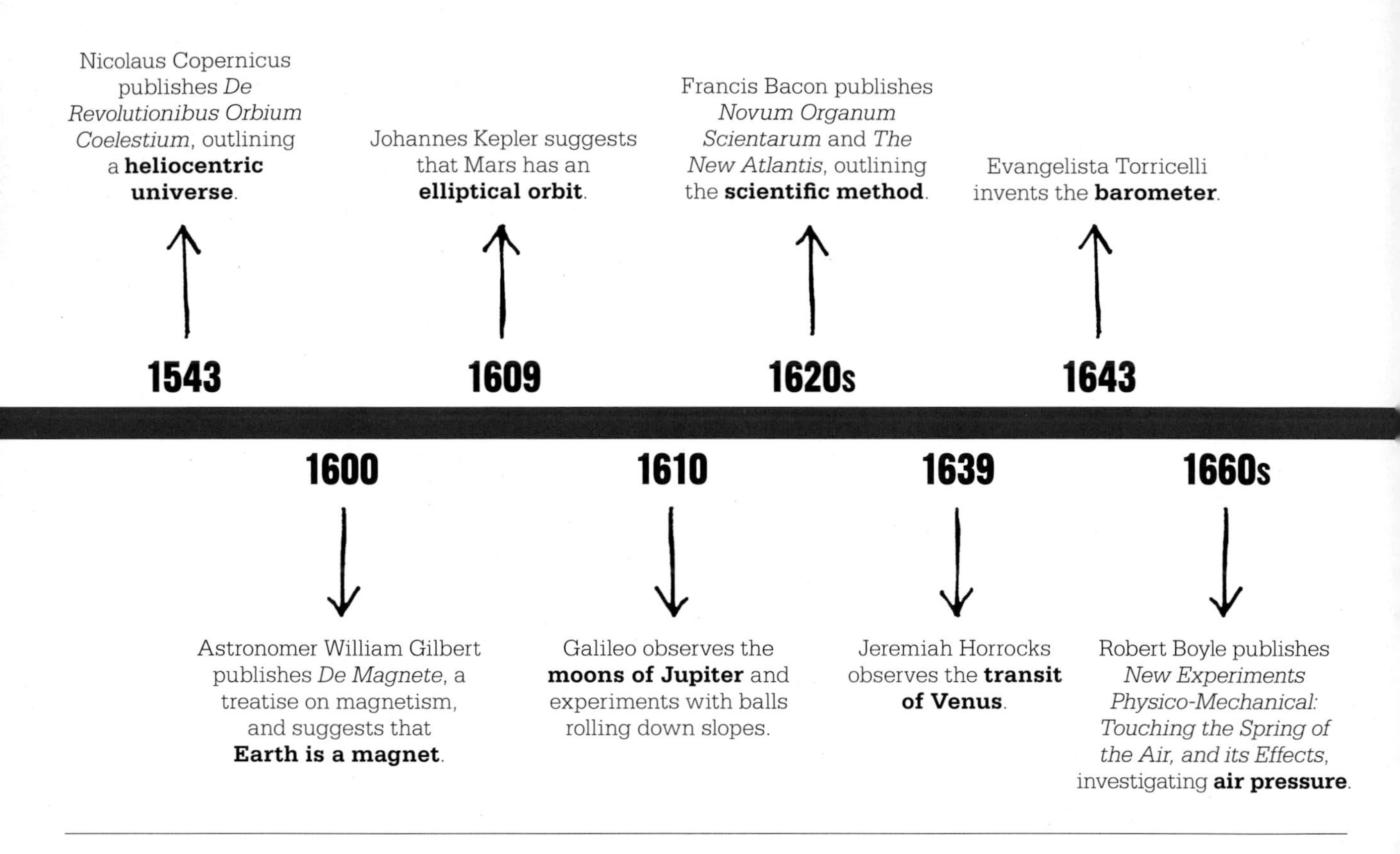

The Islamic Golden Age was a great flowering of the sciences and arts that began in the capital of the Abbasid Caliphate, Baghdad, in the mid-8th century and lasted for about 500 years. It laid the foundations for experimentation and the modern scientific method. In the same period in Europe, however, several hundred years were to pass before scientific thought was to overcome the restrictions of religious dogma.

Dangerous thinking

For centuries, the Catholic Church's view of the universe was based on Aristotle's idea that Earth was at the orbital center of all celestial bodies. Then, in about 1532, after years of struggling with its complex mathematics, Polish physician Nicolaus Copernicus completed his heretical model of the universe that had the Sun at its center. Aware of the heresy, he was careful to state that it was only a mathematical model, and he waited until he was on the point of death before publishing, but the Copernican model quickly won many advocates. German astrologer Johannes Kepler refined Copernicus's theory using observations by his Danish mentor Tycho Brahe, and calculated that the orbits of Mars and, by inference, the other planets were ellipses. Improved telescopes allowed Italian polymath Galileo Galilei to identify four moons of Jupiter in 1610. The new cosmology's explanatory power was becoming undeniable.

Galileo also demonstrated the power of scientific experiment, investigating the physics of falling objects and devising the pendulum as an effective timekeeper, which Dutchman Christiaan Huygens used to build the first pendulum clock in 1657. English philosopher Francis Bacon wrote two books laying out his ideas for a scientific method, and the theoretical groundwork for modern science, based on experiment, observation, and measurement, was developed.

New discoveries followed thick and fast. Robert Boyle used an air pump to investigate the properties of air, while Huygens and English physicist Isaac Newton came up with opposing theories of how light travels, establishing the science of optics. Danish astronomer Ole Rømer noted discrepancies in the timetable of eclipses of the moons of Jupiter, and used these to calculate an approximate value

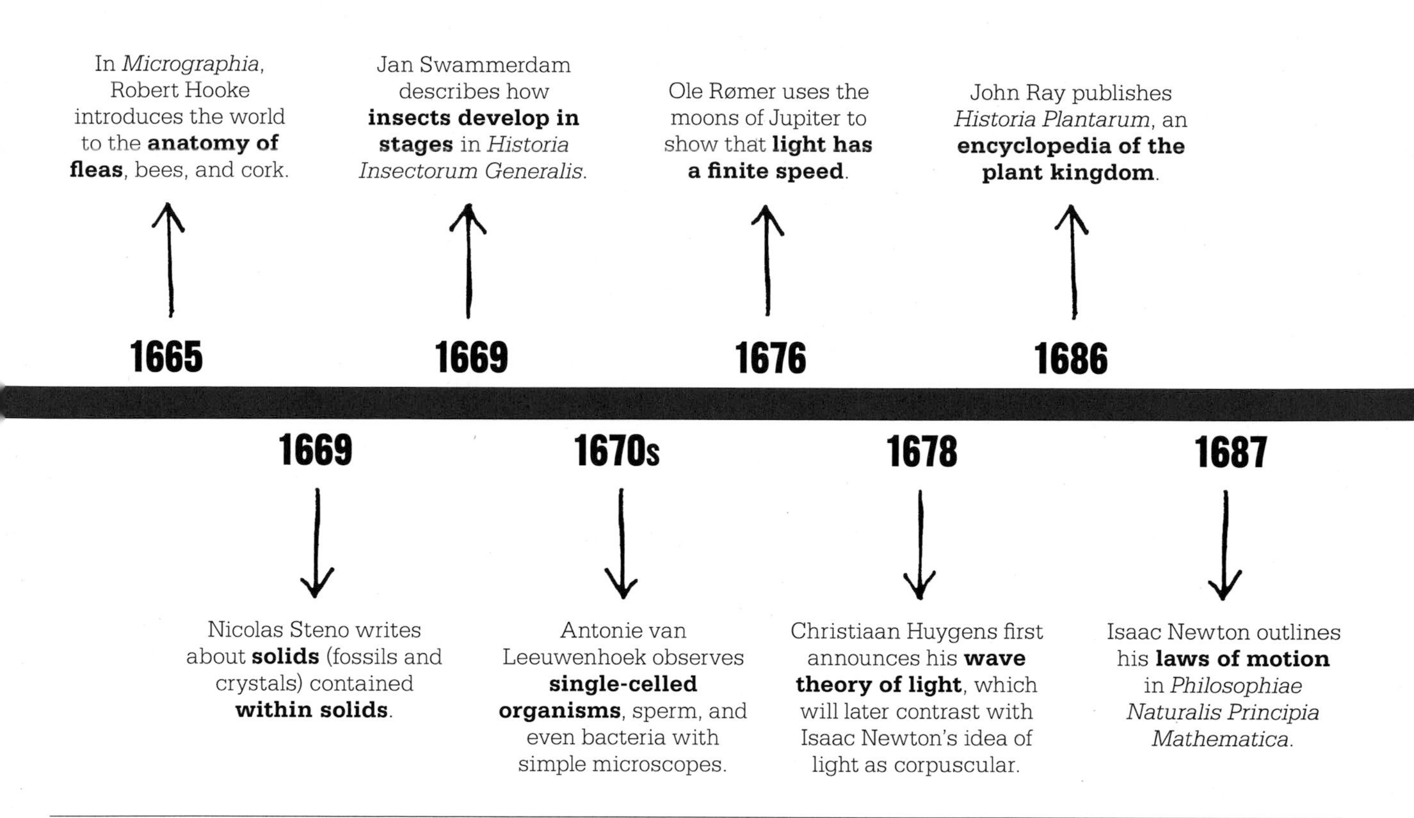

for the speed of light. Rømer's compatriot, Bishop Nicolas Steno, was sceptical of much ancient wisdom, and developed his own ideas in both anatomy and geology. He laid down the principles of stratigraphy (the study of rock layers), establishing a new scientific basis for geology.

Microworlds

Throughout the 17th century, developments in technology drove scientific discovery at the smallest scale. In the early 1600s, Dutch eyeglasses-makers developed the first microscopes, and, later that century, Robert Hooke built his own and made beautiful drawings of his findings, revealing the intricate structure of tiny bugs such as fleas for the first time. Dutch fabric-store owner Antonie van Leeuwenhoek, perhaps inspired by Hooke's drawings, made hundreds of his own microscopes and found tiny life forms in places where no one had thought of looking before, such as water. Leeuwenhoek had discovered single-celled life forms such as protists and bacteria, which he called "animalcules." When he reported his findings to the British Royal Society, they sent three priests to certify that he had really seen such things. Dutch microscopist Jan Swammerdam showed that egg, larva, pupa, and adult are all stages in the development of an insect, and not separate animals created by God. Old ideas dating back to Aristotle were swept away by these new discoveries. Meanwhile, English biologist John Ray compiled an enormous encyclopedia of plants, which marked the first serious attempt at systematic classification.

Mathematical analysis

Heralding the Enlightenment, these discoveries laid the groundwork for the modern scientific disciplines of astronomy, chemistry, geology, physics, and biology. The century's crowning achievement came with Newton's treatise *Philosophiæ Naturalis Principia Mathematica*, which laid out his laws of motion and gravity. Newtonian physics was to remain the best description of the physical world for more than two centuries, and together with the analytical techniques of calculus developed independently by Newton and Gottfried Wilhelm Leibniz, it would provide a powerful tool for future scientific study. ■

AT THE CENTER OF EVERYTHING IS THE SUN

NICOLAUS COPERNICUS (1473–1543)

IN CONTEXT

BRANCH
Astronomy

BEFORE
3rd century BCE In a work called *The Sand Reckoner*, Archimedes reports the ideas of Aristarchus of Samos, who proposed that the universe was much larger than commonly believed, and that the Sun was at its center.

150 CE Ptolemy of Alexandria uses mathematics to describe a geocentric (Earth-centered) model of the universe.

AFTER
1609 Johannes Kepler resolves the outstanding conflicts in the heliocentric (Sun-centered) model of the solar system by proposing elliptical orbits.

1610 After observing the moons of Jupiter, Galileo becomes convinced that Copernicus was right.

Throughout its early history, Western thought was shaped by an idea of the universe that placed Earth at the center of everything. This "geocentric model" seemed at first to be rooted in everyday observations and common sense—we do not feel any motion of the ground on which we stand, and superficially there seems to be no observational evidence that our planet is in motion either. Surely the simplest explanation was that the Sun, Moon, planets and stars were all spinning around Earth at different rates? This system appears to have been widely accepted in the ancient world, and became entrenched in classical philosophy through the works of Plato and Aristotle in the 4th century BCE.

However, when the ancient Greeks measured the movements of the planets, it became clear that the geocentric system had problems. The orbits of the known planets—five wandering lights in the sky—followed complex paths. Mercury and Venus were always seen in the morning and evening skies, describing tight loops around the Sun. Mars, Jupiter, and Saturn, meanwhile, took 780 days, 12 years, and 30 years respectively to circle against the background stars, their motion complicated by "retrograde" loops in which they slowed and temporarily reversed the general direction of their motion.

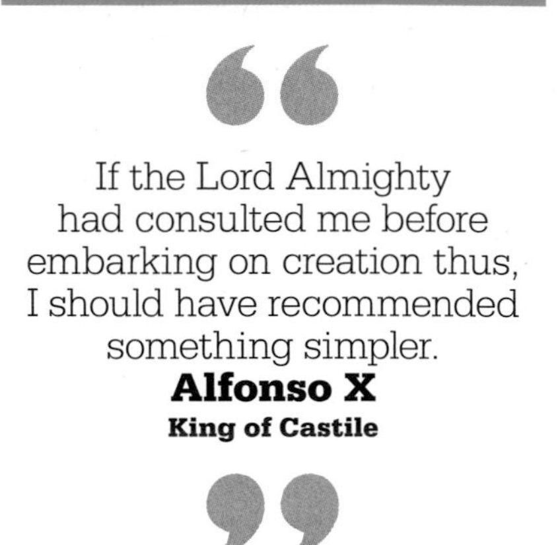

Ptolemaic system

To explain these complications, Greek astronomers introduced the idea of epicycles—"sub-orbits" around which the planets circled as the central "pivot" points of the

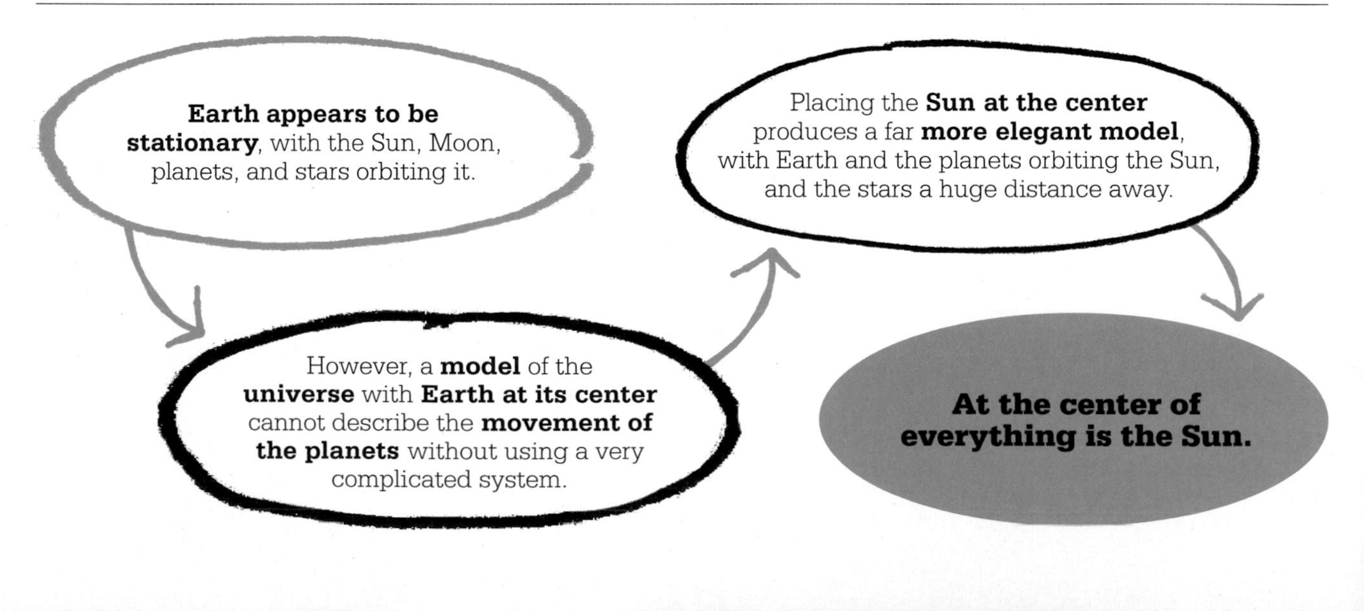

See also: Zhang Heng 26–27 ▪ Johannes Kepler 40–41 ▪ Galileo Galilei 42–43 ▪ William Herschel 86–87 ▪ Edwin Hubble 236–41

sub-orbits were carried around the Sun. This system was best refined by the great Greco-Roman astronomer and geographer Ptolemy of Alexandria in the 2nd century CE.

Even in the classical world, however, there were differences of opinion—the Greek thinker Aristarchus of Samos, for instance, used ingenious trigonometric measurements to calculate the relative distances of the Sun and Moon in the 3rd century BCE. He found that the Sun was huge, and this inspired him to suggest that the Sun was a more likely pivot point for the motion of the cosmos.

However, the Ptolemaic system ultimately won out over rival theories, with far-reaching implications. While the Roman Empire dwindled in subsequent centuries, the Christian Church inherited many of its assumptions. The idea that Earth was the center of everything, and that man was the pinnacle of God's creation, with dominion over Earth, became a central tenet of Christianity and held sway in Europe until the 16th century.

However, this does not mean that astronomy stagnated for a millennium and a half after Ptolemy. The ability to accurately predict the movements of the planets was not only a scientific and philosophical puzzle, but also had supposed practical purposes thanks to the superstitions of astrology. Stargazers of all persuasions had good reason to attempt ever more accurate measurements of the motions of the planets.

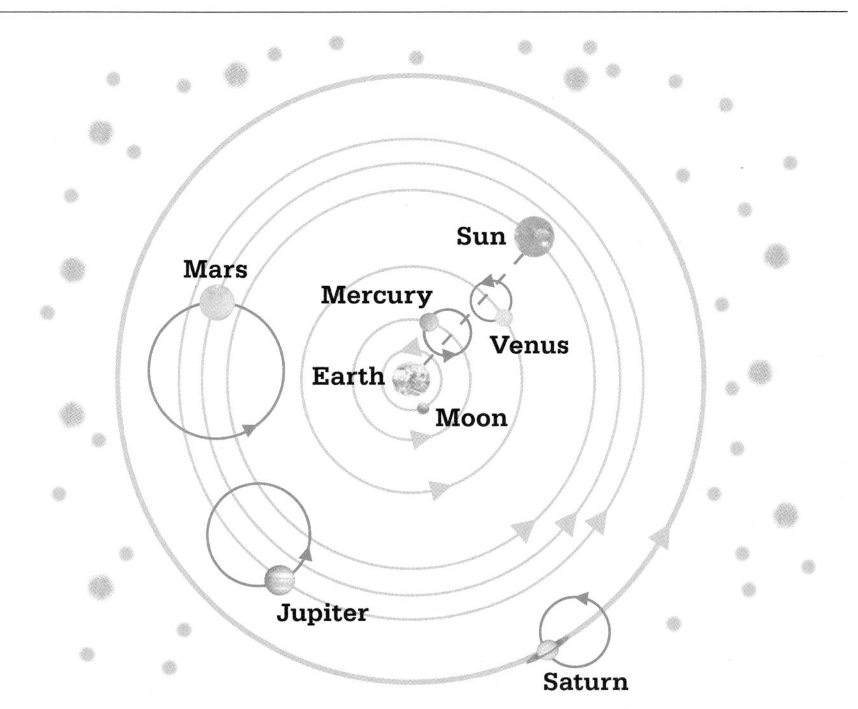

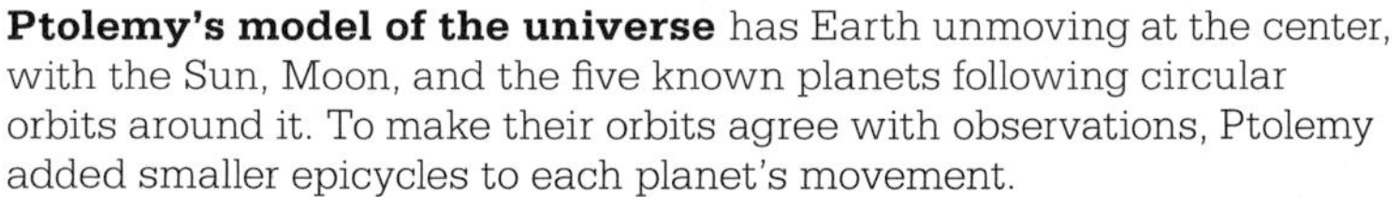
Ptolemy's model of the universe has Earth unmoving at the center, with the Sun, Moon, and the five known planets following circular orbits around it. To make their orbits agree with observations, Ptolemy added smaller epicycles to each planet's movement.

Arabic scholarship

The later centuries of the first millennium corresponded with the first great flowering of Arabic science. The rapid spread of Islam across the Middle East and North Africa from the 7th century brought Arab thinkers into contact with classical texts, including the astronomical writings of Ptolemy and others.

The practice of "positional astronomy"—calculating the positions of heavenly bodies—reached its apogee in Spain, which had become a dynamic melting pot of Islamic, Jewish, and Christian thought. In the late 13th century, King Alfonso X of Castile sponsored the compilation of the Alfonsine Tables, which combined new observations with centuries of Islamic records to bring new precision to the Ptolemaic system and provide the data that would be used to calculate planetary positions until the early 17th century.

Questioning Ptolemy

However, by this point the Ptolemaic model was becoming absurdly complicated, with yet more epicycles added to keep prediction in line with observation. In 1377, French philosopher Nicole Oresme, Bishop of Lisieux, addressed this problem head-on in the work *Livre du Ciel et du Monde* (*Book of the Heavens and the Earth*). He demonstrated the lack of observational proof that Earth was static, and argued that there was no reason to suppose that it » was not in motion. Yet, despite his demolition of the evidence for the Ptolemaic system, Oresme concluded that he did not himself believe in a moving Earth.

By the beginning of the 16th century, the situation had become very different. The twin forces of the Renaissance and the Protestant Reformation saw many old religious dogmas opened up to question. It was in this context that Nicolaus Copernicus, a Polish Catholic canon from the province of Warmia, put forward the first modern heliocentric theory, shifting the center of the universe from Earth to the Sun.

Copernicus first published his ideas in a short pamphlet known as the *Commentariolus*, circulated among friends from around 1514. His theory was similar in essence to the system proposed by Aristarchus, and while it overcame many of the earlier model's failings, it remained deeply attached to certain pillars of Ptolemaic thought—most significantly the idea that the orbits of celestial objects were mounted on crystalline spheres that rotated in perfect circular motion. As a result, Copernicus had to introduce "epicycles" of his own in order to regulate the speed of planetary motions on certain parts of their orbits. One important implication of his model was that it vastly increased the size of the universe. If Earth was moving around the Sun, then this should give itself away through parallax effects caused by our changing point of view: the stars should appear to shift back and forth across the sky throughout the year. Because they do not do so, they must be very far away indeed.

Since the Sun remains stationary, whatever appears as a motion of the Sun is due to the motion of the Earth.
Nicolaus Copernicus

This 17th-century illustration of the Copernican system shows the planets in circular orbits around the Sun. Copernicus believed that the planets were attached to heavenly spheres.

The Copernican model soon proved itself far more accurate than any refinement of the old Ptolemaic system, and word spread among intellectual circles across Europe. Notice even reached Rome, where, contrary to popular belief, the model was at first welcomed in some Catholic circles. The new model caused enough of a stir for German mathematician Georg Joachim Rheticus to travel to Warmia and become Copernicus's pupil and assistant from 1539.

It was Rheticus who published the first widely circulated account of the Copernican system, known as the *Narratio Prima*, in 1540. Rheticus urged the aging priest to publish his own work in full—something that Copernicus had contemplated for many years, but only conceded to in 1543 as he lay on his deathbed.

Mathematical tool

Published posthumously, *De Revolutionibus Orbium Coelestium* (*On the Revolutions of the Heavenly Spheres*) was not initially greeted with outrage, even though any suggestion that Earth was in motion directly contradicted several passages of Scripture and was therefore regarded as heretical by both Catholic and Protestant theologians. To sidestep the issue, a preface had been inserted that explained the heliocentric model as purely a mathematical tool for prediction, not a description of the physical universe. In his life, however, Copernicus himself had shown no such reservations. Despite its heretical implications, the Copernican model was used for the calculations involved in the great calendar reform introduced by Pope Gregory XIII in 1582.

However, new problems with the model's predictive accuracy soon began to emerge, thanks to the meticulous observations of the Danish astronomer Tycho Brahe (1546–1601), which showed that the Copernican model did not adequately describe planetary motions. Brahe attempted to resolve these contradictions with a model of his own in which the planets went around the Sun but the Sun and Moon remained in orbit around Earth. The real solution—that of elliptical orbits—would only be found by his pupil Johannes Kepler.

It would be six decades before Copernicanism became truly emblematic of the split caused in Europe by the Reformation of the Church, thanks largely to the controversy surrounding Italian scientist Galileo Galilei. Galileo's 1610 observations of the phases displayed by Venus and the presence of moons orbiting Jupiter convinced him that the heliocentric theory was correct, and his ardent support for it, from the heart of Catholic Italy, was ultimately expressed in his *Dialogue Concerning the Two Chief World Systems* (1632). This led Galileo into conflict with the papacy, one result of which was the retrospective censorship of controversial passages in *De Revolutionibus* in 1616. This prohibition would not be lifted for more than two centuries. ■

> As though seated on a royal throne, the Sun governs the family of planets revolving around it.
> **Nicolaus Copernicus**

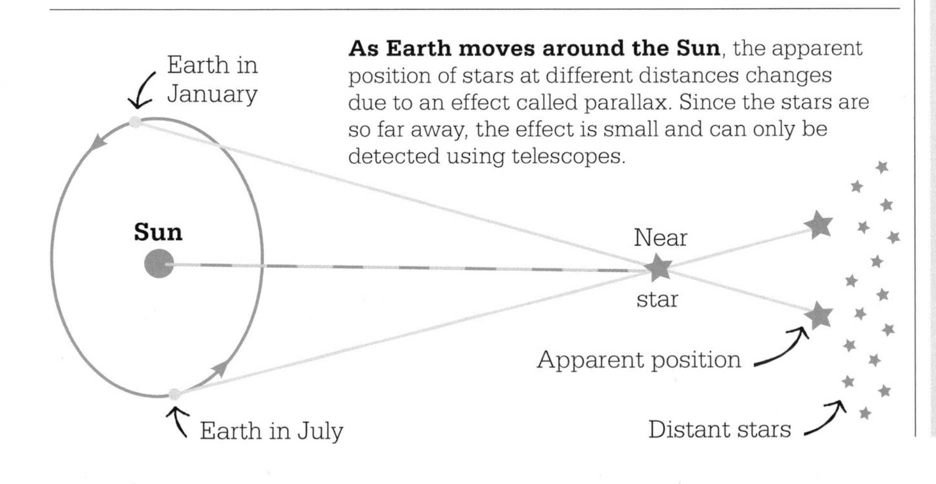

As Earth moves around the Sun, the apparent position of stars at different distances changes due to an effect called parallax. Since the stars are so far away, the effect is small and can only be detected using telescopes.

Nicolaus Copernicus

Born in the Polish city of Torun in 1473, Nicolaus Copernicus was the youngest of four children of a wealthy merchant. His father died when Nicolaus was 10. An uncle took him under his wing and oversaw his education at the University of Krakow. He spent several years in Italy studying medicine and law, returning in 1503 to Poland, where he joined the canonry under his uncle, who was now Prince-Bishop of Warmia.

Copernicus was a master of both languages and mathematics, translating several important works and developing ideas about economics, as well as working on his astronomical theories. The theory he outlined in *De Revolutionibus* was daunting in its mathematical complexity, so while many recognized its significance, it was not widely adopted by astronomers for practical everyday use.

Key works

1514 *Commentariolus*
1543 *De Revolutionibus Orbium Coelestium* (*On the Revolutions of the Heavenly Spheres*)

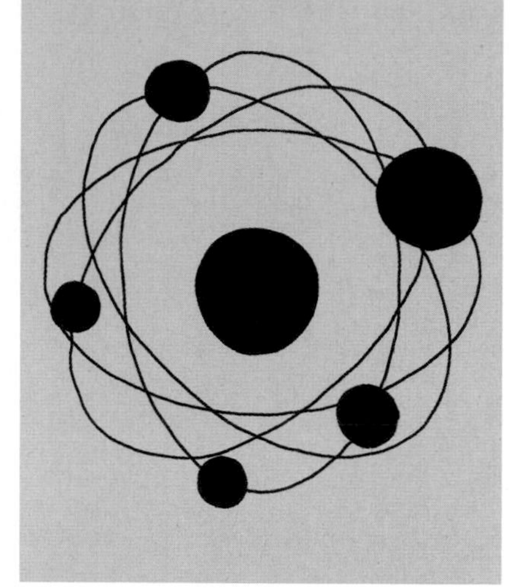

THE ORBIT OF EVERY PLANET IS AN ELLIPSE

JOHANNES KEPLER (1571–1630)

IN CONTEXT

BRANCH
Astronomy

BEFORE
150 CE Ptolemy of Alexandria publishes the *Algamest*, a model of the universe built on the assumption that Earth lies at its center and the Sun, Moon, planets and stars revolve around it in circular orbits on fixed celestial spheres.

16th century The idea of a Sun-centered cosmology begins to gain followers through the ideas of Nicolaus Copernicus.

AFTER
1639 Jeremiah Horrocks uses Kepler's ideas to predict and view a transit of Venus across the face of the Sun.

1687 Isaac Newton's laws of motion and gravitation reveal the physical principles that give rise to Kepler's laws.

While the work of Nicolaus Copernicus on celestial orbits, published in 1543, made a convincing case for a heliocentric (Sun-centered) model of the universe, his system suffered from significant problems. Unable to break free from ancient ideas that heavenly bodies were mounted on crystal spheres, Copernicus had stated that the planets orbited the Sun on perfect circular paths, and was forced to introduce a variety of complications to his model to account for their irregularities.

Supernova and comets

In the latter half of the 16th century, Danish nobleman Tycho Brahe (1546–1601) made observations that

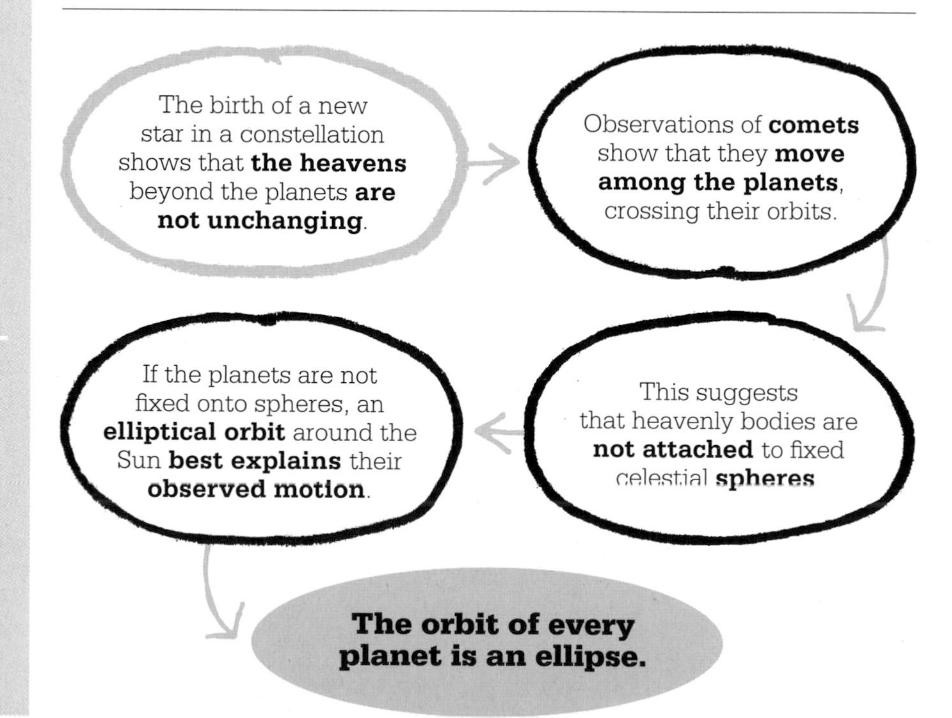

See also: Nicolaus Copernicus 34–39 ▪ Jeremiah Horrocks 52 ▪ Isaac Newton 62–69

would prove vital to resolving the problems. A bright supernova explosion seen in the constellation of Cassiopeia in 1572 undermined the Copernican idea that the universe beyond the planets was unchanging. In 1577, Brahe plotted the motion of a comet. Comets had been thought of as local phenomena, closer than the Moon, but Brahe's observations showed that the comet must lie well beyond the Moon, and was in fact moving among the planets. In one stroke, this evidence demolished the idea of "heavenly spheres." However, Brahe remained wedded to the idea of circular orbits in his geocentric (Earth-centered) model.

In 1597, Brahe was invited to Prague, where he spent his last years as Imperial Mathematician to Emperor Rudolph II. Here he was joined by German astrologer Johannes Kepler, who continued Brahe's work after his death.

Breaking with circles

Kepler had already begun to calculate a new orbit for Mars from Brahe's observations, and around this time concluded that its orbit must be ovoid (egg-shaped) rather than truly circular. Kepler formulated a heliocentric model with ovoid orbits, but this still did not match the observational data. In 1605, he concluded that Mars must instead orbit the Sun in an ellipse—a "stretched circle" with the Sun as one of two focus points. In his *Astronomia Nova* (*New Astronomy*) of 1609, he outlined two laws of planetary motion. The first law stated that the orbit of every planet is an ellipse. The second law stated that a line joining a planet to the Sun sweeps across equal areas during equal periods of time. This means that the speed of the planets increases the closer they are to the Sun. A third law, in 1619, described the relationship of a planet's year to its distance from the Sun: the square of a planet's orbital period (year) is proportional to the cube of its distance from the Sun. So a planet that is twice the distance from the Sun than another planet will have a year that is almost three times as long.

The nature of the force keeping the planets in orbit was unknown. Kepler believed it was magnetic, but it would be 1687 before Newton showed that it was gravity. ■

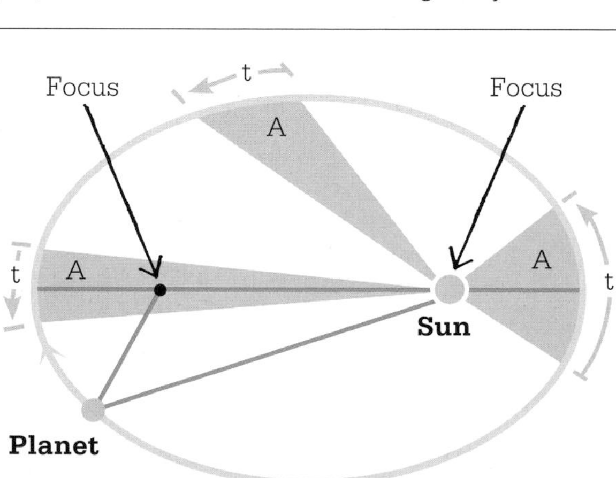

Kepler's laws state that planets follow elliptical orbits with the Sun as one of the two foci of the ellipse. In any given time, *t*, a line joining the planets to the Sun sweeps across equal areas (A) in the ellipse.

Johannes Kepler

Born in the city of Weil der Stadt near Stuttgart, southern Germany, in 1571, Johannes Kepler witnessed the Great Comet of 1577 as a small child, marking the start of his fascination with the heavens. While studying at the University of Tübingen, he developed a reputation as a brilliant mathematician and astrologer. He corresponded with various leading astronomers of the time, including Tycho Brahe, ultimately moving to Prague in 1600 to become Brahe's student and academic heir.

Following Brahe's death in 1601, Kepler took on the post of Imperial Mathematician, with a royal commission to complete Brahe's work on the so-called *Rudolphine Tables* for predicting the movements of the planets. He completed this work in Linz, Austria, where he worked from 1612 until his death in 1630.

Key works

1596 *The Cosmic Mystery*
1609 *Astronomia Nova* (*New Astronomy*)
1619 *The Harmony of the World*
1627 *Rudolphine Tables*

A FALLING BODY ACCELERATES UNIFORMLY

GALILEO GALILEI (1564–1642)

IN CONTEXT

BRANCH
Physics

BEFORE
4th century BCE Aristotle develops ideas about forces and motion, but does not test them experimentally.

1020 Persian scholar Ibn Sina (Avicenna) writes that moving objects have innate "impetus," slowed only by external factors such as air resistance.

1586 Flemish engineer Simon Stevin drops two lead balls of unequal weight from a church tower in Delft to show that they fall at the same speed.

AFTER
1687 Isaac Newton's *Principia* formulates his laws of motion.

1971 US astronaut Dave Scott demonstrates Galileo's ideas about falling bodies by showing that a hammer and a feather fall at the same rate on the Moon, which has almost no atmosphere to cause drag.

For 2,000 years, few people challenged Aristotle's assertion that an external force keeps things moving and that heavy objects fall faster than lighter ones. Only in the 17th century did the Italian astronomer and mathematician Galileo Galilei insist that the ideas had to be tested. He devised experiments to test how and why objects move and stop moving, and was the first to figure out the principle of inertia—that objects resist a change in motion and need a force to start moving, speed up, or slow down. By timing objects falling, Galileo showed that the rate of fall is the same for all objects, and came to realize the part played by friction in slowing them down.

With the equipment available during the 1630s, Galileo could not directly measure the speed or acceleration of freely falling objects. By rolling balls down one ramp and up another, he showed that the speed of a ball at the bottom of the ramp depended on its starting height, not on the steepness of the ramp, and that a ball would always roll up to the same height it had started from, no matter how steep or shallow the inclines were.

Galileo carried out his remaining experiments with a ramp 16 ft (5 m) long, lined with a smooth material to reduce friction. For timing, he used a large container of water with a small pipe in the bottom. He collected the water during the interval he was measuring, and weighed the water

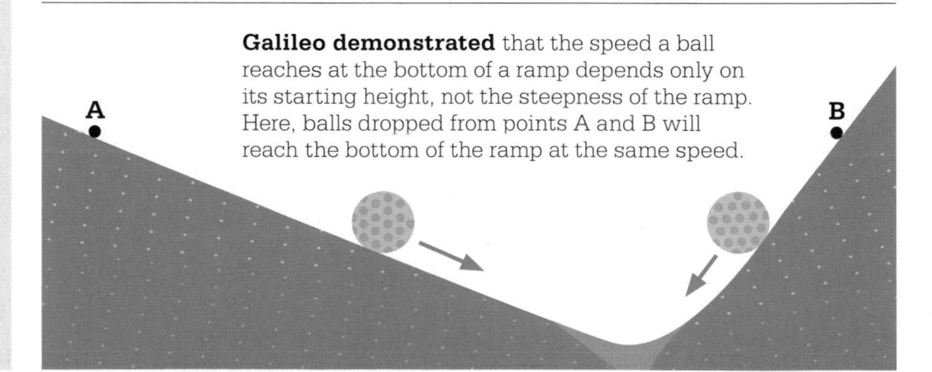

Galileo demonstrated that the speed a ball reaches at the bottom of a ramp depends only on its starting height, not the steepness of the ramp. Here, balls dropped from points A and B will reach the bottom of the ramp at the same speed.

See also: Nicolaus Copernicus 34–39 ▪ Isaac Newton 62–69

Count what is countable, measure what is measurable, and what is not measurable, make it measurable.
Galileo Galilei

collected. By letting the ball go at different points on the ramp, he showed that the distance traveled depended on the square of the time taken—in other words, the ball accelerated down the ramp.

The law of falling bodies

Galileo's conclusion was that bodies all fall at the same speed in a vacuum, an idea later developed further by Isaac Newton. There is a greater force from gravity on a larger mass, but the larger mass also needs a bigger force to make it accelerate. The two effects cancel each other out, so in the absence of any other forces, all falling objects will accelerate at the same rate. We see things falling at different rates in everyday life because of the effect of air resistance, which slows objects down at different rates depending on their size and shape. A beach ball and a bowling ball of the same size will initially accelerate at the same rate. Once they are moving, the same amount of air resistance will act on them, but the size of this force will be a much greater proportion of the downward force on the beach ball than the bowling ball, and so the beach ball will slow down more.

Galileo's insistence on testing theories with careful observation and measurable experiments marks him, like Alhazen, as one of the founders of modern science. His ideas on forces and motion paved the way for Newton's laws of motion 50 years later and underpin our understanding of movement in the universe, from atoms to galaxies. ■

Objects of different masses **appear to fall** at **different rates**.

↓

All moving objects are affected by **air resistance**.

↓

Without air resistance, all objects would fall at **the same rate**.

↓

A falling body accelerates uniformly.

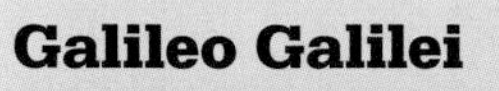

Galileo Galilei

Galileo was born in Pisa, but later moved with his family to Florence. In 1581, he enrolled in the University of Pisa to study medicine, then switched to mathematics and natural philosophy. He investigated many areas of science, and is perhaps most famous for his discovery of the four largest moons of Jupiter (still called the Galilean moons). Galileo's observations led him to support the Sun-centered model of the solar system, which at the time was in opposition to the teachings of the Roman Catholic Church. In 1633, he was tried and made to recant this and other ideas. He was sentenced to house arrest, which lasted the rest of his life. During his confinement, he wrote a book summarizing his work on kinematics (the science of movement).

Key works

1623 *The Assayer*
1632 *Dialogue Concerning the Two Chief World Systems*
1638 *Discourses and Mathematical Demonstrations Relating to Two New Sciences*

THE GLOBE OF THE EARTH IS A MAGNET

WILLIAM GILBERT (1544–1603)

IN CONTEXT

BRANCH
Geology

BEFORE
6th century BCE The Greek thinker Thales of Miletus notes magnetic rocks, or lodestones.

1st century CE Chinese diviners make primitive compasses with iron ladles that swivel to point south.

1269 French scholar Pierre de Maricourt sets out the basic laws of magnetic attraction, repulsion, and poles.

AFTER
1824 French mathematician Siméon Poisson models the forces in a magnetic field.

1940s American physicist Walter Maurice Elsasser attributes Earth's magnetic field to iron swirling in its outer core as the planet rotates.

1958 Explorer 1 space mission shows Earth's magnetic field extending far out into space.

By the late 1500s, ships' captains already relied on magnetic compasses to maintain their course across the oceans. Yet no one knew how they worked. Some thought the compass needle was attracted to the North Star, others that it was drawn to magnetic mountains in the Arctic. It was English physician William Gilbert who discovered that Earth itself is magnetic.

Stronger reasons are obtained from sure experiments and demonstrated arguments than from probable conjectures and the opinions of philosophical speculators.
William Gilbert

Gilbert's breakthrough came not from a flash of inspiration, but from 17 years of meticulous experiment. He learned all he could from ships' captains and compass makers, and then he made a model globe, or "terrella," out of the magnetic rock lodestone and tested compass needles against it. The needles reacted around the terrella just as ships' compasses did on a larger scale—showing the same patterns of declination (pointing slightly away from true north at the geographic pole, which differs from magnetic north) and inclination (tilting down from the horizontal toward the globe).

Gilbert concluded, rightly, that the entire planet is a magnet and has a core of iron. He published his ideas in the book *De Magnete* (*On the Magnet*) in 1600, causing a sensation. Johannes Kepler and Galileo, in particular, were inspired by his suggestion that Earth is not fixed to rotating celestial spheres, as most people still thought, but is made to spin by the invisible force of its own magnetism. ■

See also: Thales of Miletus 20 ■ Johannes Kepler 40–41 ■ Galileo Galilei 42–43 ■ Hans Christian Ørsted 120 ■ James Clerk Maxwell 180–85

NOT BY ARGUING, BUT BY TRYING

FRANCIS BACON (1561–1626)

IN CONTEXT

BRANCH
Experimental science

BEFORE
4th century BCE Aristotle deduces, argues, and writes, but does not test with experiments—his methods persist for the next millennium.

c.750–1250 CE Arab scientists conduct experiments during the Golden Age of Islam.

AFTER
1630s Galileo experiments with falling bodies.

1637 French philosopher René Descartes insists on rigorous scepticism and inquiry in his *Discourse on Method.*

1665 Isaac Newton uses a prism to investigate light.

1963 In *Conjectures and Refutations*, the Austrian philosopher Karl Popper insists that a theory may be tested and proved false, but cannot conclusively be proved correct.

The English philosopher, statesman, and scientist Francis Bacon was not the first to conduct experiments—Alhazen and other Arab scientists conducted them 600 years earlier—but he was the first to explain the methods of inductive reasoning and set out the scientific method. He also saw science as a "spring of a progeny of inventions, which shall overcome, to some extent, and subdue our needs and miseries."

Evidence from experiment

According to the Greek philosopher Plato, truth was found by authority and argument—if enough intelligent men discussed something for long enough, the truth would result. His student, Aristotle, saw no need for experiments. Bacon parodied such "authorities" as spiders, spinning webs from their own substance. He insisted on evidence from the real world, particularly from experiment.

Two key works by Bacon laid out the future of scientific inquiry. In *Novum Organum* (1620), he sets out his three fundamentals for the scientific method: observation, deduction to formulate a theory that might explain what has been observed, and experiment to test whether the theory is correct. In *The New Atlantis* (1623), Bacon describes a fictitious island and its House of Salomon—a research institution where scholars conduct pure research centered on experiment and make inventions. Sharing those goals, the Royal Society was founded in 1660 in London, with Robert Hooke as its first Curator of Experiments. ■

> Whether or no anything can be known, can be settled not by arguing, but by trying.
> **Francis Bacon**

See also: Alhazen 28–29 ▪ Galileo Galilei 42–43 ▪ William Gilbert 44 ▪ Robert Hooke 54 ▪ Isaac Newton 62–69

TOUCHING THE SPRING OF THE AIR

ROBERT BOYLE (1627–1691)

IN CONTEXT

BRANCH
Physics

BEFORE
1643 Evangelista Torricelli invents the barometer using a tube of mercury.

1648 Blaise Pascal and his brother-in-law demonstrate that air pressure decreases with altitude.

1650 Otto von Guericke performs experiments on air and vacuums, first published in 1657.

AFTER
1738 Swiss physicist Daniel Bernoulli publishes *Hydrodynamica*, describing a kinetic theory of gases.

1827 Scottish botanist Robert Brown explains the motion of pollen in water as a result of collisions with water molecules moving in random directions.

In the 17th century, several scientists across Europe investigated the properties of air, and their work was to lead Anglo-Irish scientist Robert Boyle to produce his mathematical laws describing pressure in a gas. This work was tied in to a wider debate about the nature of the space between stars and planets. The "atomists" held that there was empty space between celestial bodies, whereas the Cartesians (followers of the French philosopher René Descartes) held that the space between particles was filled with an unknown substance called the ether, and that it was impossible to produce a vacuum.

See also: Isaac Newton 62–69 ▪ John Dalton 112–13 ▪ Robert FitzRoy 150–55

We live submerged at the bottom of an ocean of the element air, that by unquestioned experiments is known to have weight.
Evangelista Torricelli

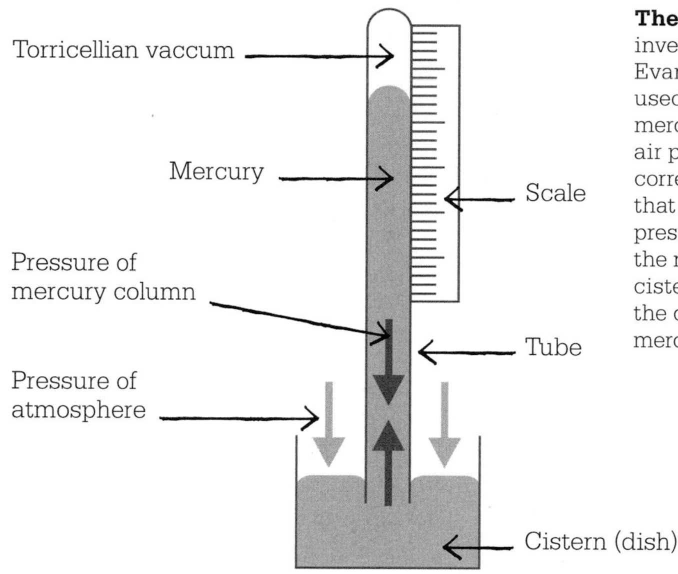

The barometer invented by Evangelista Torricelli used a column of mercury to measure air pressure. Torricelli correctly reasoned that it was the air pressing down on the mercury in the cistern that balanced the column of mercury in the tube.

Barometers

In Italy, the mathematician Gasparo Berti performed experiments designed to figure out why a suction pump could not raise water more than 33 ft (10 m) high. Berti took a long tube, sealed it at one end and filled it with water. He then inverted the tube with its mouth in a tub of water. The level of water in the tube fell until the column was about 30 ft (10 m) high. In 1642, fellow Italian Evangelista Torricelli, hearing of Berti's work, constructed a similar apparatus but used mercury instead of water. Mercury is more than 13 times denser than water, so his column of liquid was only about 30 in (76 cm) high. Torricelli's explanation for this was that the weight of the air above the mercury in the dish was pressing down on it, and that this balanced the weight of the mercury inside the column. He said that the space in the tube above the mercury was a vacuum. This is explained today in terms of pressure (force on a certain area), but the basic idea is the same. Torricelli had invented the first mercury barometer.

French scientist Blaise Pascal heard of Torricelli's barometer in 1646, prompting him to start some experiments of his own. One of these, performed by his brother-in-law Florin Périer, was to demonstrate that air pressure changed depending on altitude. One barometer was set up on the grounds of a monastery in Clermont, and observed by a monk during the day. Périer carried the other to the top of Puy de Dôme, about 3,200 ft (1,000 m) above the town. The column of mercury was more than 3 in (8 cm) shorter at the top of the mountain than in the monastery garden. Since there is less air above a mountain than there is above the valley below it, this showed that it was indeed the weight of the air that held the liquid in the tubes of mercury or water. For this, and other work, the modern unit of pressure is named after Pascal.

Blaise Pascal's experiments with barometers showed how air pressure varied with altitude. In addition to physics, Pascal also made significant contributions to mathematics.

Air pumps

The next important breakthrough was made by Prussian scientist Otto von Guericke, who made a pump that was capable of pumping some of the air out of a container. He performed his most famous »

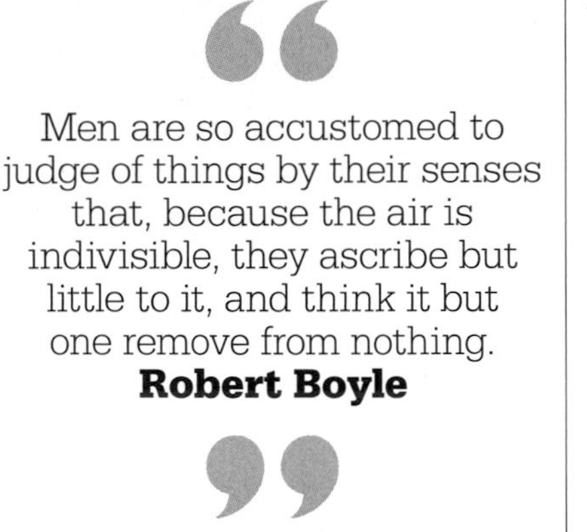

> Men are so accustomed to judge of things by their senses that, because the air is indivisible, they ascribe but little to it, and think it but one remove from nothing.
> **Robert Boyle**

demonstration in 1654, when he put two metal hemispheres together with an airtight seal between them and pumped the air out of them—two teams of horses were unable to pull the hemispheres apart. Before the air was pumped out, the air pressure inside the sealed hemispheres was the same as the air pressure outside. Without the air inside, pressure from the outside air held the hemispheres together.

Robert Boyle learned of von Guericke's experiments when they were published in 1657. To do experiments of his own, Boyle commissioned Robert Hooke (p.54) to design and build an air pump. Hooke's air pump consisted of a glass "receiver" (container) whose diameter was nearly 16 in (40 cm), a cylinder with a piston below it, and an arrangement of plugs and valves between them. Successive movements of the piston drew more and more air out of the receiver. Due to slow leaks in the seals of the equipment, the near-vacuum inside the receiver could only be maintained for a short time. Nevertheless, the machine was a great improvement on anything made previously, an example of the importance of technology to the furthering of scientific investigation.

Experimental results

Boyle performed a number of different experiments with the air pump, which he described in his 1660 book *New Experiments Physico-Mechanical*. In the book, he was intent on pointing out that the results described are all from experiments, since at the time even such noted experimentalists as Galileo often also reported the results of "thought experiments."

Many of Boyle's experiments were directly connected to air pressure. The receiver could be modified to hold a Torricelli barometer, with the tube sticking

Otto von Guericke built the first air pump. His experiments with the pump provided evidence against Aristotle's idea that "Nature abhors a vacuum."

Robert Boyle

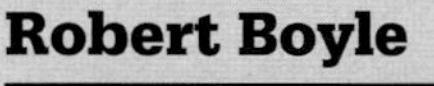

Robert Boyle was born in Ireland, the 14th child of the Earl of Cork. He was tutored at home before attending Eton College in England and then touring Europe. His father died in 1643, leaving him enough money to indulge his interest in science full time. Boyle moved back to Ireland for a couple of years, but lived in Oxford from 1654 to 1668 so that he could do his work more easily, and then moved to London.

Boyle was part of a group of men studying scientific subjects called the "Invisible College," who met in London and Oxford to discuss their ideas. This group became the Royal Society in 1663, and Boyle was one of the first council members. In addition to his interests in science, Boyle performed experiments in alchemy and wrote about theology and the origin of different human races.

Key works

1660 *New Experiments Physico-Mechanical: Touching the Spring of the Air and their Effects*
1661 *The Sceptical Chymist*

out of the top of the receiver and sealed in place with cement. As the pressure in the receiver was reduced, the level of the mercury fell. He also performed the opposite experiment, and found that raising the pressure inside the receiver made the level of the mercury rise. This confirmed the previous findings of Torricelli and Pascal.

Boyle noted that it became harder and harder to pump air out of the receiver as the amount of air left decreased, and also showed that a half-inflated bladder in the receiver increased in volume as the air surrounding it was removed. A similar effect on the bladder could be achieved by holding it in front of a fire. He gave two possible explanations for the "spring" of the air that caused these effects: each particle of the air was compressible like a spring and the whole mass of air resembled fleece, or the air consisted of particles moving randomly.

This was similar to the view of the Cartesians, although Boyle did not agree with the idea of the ether, but suggested that the "corpuscles" were moving in empty space. His explanation is remarkably similar to the modern kinetic theory, which describes the properties of matter in terms of moving particles.

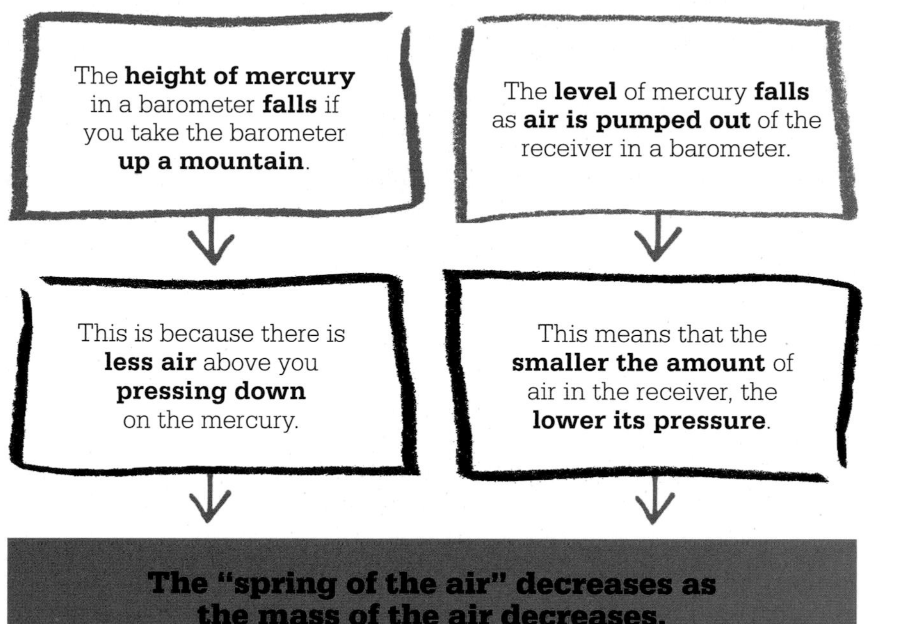

If the height of the mercury column is less on the top of a mountain than at the foot of it, it follows that the weight of the air must be the sole cause of the phenomenon.
Blaise Pascal

Some of Boyle's experiments were physiological, investigating the effects on birds and mice of reducing the pressure of the air, and speculating on how air is moved in and out of lungs.

Boyle's law

Boyle's law states that the pressure of a gas multiplied by its volume is a constant, as long as the amount of gas and the temperature are kept the same. In other words, if you decrease the volume of a gas, its pressure increases. It is this increased pressure that produces the spring of the air. You can feel this effect using a bicycle pump by covering the end with a finger and pushing the handle in.

Although it bears his name, this law was first proposed not by Boyle, but by English scientists Richard Towneley and Henry Power, who performed a series of experiments with a Torricelli barometer and published their results in 1663. Boyle saw an early draft of the book and discussed the results with Towneley. He confirmed them by experiment and published "Mr Towneley's hypothesis" in 1662 as part of a response to criticism of his original experiments.

Boyle's work on gases was particularly significant because of his careful experimental technique, and also his full reporting of all his experiments and their possible sources of error, whether or not they gave the expected results. This led many to seek to extend his work. Today, Boyle's law has been combined with laws figured out by other scientists to form the "ideal-gas law," which approximates to the behavior of real gases under changes of temperature, pressure, or volume. His ideas would also eventually lead to the development of the kinetic theory. ■

IS LIGHT A PARTICLE OR A WAVE?

CHRISTIAAN HUYGENS (1629–1695)

IN CONTEXT

BRANCH
Physics

BEFORE
11th century Alhazen shows that light travels in straight lines.

1630 René Descartes proposes a wave description of light.

1660 Robert Hooke states that light is a vibration of the medium through which it propagates.

AFTER
1803 Thomas Young describes experiments that demonstrate how light behaves as a wave.

1864 James Clerk Maxwell predicts the speed of light and concludes that light is a form of electromagnetic wave.

1900s Albert Einstein and Max Planck show that light is both a particle and a wave. The quanta of electromagnetic radiation they recognize become known as "photons."

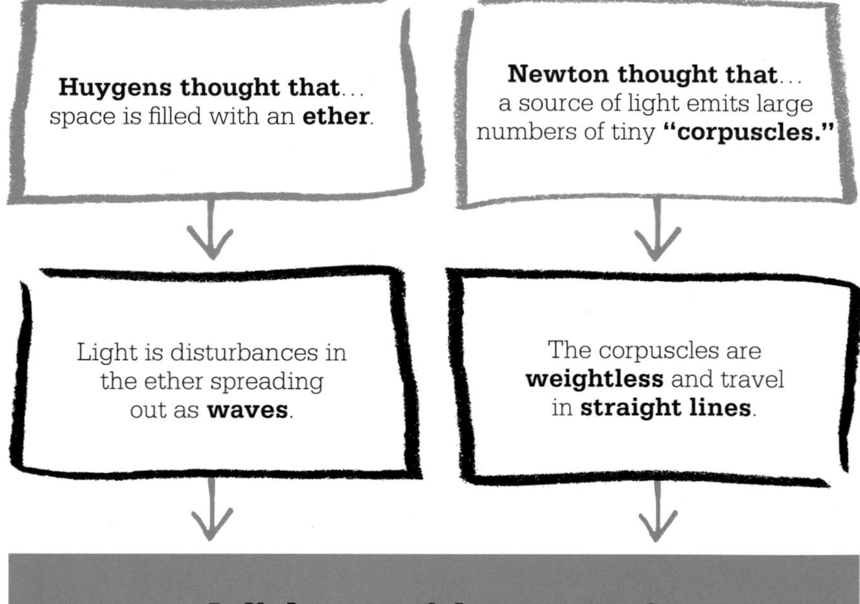

In the 17th century, Isaac Newton and the Dutch astronomer Christiaan Huygens both pondered the true nature of light, and reached very different conclusions. The problem they faced was that any theory about the nature of light had to explain reflection, refraction, diffraction, and color. Refraction is the bending of light as it passes from one substance to another, and is the reason that lenses can focus light. Diffraction is the spreading out of light when it passes through a very narrow gap.

Before Newton's experiments, it was widely accepted that light gained its quality of color by interacting with matter—that

See also: Alhazen 28–29 ▪ Robert Hooke 54 ▪ Isaac Newton 62–69 ▪ Thomas Young 110–11 ▪ James Clerk Maxwell 180–85 ▪ Albert Einstein 214–21

the "rainbow" effect seen when light passes through a prism is produced because the prism has somehow stained the light. Newton demonstrated that the "white" light that we see is actually a mixture of different colors of light, and these are split up by a prism because they are all refracted by slightly different amounts.

As with many natural philosophers of the time, Newton held that light was made up of a stream of particles, or "corpuscles." This idea explained how light traveled in straight lines and "bounced" off reflective surfaces. It also explained refraction in terms of forces at the boundaries between different materials.

Partial reflection

However, Newton's theory could not explain how, when light hits many surfaces, some is reflected and some is refracted. In 1678, Huygens argued that space was filled with weightless particles (the ether), and that light caused disturbances in the ether that spread out in spherical waves. Refraction was thus explained if different materials (be they ether, water, or glass) caused light waves to travel at different speeds. Huygens' theory could explain why both reflection and refraction can occur at a surface. It could also explain diffraction.

Huygens' ideas made little impact at the time. This was in part due to Newton's already giant stature as a scientist. However, a century later, in 1803, Thomas Young showed that light does indeed behave as a wave, and experiments in the 20th century have shown that it behaves both like a wave and a particle, although there are big differences between Huygens' "spherical waves" and our modern models of light. Huygens said that light waves were longitudinal as they passed through a substance—the ether. Sound waves are also longitudinal waves, in which the particles of the substance the wave is passing through vibrate in the same direction as the wave is traveling. Our modern view of light waves is that they are transverse waves that behave more like waves of water. They do not need matter to propagate (transmit), while particles vibrate at right angles (up and down) to the wave's direction. ■

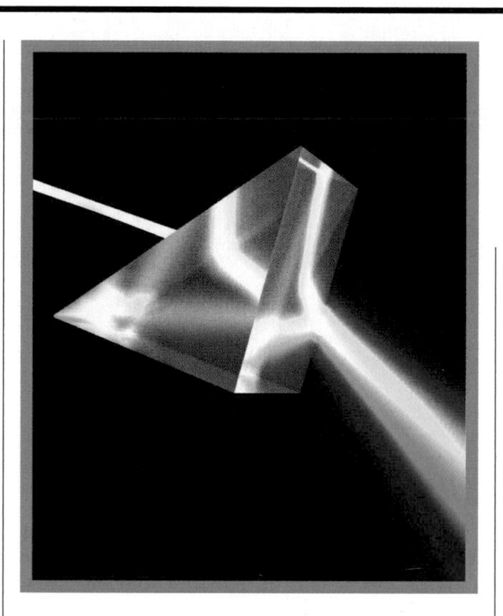

When white light passes through a prism, it is refracted into its component parts. Huygens explained that this is due to light waves traveling at different speeds through different materials.

Christiaan Huygens

Dutch mathematician and astronomer Christiaan Huygens was born in The Hague in 1629. He studied law and mathematics at his university, then devoted some time to his own research, initially in mathematics but then also in optics, working on telescopes and grinding his own lenses.

Huygens visited England several times, and met Isaac Newton in 1689. In addition to his work on light, Huygens had studied forces and motion, but he did not accept Newton's idea of "action at a distance" to describe the force of gravity. Huygens' wide-ranging achievements included some of the most accurate clocks of his time, the result of his work on pendulums. His astronomical work, carried out using his own telescopes, included the discovery of Titan, the largest of Saturn's moons, and the first correct description of Saturn's rings.

Key works

1656 *De Saturni Luna Observatio Nova*
1690 *Treatise on Light*

THE FIRST OBSERVATION OF A TRANSIT OF VENUS

JEREMIAH HORROCKS (1618–1641)

IN CONTEXT

BRANCH
Astronomy

BEFORE
1543 Nicolaus Copernicus makes the first complete argument for a Sun-centered (heliocentric) universe.

1609 Johannes Kepler proposes a system of elliptical orbits—the first complete description of planetary motion.

AFTER
1663 Scottish mathematician James Gregory devises a way to measure the exact distance from Earth to the Sun using observations of the transits of Venus in 1631 and 1639.

1769 British explorer Captain James Cook observes and records the transit of Venus in Tahiti in the South Pacific.

2012 Astronomers observe the last transit of Venus of the 21st century.

Planetary transits offered an opportunity to test the first of Johannes Kepler's three laws of planetary motion—that the planets orbit the Sun in an elliptical path. The brief passages by Venus and Mercury across the disk of the Sun—at the times predicted by Kepler's Rudolphine Tables—would reveal whether the underlying theory was correct.

The first test—a 1631 transit of Mercury observed by French astronomer Pierre Gassendi—proved encouraging. However, his attempt to spot the transit of Venus a month later failed due to inaccuracies in Kepler's figures. These same figures predicted a "near miss" for Venus and the Sun in 1639, but English astronomer Jeremiah Horrocks calculated that a transit would in fact occur.

At sunrise on December 4, 1639, Horrocks set up his best telescope, focusing the Sun's disk onto a piece of card. Around 3:15 pm, the clouds cleared, revealing a "spot of unusual magnitude"—Venus—edging across the Sun. While Horrocks marked its progress on the card, timing each interval, a friend measured the transit in another location. By using the two sets of measurements from the different viewpoints, and by recalculating the diameter of Venus relative to the Sun, Horrocks could then estimate Earth's distance from the Sun more accurately than ever before. ■

> I received my first intimation of the remarkable conjunction of Venus and the Sun…it induced me, in expectation of so grand a spectacle, to observe with increased attention.
> **Jeremiah Horrocks**

See also: Nicolaus Copernicus 34–39 ▪ Johannes Kepler 40–41

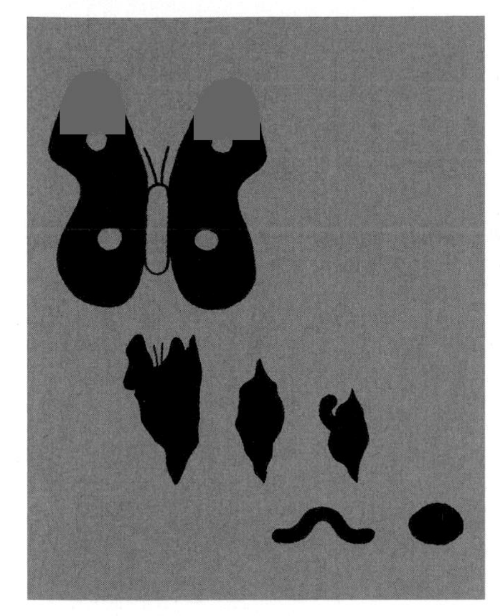

ORGANISMS DEVELOP IN A SERIES OF STEPS

JAN SWAMMERDAM (1637–1680)

IN CONTEXT

BRANCH
Biology

BEFORE
c.320 BCE Aristotle declares that worms and insects arise by spontaneous generation.

1651 English physician William Harvey considers the insect larva a "crawling egg" and the pupa a "second egg" with little internal development.

1668 Italian Francesco Redi provides early evidence to refute spontaneous generation.

AFTER
1859 Charles Darwin explains how each stage of an insect's life is adapted to its activity and environment at that stage.

1913 Italian zoologist Antonio Berlese proposes that an insect larva hatches at a premature stage of embryo development.

1930s British entomologist Vincent Wigglesworth finds hormones control life cycles.

The metamorphosis of a butterfly from egg to caterpillar to chrysalis to adult is a familiar process to us today, but in the 17th century, reproduction was viewed very differently. Following the Greek philosopher Aristotle, most people believed that life—especially "lower" creatures such as insects—arose by spontaneous generation from nonliving matter. The theory of "preformism" held that a "higher" organism took its fully mature form in its miniscule beginning, but that "lower" animals were too simple to have complex innards. In 1669, pioneering Dutch microscopist Jan Swammerdam disproved Aristotle by dissecting insects under the microscope, including butterflies, dragonflies, bees, wasps, and ants.

In the anatomy of a louse, you will find miracles heaped on miracles and will see the wisdom of God clearly manifested in a minute point.
Jan Swammerdam

A new metamorphosis

The term "metamorphosis" had once meant the death of one individual followed by another's appearance from its remains. Swammerdam showed that the stages in an insect's life cycle—adult female, egg, larva and pupa (or nymph), adult—are different forms of the same creature. Each life stage has its own fully formed internal organs, as well as early versions of the organs for later stages. Seen in this new light, insects clearly warranted further scientific study. Swammerdam went on to pioneer the classification of insects based on their reproduction and development, before dying of malaria at 43. ■

See also: Robert Hooke 54 ▪ Antonie van Leeuwenhoek 56–57 ▪ John Ray 60–61 ▪ Carl Linnaeus 74–75 ▪ Louis Pasteur 156–59

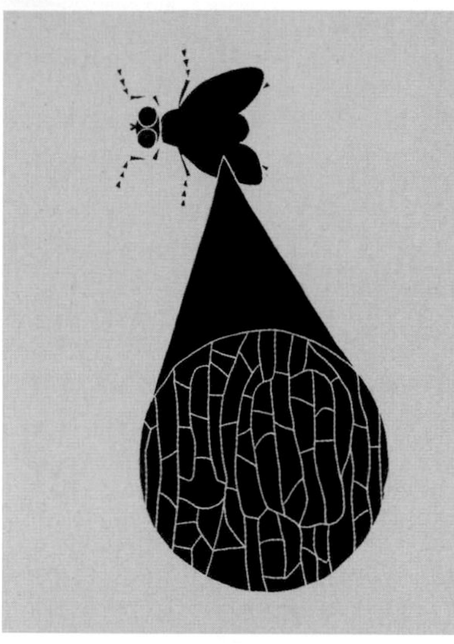

ALL LIVING THINGS ARE COMPOSED OF CELLS

ROBERT HOOKE (1635–1703)

IN CONTEXT

BRANCH
Biology

BEFORE
c.1600 The first compound microscope is developed in the Netherlands, probably by either Hans Lippershey or Hans and Zacharius Janssen.

1644 Italian priest and self-taught scientist Giovanni Battista Odierna produces the first description of living tissue, using a microscope.

AFTER
1674 Antonie van Leeuwenhoek is the first to see single-celled organisms under the microscope.

1682 Van Leeuwenhoek observes the nuclei inside the red blood cells of salmon.

1931 The invention of the electron microscope by Hungarian physicist Leó Szilárd allows much higher resolution images to be made.

The development of the compound microscope in the 17th century opened up a whole new world of previously unseen structures. A simple microscope consists of just one lens, while the compound microscope, developed by Dutch eyeglasses makers, uses two or more lenses, and generally provides greater magnification.

English scientist Robert Hooke was not the first to observe living things using a microscope. However, with the publication of his *Micrographia* in 1665, he became the first best-selling popular science author, stunning his readers with the new science of microscopy. Accurate copperplate drawings made by Hooke himself showed objects the public had never seen before—the detailed anatomies of lice and fleas; the compound eyes of a fly; the delicate wings of a gnat. He also drew some man-made objects—the sharp point of a needle appeared blunt under the microscope—and used his observations to explain how crystals form and what happens when water freezes. The English diarist Samuel Pepys called *Micrographia* "the most ingenious book that I ever read in my life."

Describing cells

One of Hooke's drawings was of a thin slice of cork. In the structure of the cork, he noted what looked like the walls dividing monks' cells in a monastery. These were the first recorded descriptions and drawings of cells, the basic units from which all living things are made. ■

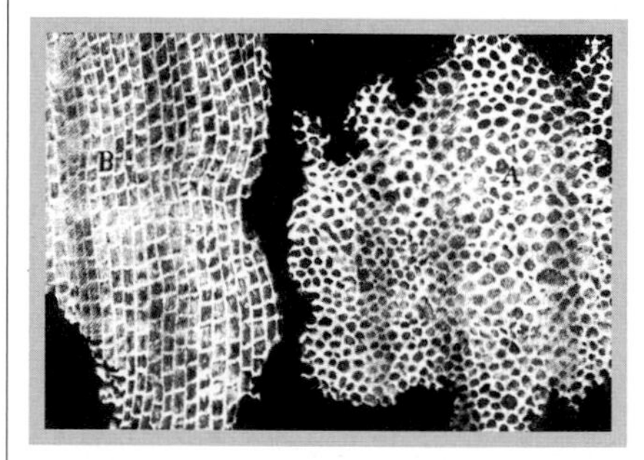

Hooke's drawings of dead cork cells show empty spaces between the cell walls—living cells contain protoplasm. He calculated that there were more than a billion cells in 1 in^3 (16 cm^3) of cork.

See also: Antonie van Leeuwenhoek 56–57 ▪ Isaac Newton 62–69 ▪ Lynn Margulis 300–01

LAYERS OF ROCK FORM ON TOP OF ONE ANOTHER

NICOLAS STENO (1638–1686)

IN CONTEXT

BRANCH
Geology

BEFORE
Late 15th century Leonardo da Vinci writes about his observations of the erosional and depositional action of wind and water on landscapes and surface materials.

AFTER
1780s James Hutton refers Steno's principles to a continuing and cyclical geological process stretching back in time.

1810s Georges Cuvier and Alexandre Brongniart in France and William Smith in Britain apply Steno's principles of stratigraphy to geological mapping.

1878 The first International Geological Congress in Paris sets out procedures for the production of a standard stratigraphic scale.

The sedimentary strata of rocks that make up much of Earth's surface also form the basis for Earth's geological history, which is normally depicted as a column of layers with the oldest strata at the bottom and the youngest at the top. The process of deposition of rock by water and gravity had been known for centuries, but Danish bishop and scientist Niels Stensius, also known as Nicolas Steno, was the first to describe the principles that underlie the process. His conclusions, published in 1669, were drawn from his observations of geological strata in Tuscany, Italy.

Steno's Law of Superposition states that any single sedimentary deposit, or stratum, is younger than the sequence of strata upon which it rests, and older than the strata that rest upon it. Steno's principles of original horizontality and lateral continuity state that strata are deposited as horizontal and continuous layers, and if they are found tilted, folded, or broken, they must have experienced such disturbance after their deposition. Finally, his principle of crosscutting relationships states that "if a body or discontinuity cuts across a stratum, it must have formed after that stratum".

Rock strata, as Steno realized, all start life as horizontal layers, which are subsequently deformed and twisted over time by huge forces acting on them.

Steno's insights allowed the later mapping of geological strata by the likes of William Smith in Britain and Georges Cuvier and Alexandre Brongniart in France. They also allowed the subdivision of strata into time-related units, which could be correlated with each other across the world. ■

See also: James Hutton 96–101 ▪ William Smith 115

MICROSCOPIC OBSERVATIONS OF ANIMALCULES

ANTONIE VAN LEEUWENHOEK (1632–1723)

IN CONTEXT

BRANCH
Biology

BEFORE
2000 BCE Chinese scientists make a water microscope with a glass lens and a water-filled tube to see very small things.

1267 English philosopher Roger Bacon suggests the idea of the telescope and the microscope.

c.1600 The microscope is invented in the Netherlands.

1665 Robert Hooke observes living cells and publishes *Micrographia*.

AFTER
1841 Swiss anatomist Albert von Kölliker finds that each sperm and egg is a cell with a nucleus.

1951 German physicist Erwin Wilhelm Müller invents the field ion microscope and sees atoms for the first time.

Antonie van Leeuwenhoek rarely ventured far from his home above a cloth store in Delft in the Netherlands. But working on his own in his back room, he discovered an entirely new world—the world of previously unseen microscopic life, including human sperm, blood cells, and, most dramatically of all, bacteria.

Before the 17th century, no one suspected there was life too small to see with the naked eye. Fleas were thought to be the smallest possible form of life. Then, in about 1600, the microscope was invented by Dutch eyeglasses makers who put two glass lenses together to boost their magnification (p.54). In 1665, English scientist Robert Hooke made the first drawing of tiny living cells that he had seen in a slice of cork through a microscope.

It never occurred to Hooke or any other microscopist of the time to look for life anywhere they could not already see it with their own eyes. Van Leeuwenhoek, by contrast, turned his lenses on places where there appeared to be no life at all, particularly in liquids. He studied raindrops, tooth plaque, dung, sperm, blood, and much more. It was here, in these

When van Leeuwenhoek's drawings of human sperm were first published in 1719, many people did not accept that such tiny swimming "animalcules" could exist in semen.

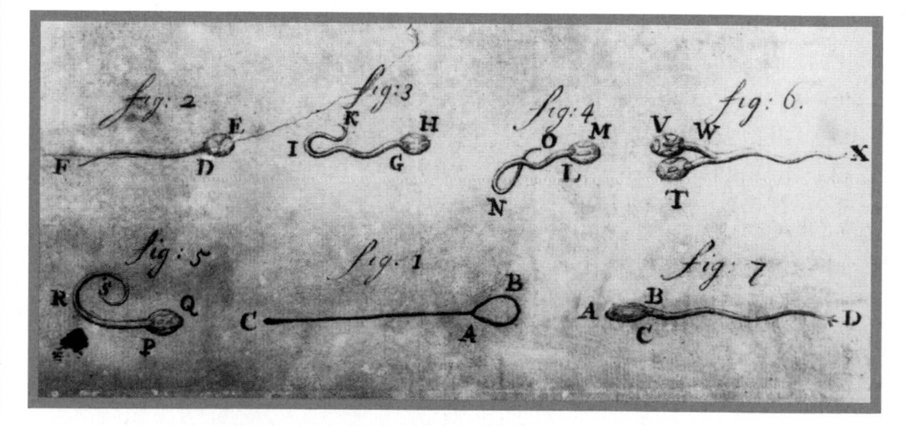

See also: Robert Hooke 54 ▪ Louis Pasteur 156–59 ▪ Martinus Beijerinck 196–97 ▪ Lynn Margulis 300–01

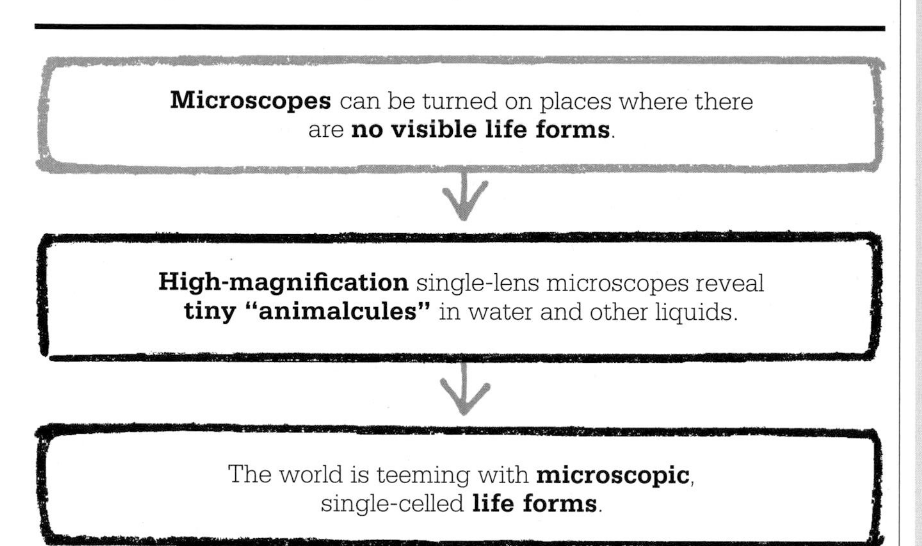

apparently lifeless substances, that van Leeuwenhoek discovered the richness of microscopic life.

Unlike Hooke, van Leeuwenhoek did not use a two-lens "compound" microscope, but a single, high-quality lens—really a magnifying glass. At the time, it was in fact easier to produce a clear picture with such simple microscopics. A magnification greater than 30 times was impossible with compound microscopes since the image became blurred. Van Leeuwenhoek ground his own single lens microscopes, and after years of honing his technique, managed a magnification of more than 200 times. His microscopes were small devices with tiny lenses just fractions of an inch (a few millimeters) wide. The sample was placed on a pin on one side of the lens, and van Leeuwenhoek held one eye up close to the other side.

Single-celled life

At first, van Leeuwenhoek found nothing unusual, but then, in 1674, he reported seeing tiny creatures thinner than a human hair in a sample of lake water. These were the green algae *Spirogyra*, an example of the simple life forms that are now known as protists. Van Leeuwenhoek called these tiny creatures "animalcules." In October 1676, he discovered even smaller single-celled bacteria in drops of water. In the following year, he described how his own semen was swarming with the little creatures we now call sperm. Unlike the creatures he had found in water, the animalcules in semen were all identical. Each of the many thousands he looked at had the same tiny tail and the same tiny head, and nothing else, and he could see them swimming like tadpoles in the semen.

Van Leeuwenhoek reported his findings in a series of hundreds of letters to the Royal Society in London. While he published his findings, he kept his lens-making techniques secret. It is probable that he made his tiny lenses by fusing thin glass threads, but we do not know for sure. ■

Antonie van Leeuwenhoek

The son of a basket maker, Antonie van Leeuwenhoek was born in Delft in 1632. After working in his uncle's linen business, he established his own fabric store at 20 years old and remained there for the rest of his long life.

Van Leeuwenhoek's business allowed him time to pursue his hobby—microscopy. He began in about 1668 after a visit to London, where he may have seen a copy of Robert Hooke's *Micrographia*. From 1673 onward, he reported his findings in letters to the Royal Society in London, writing more reports to them than any scientist in history. The Royal Society was initially sceptical of the amateur's reports, but Hooke repeated many of his experiments and confirmed his discoveries. Van Leeuwenhoek made over 500 microscopes, many designed to view specific objects.

Key works

1673 *Letter 1, van Leeuwenhoek's first letter to the Royal Society*
1676 *Letter 18, revealing his discovery of bacteria*

MEASURING THE SPEED OF LIGHT

OLE RØMER (1644–1710)

IN CONTEXT

BRANCH
Astronomy and physics

BEFORE
1610 Galileo Galilei discovers the four largest moons of Jupiter.

1668 Giovanni Cassini publishes the first accurate tables predicting eclipses of the moons of Jupiter.

AFTER
1729 James Bradley calculates a speed of light of 185,000 miles/s (301,000 km/s) based on variations in the positions of stars.

1809 Jean-Baptiste Delambre uses 150 years' worth of observations of Jupiter's moons to calculate a speed of light of 186,600 miles/s (300,300 km/s).

1849 Hippolyte Fizeau measures the speed of light in a laboratory, rather than using astronomical data.

Eclipses of Jupiter's moons **do not** always **match predictions**.

The **distance** between Earth and Jupiter **changes as the planets orbit the Sun**.

If **light does not propagate instantaneously**, this explains the discrepancies.

The speed of light can be calculated from the time differences and distances in the solar system.

Jupiter has many moons, but only the four largest (Io, Europa, Ganymede, and Callisto) were visible through a telescope at the time that Ole Rømer was observing the skies of northern Europe, in the late 17th century. These moons are eclipsed as they pass through the shadow cast by Jupiter and at certain times they can be observed either entering or leaving the shadow, depending on the relative positions of Earth and Jupiter around the Sun. For nearly half of the year, the eclipses of the moons cannot be observed at all, because the Sun is between Earth and Jupiter.

Giovanni Cassini, the director of the Royal Observatory in Paris when Rømer started work there in the late 1660s, published a set of tables predicting the moons' eclipses. Knowing the times of these eclipses provided a new way to figure out longitude. The measurement of longitude depends on knowing the difference between the time at a given location and the time at a reference meridian (in this case, Paris). On land at least, it was now possible to calculate longitude by observing the time of an eclipse

See also: Galileo Galilei 42–43 ▪ John Michell 88–89 ▪ Léon Foucault 136–37

of one of Jupiter's moons and comparing it to the predicted time of the eclipse in Paris. It was not possible to hold a telescope steadily enough onboard ship to observe the eclipses, and measuring longitude at sea remained impossible until John Harrison built the first marine chronometers—clocks that could keep time at sea—in the 1730s.

Finite or infinite speed?

Rømer studied observations of the eclipses of the moon Io taken over a period of two years and compared these to the times predicted by Cassini's tables. He found a discrepancy of 11 minutes between observations taken when Earth was closest to Jupiter and those taken when it was farthest away. This discrepancy could not be explained by any of the known irregularities in the orbits of Earth, Jupiter, or Io. It had to be the time it took for light to travel the diameter of Earth's orbit. Knowing the diameter of Earth's orbit, Rømer could now calculate the speed of light. He produced a figure of 133,000 miles/s (214,000 km/s). The current value is 186,282 miles/s (299,792 km/s), so Rømer's calculation was off by about 25 percent. Nevertheless, this was an excellent first approximation, and it solved the previously open question as to whether light had a finite speed.

In England, Isaac Newton readily accepted Rømer's hypothesis that light did not travel instantaneously. However, not everyone agreed with Rømer's reasoning. Cassini pointed out that discrepancies in the observations of the other moons were still not accounted for. Rømer's findings were not universally accepted until English astronomer James Bradley produced his more accurate figure for the speed of light in 1729 by measuring the parallax of stars (p.39). ■

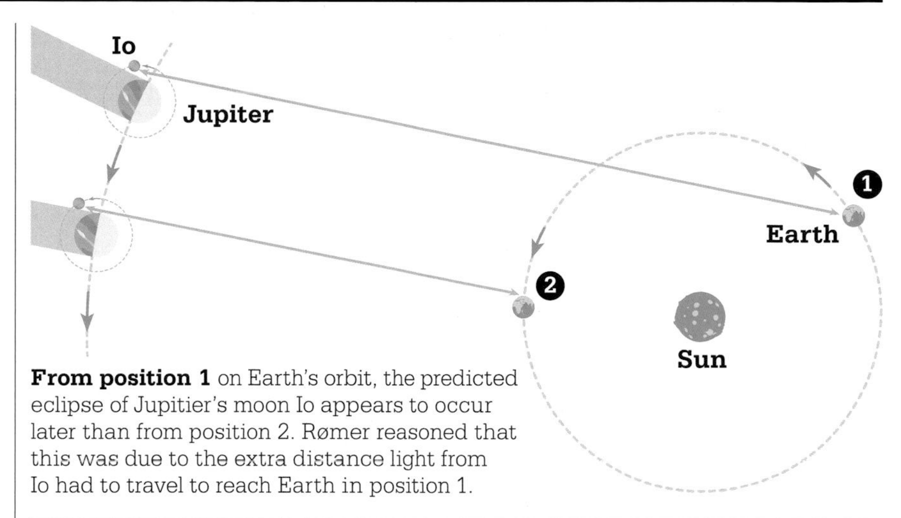

From position 1 on Earth's orbit, the predicted eclipse of Jupitier's moon Io appears to occur later than from position 2. Rømer reasoned that this was due to the extra distance light from Io had to travel to reach Earth in position 1.

For the distance of about 3,000 leagues, which is nearly equal to the diameter of the Earth, light needs not one second of time.
Ole Rømer

Ole Rømer

Born in the Danish city of Aarhus in 1644, Ole Rømer studied at the University of Copenhagen. On leaving the university, he helped to prepare the astronomical observations of Tycho Brahe for publication. Rømer also made his own observations, recording the times of the eclipses of Jupiter's moons from Brahe's old observatory at Uraniborg, near Copenhagen. From there, he moved to Paris, where he worked at the Royal Observatory under Giovanni Cassini. In 1679, he visited England and met Isaac Newton.

Returning to the University of Copenhagen in 1681, Rømer became professor of astronomy. He was involved in modernizing weights and measures, the calendar, and building codes, and even the water supplies. Unfortunately, his astronomical observations were destroyed in a fire in in 1728.

Key work

1677 *On the Motion of Light*

ONE SPECIES NEVER SPRINGS FROM THE SEED OF ANOTHER

JOHN RAY (1627–1705)

IN CONTEXT

BRANCH
Biology

BEFORE
4th century BCE The Greeks use the terms "genus" and "species" to describe groups of similar things.

1583 Italian botanist Andrea Cesalpino classifies plants based on seeds and fruits.

1623 Swiss botanist Caspar Bauhin classifies more than 6,000 plants in his *Illustrated Exposition of Plants*.

AFTER
1690 English philosopher John Locke argues that species are artificial constructs.

1735 Carl Linnaeus publishes *Systema Naturae*, the first of his many works classifying plants and animals.

1859 Charles Darwin proposes the evolution of species by natural selection in *On the Origin of Species*.

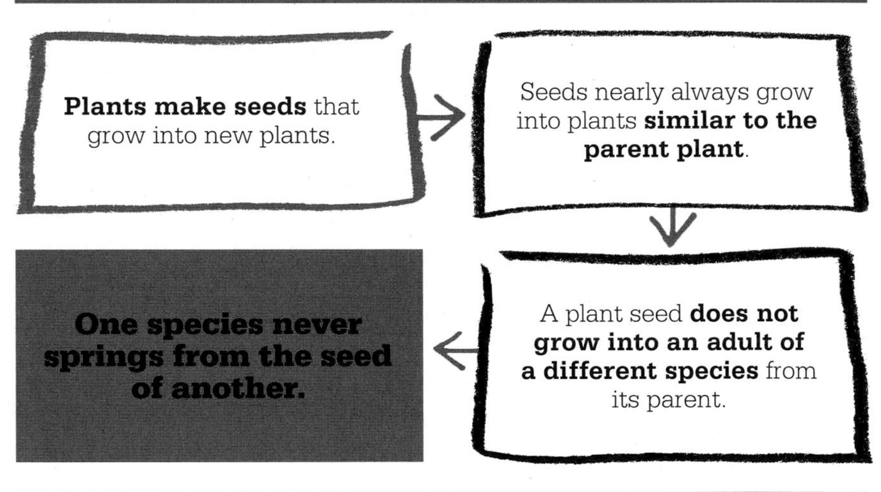

The modern concept of a plant or animal species is based on reproduction. A species includes all individuals that can actually or potentially breed together to produce offspring, which in turn can do the same. This concept, first introduced by English natural historian John Ray in 1686, still underpins taxonomy—the science of classification, in which genetics now plays a major role.

Metaphysical approach

During this period, the term "species" was in common usage, but intricately connected with religion and metaphysics—an approach persisting from ancient Greece. The Greek philosophers Plato, Aristotle, and Theophrastus had discussed classification and used terms such as "genus" and "species" to describe groups and subgroups of all manner of things, living or inanimate. In doing so, they had invoked vague qualities such as "essence" and "soul." So members belonged to a species because they shared the same "essence," rather than sharing the same appearance or the ability to breed with one another.

By the 17th century, myriad classifications existed. Many were organized in alphabetical order, or

See also: Jan Swammerdam 53 ▪ Carl Linnaeus 74–75 ▪ Christian Sprengel 104 ▪ Charles Darwin 142–49 ▪ Michael Syvanen 318–19

Nothing is invented and perfected at the same time.
John Ray

by groups derived from folklore, such as grouping plants according to which illnesses they could treat. In 1666, Ray returned from a three-year European tour with a large collection of plants and animals that he and his colleague Francis Willughby intended to classify along more scientific lines.

Practical nature

Ray introduced a novel practical, observational approach. He examined all parts of the plants, from roots to stem tips and flowers. He encouraged the terms "petal" and "pollen" into general usage and decided that floral type should be an important feature for classification, as should seed type. He also introduced the distinction between monocotyledons (plants with a single seed leaf) and dicotyledons (plants with two seed leaves). However, he recommended a limit to the number of features used for classification, to prevent species numbers multiplying to unworkable proportions. His major work, *Historia Plantarum* (*Treatise on Plants*), published in three volumes in 1686, 1688, and 1704, contains more than 18,000 entries.

For Ray, reproduction was the key to defining a species. His own definition came from his experience gathering specimens, sowing seeds, and observing their germination: "no surer criterion for determining [plant] species has occurred to me than the distinguishing features that perpetuate themselves in propagation from seed…Animals likewise that differ specifically preserve their distinct species permanently; one species never springs from the seed of another nor vice versa." Ray established the basis of a true-breeding group by which a species is still defined today. In so doing, he made botany and zoology scientific pursuits. Devoutly religious, Ray saw his work as a means of displaying the wonders of God. ■

Wheat is a monocotyledon (a plant whose seed contains a single leaf) as defined by Ray. Around 30 species of this major food crop have evolved from 10,000 years of cultivation, and all of them belong to the genus *Triticum*.

John Ray

Born in 1627 in Black Notley, Essex, England, John Ray was the son of the village blacksmith and the local herbalist. At 16, he went to Cambridge University, where he studied widely and lectured on topics from Greek to mathematics, before joining the priesthood in 1660. To recuperate from an illness in 1650, he had taken to nature walks and developed an interest in botany.

Accompanied by his wealthy student and supporter Francis Willughby, Ray toured Britain and Europe in the 1660s, studying and collecting plants and animals. He married Margaret Oakley in 1673 and, after leaving Willughby's household, lived quietly in Black Notley to the age of 77. He spent his later years studying specimens in order to assemble ever-more ambitious plant and animal catalogues. He wrote more than 20 works on plants and animals and their taxonomy, form, and function, and on theology and his travels.

Key work

1686–1704 *Historia Plantarum*

GRAVITY AFFECTS EVERYTHING IN THE UNIVERSE

ISAAC NEWTON (1642–1727)

IN CONTEXT

BRANCH
Physics

BEFORE
1543 Nicolaus Copernicus argues that the planets orbit the Sun, not Earth.

1609 Johannes Kepler argues that the planets move freely in elliptical orbits around the Sun.

1610 Galileo's astronomical observations support Copernicus's views.

AFTER
1846 Johann Galle discovers Neptune after French mathematician Urbain Le Verrier uses Newton's laws to calculate where it should be.

1859 Le Verrier reports that Mercury's orbit is not explained by Newtonian mechanics.

1915 With his general theory of relativity, Albert Einstein explains gravity in terms of the curvature of space-time.

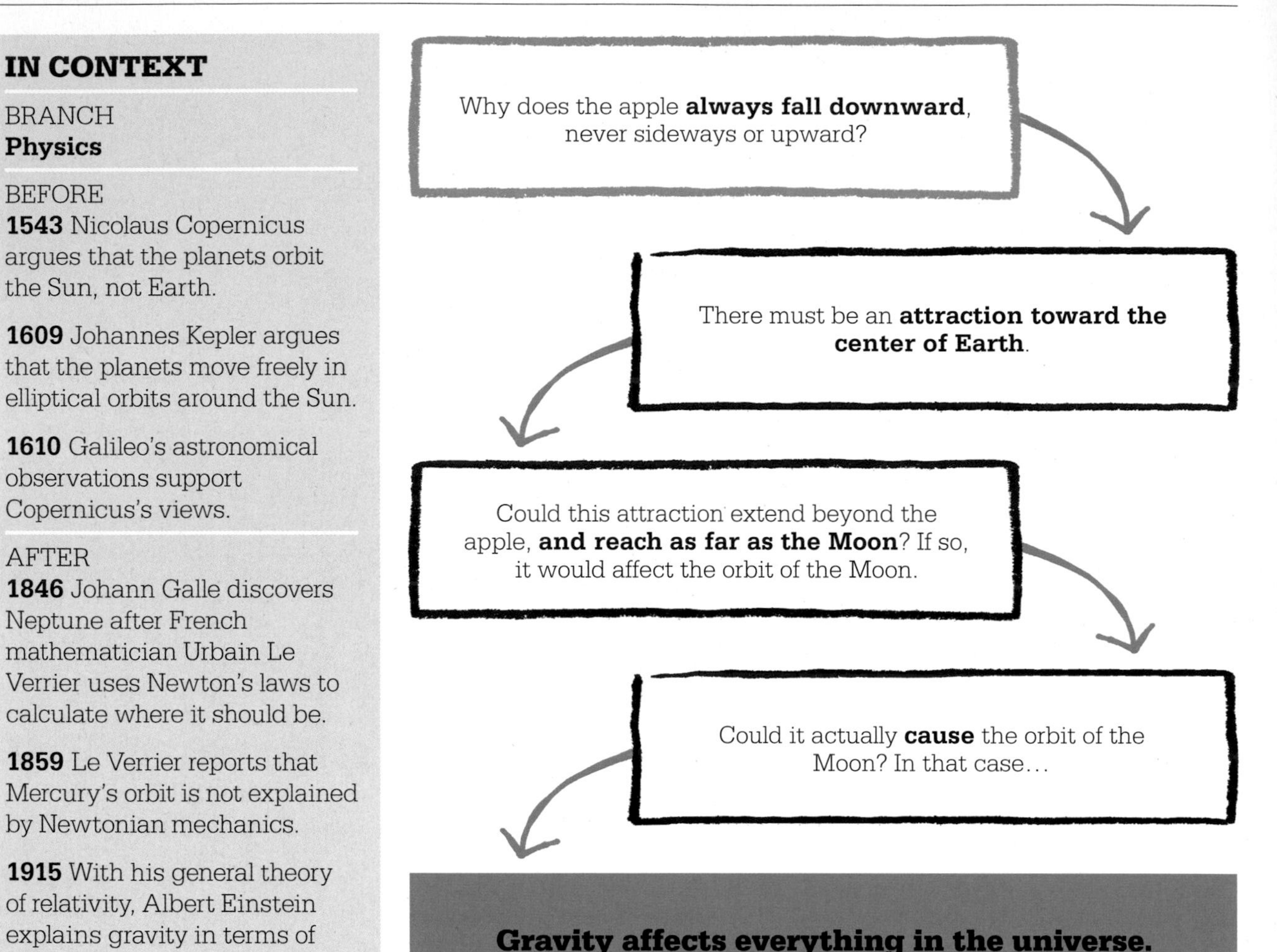

At the time Isaac Newton was born, the heliocentric model of the universe, in which Earth and the other planets orbit the Sun, was the accepted explanation for the observed movements of the Sun, Moon, and planets. This model was not new, but had returned to prominence when Nicolaus Copernicus published his ideas at the end of his life in 1543. In Copernicus's model, the Moon and each of the planets revolved in its own crystalline sphere around the Sun, with an outer sphere holding the "fixed" stars. This model was superseded when Johannes Kepler published his laws of planetary motion in 1609. Kepler dispensed with Copernicus's crystalline spheres, and showed that the orbits of the planets were ellipses, with the Sun at one focus of each ellipse. He also described how the speed of a planet changes as it moves.

What all these models of the universe lacked was an explanation of why the planets moved in the way they did. This is where Newton came in. He realized that the force that pulled an apple toward the center of Earth was the same force that kept the planets in their orbits around the Sun, and demonstrated mathematically how this force changed with distance. The mathematics he used involved Newton's three Laws of Motion and his Law of Universal Gravitation.

Changing ideas

For centuries, scientific thinking had been dominated by the ideas of Aristotle, who reached his conclusions without carrying out experiments to test them. Aristotle

See also: Nicolaus Copernicus 34–39 ▪ Johannes Kepler 40–41 ▪ Galileo Galilei 42–43 ▪ Christiaan Huygens 50–51 ▪ William Herschel 86–87 ▪ Albert Einstein 214–21

taught that moving objects only kept moving as long as they were being pushed, and that heavy objects fell faster than lighter ones. Aristotle explained that heavy objects fell to Earth because they were moving toward their natural place. He also said that celestial bodies, being perfect, must all move in circles at constant speeds.

Galileo Galilei came up with a different set of ideas, arrived at through experiment. He observed balls running down ramps and demonstrated that objects all fall at the same rate if air resistance is minimal. He also concluded that moving objects continue to move unless a force, such as friction, acts to slow them down. Galileo's Principle of Inertia was to become part of Newton's First Law of Motion. Since friction and air resistance act on all moving objects that we encounter in daily life, the concept of friction is not immediately obvious. It was only by careful experimentation that Galileo could show that the force keeping something moving at a steady speed was only needed to counteract friction.

Laws of motion

Newton experimented in many areas of interest, but no records of his experiments on motion survive. His three laws, however, have been verified in many experiments, holding true for speeds well below the speed of light. Newton stated his first law as: "Every body perseveres in its state of rest, or of uniform motion in a right line, unless it is compelled to change that state by forces impressed thereon." In other words, a stationary object will only start to move if a force acts on it, and a moving object continues to move with constant velocity unless a force acts on it. Here, velocity means both the direction of a moving object and its speed. So an object will only change its speed or change direction if a force acts on it. The force that is important is the net force. A moving car has many forces on it, including friction and air resistance, and also the engine driving the wheels. If the forces pushing the car forward balance the forces trying to slow it down, there is no net force and the car will maintain a constant velocity.

Newton's Second Law states that the acceleration (a change of velocity) of a body depends on the size of the force acting on it, and is often written down as $F = ma$, where F is force, m is mass, and a is acceleration. This shows that the greater the force on a body, the greater the acceleration. »

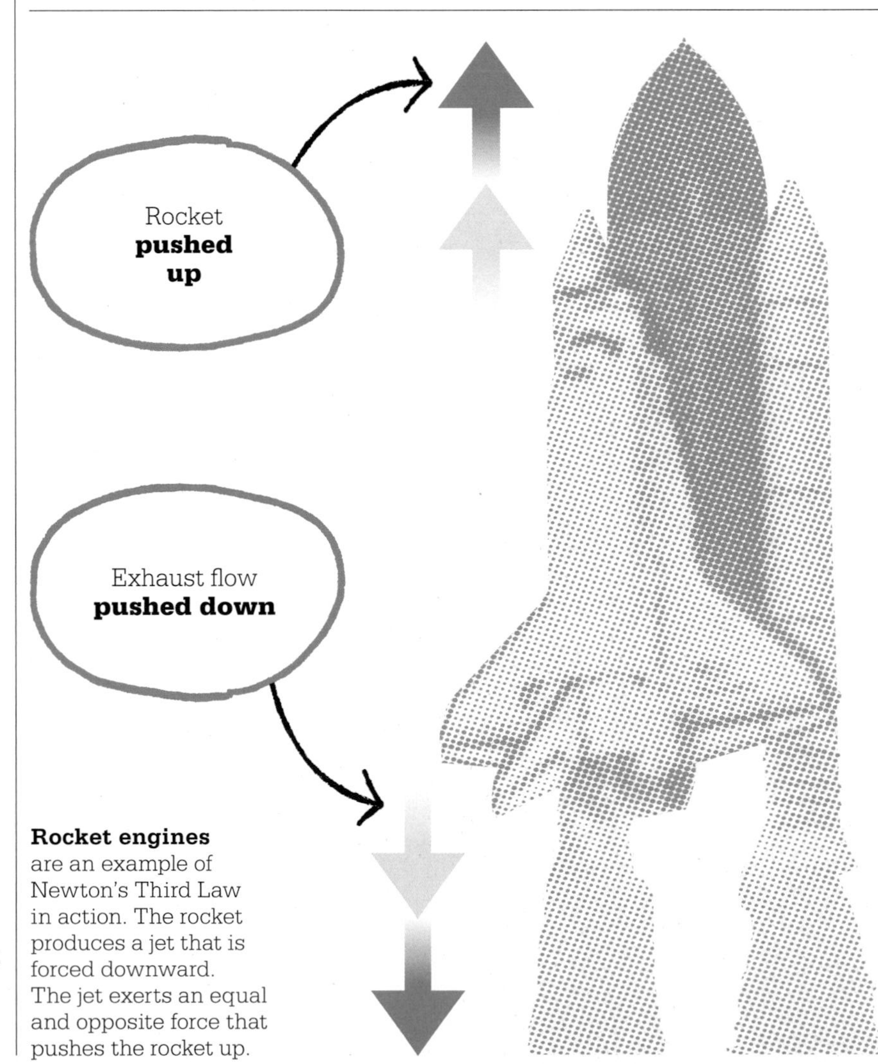

Rocket engines are an example of Newton's Third Law in action. The rocket produces a jet that is forced downward. The jet exerts an equal and opposite force that pushes the rocket up.

It also shows that the acceleration depends on the mass of a body. For a given force, a body with a small mass will accelerate faster than one with a larger mass.

The Third Law is stated as "For every action there is an equal and opposite reaction." It means that all forces exist in pairs: if one object exerts a force on a second object, then the second object simultaneously exerts a force on the first object, and both forces are equal and opposite. In spite of the term "action," movement is not required for this to be true. This is linked to Newton's ideas about gravity, since one example of his Third Law is the gravitational attraction between bodies. Not only is Earth pulling on the Moon, but the Moon is pulling on Earth with the same force.

> I have not been able to discover the cause of these properties of gravity from phenomena, and I frame no hypotheses.
> **Isaac Newton**

Universal attraction

Newton started thinking about gravity in the late 1660s, when he retired to the village of Woolsthorpe for a couple of years to avoid the plague that was ravaging Cambridge. At that time, several people had suggested that there was an attractive force from the Sun, and that the size of this force was inversely proportional to the square of the distance. In other words, if the distance between the Sun and another body is doubled, the force between them is only one quarter of the original force. However, it was not thought that this rule could be applied close to the surface of a large body such as Earth.

Newton, seeing an apple fall from a tree, reasoned that Earth must be attracting the apple and, since the apple always fell perpendicular to the ground, its direction of fall was directed to the center of Earth. So the attractive force between Earth and the apple must act as if it originated in the center of Earth. These ideas opened the way to treating the Sun and planets as small points with large masses, which made calculations much easier by measuring from their centers. Newton saw no reason to think that the force that made an apple fall was any different from the forces that kept the planets in their orbits. Gravity, then, was a universal force.

If Newton's theory of gravity is applied to falling bodies, M_1 is the mass of Earth and M_2 is the mass of the falling object. So the greater the mass of an object, the greater the force pulling it downward. However, Newton's Second Law tells us that a larger mass does not accelerate as quickly as a smaller one if the force is the same. So the greater force is needed to accelerate the greater mass, and all objects fall at the same speed, as long as there are no other forces such as air resistance to complicate matters. With no air resistance, a hammer and a feather will fall at the same speed—a fact finally demonstrated in 1971 by astronaut Dave Scott, who carried out the experiment on the surface of the Moon during the Apollo 15 mission.

Newton described a thought experiment to explain orbits in an early draft of the *Philosophiae Naturalis Principia Mathematica*. He imagined a cannon on a very high mountain, firing cannon balls horizontally at higher and higher speeds. The higher the speed at which a ball is fired, the farther

Newton's Law of Gravity produces the equation below, which shows how the force produced depends on the mass of the two objects and the square of the distance between them.

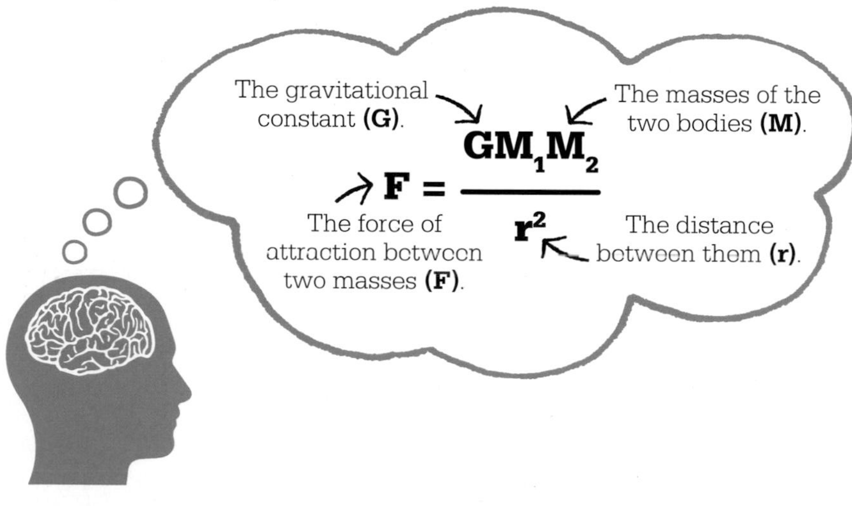

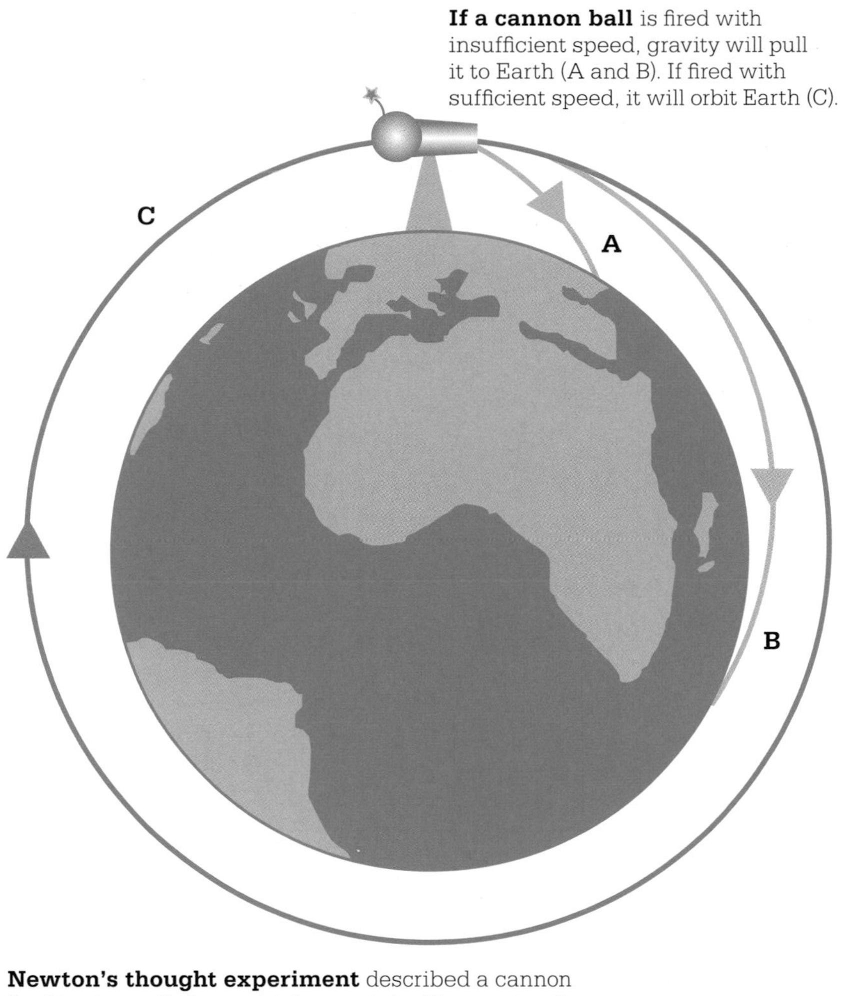

If a cannon ball is fired with insufficient speed, gravity will pull it to Earth (A and B). If fired with sufficient speed, it will orbit Earth (C).

Newton's thought experiment described a cannon fired horizontally from a high mountain. The greater the force firing the cannon ball, the farther it travels before falling to the ground. If it is fired hard enough, it will travel right around the planet back to the mountain.

which a ball is fired, the farther away it will land. If it is launched sufficiently fast, it will not land at all, but continue around Earth until it arrives back at the top of the mountain. In the same way, a satellite launched into orbit at the correct speed will continue to circle Earth. The satellite is continually being accelerated by Earth's gravity. It moves at a constant speed, but its direction is continually changing, making it circle the planet rather than whizzing off into space in a straight line. In this case, Earth's gravity only changes the direction of the satellite's velocity, not its speed.

Publishing the ideas

In 1684, Robert Hooke boasted to his friends Edmond Halley and Christopher Wren that he had discovered the laws of planetary motion. Halley was a friend of Newton, and asked him about this. Newton said that he had already solved the problem, but had lost his notes. Halley encouraged Newton to redo the work, and as a result, Newton produced *On the Motion of Bodies in an Orbit*, a short manuscript sent to the Royal Society in 1684. In this paper, Newton showed that the elliptical motion of the planets that Kepler described would result from a force pulling everything toward the Sun, where that force was inversely proportional to the distance between the bodies. Newton expanded on this work, and included other work on forces and motion, in the *Principia Mathematica,* which was published in three volumes and contained, among other things, the Law of Universal Gravitation and Newton's three Laws of Motion. The volumes were written in Latin, and it was not until 1729 that the first English translation was published, based on Newton's third edition of the *Principia Mathematica*.

> Kepler's laws ...have led to the discovery of the law of attraction of the bodies of the solar system.
> **Isaac Newton**

Hooke and Newton had already fallen out over Hooke's criticisms of Newton's theory of light. Following Newton's publication, however, much of Hooke's work on planetary motion was obscured. However, Hooke had not been the only one to suggest such a law, and he had not demonstrated that it »

Newton's laws provided the tools to calculate the orbits of heavenly bodies such as Halley's comet, shown here on the Bayeux Tapestry after its appearance in 1066.

worked. Newton had shown that his Law of Universal Gravitation and laws of motion could be used mathematically to describe the orbits of planets and comets, and that these descriptions matched observations.

Sceptical reception

Newton's ideas on gravity were not welcomed everywhere. The "action at a distance" of Newton's force of gravity, with no way of explaining how or why it happened, was seen as an "occult" idea. Newton himself refused to speculate on the nature of gravity. For him, it was enough that he had shown that the idea of an inverse-square attraction could explain planetary motions, so the mathematics was correct. However, Newton's laws described so many phenomena that they soon came to be widely accepted, and today the internationally used unit of force is named after him.

> Why should that apple always descend perpendicularly to the ground, thought he to himself...
> **William Stukeley**

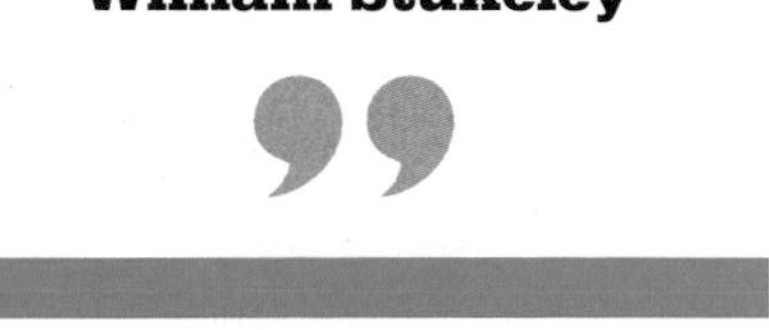

Using the equations

Edmond Halley used Newton's equations to calculate the orbit of a comet seen in 1682, and showed that it was the same comet as that observed in 1531 and 1607. The comet is now called Halley's comet. Halley successfully predicted that it would return in 1758, which was 16 years after his death. This was the first time that comets had been shown to orbit the Sun. Halley's comet passes close to Earth every 75–76 years, and was the same comet as that seen in 1066 before the Battle of Hastings in southern England.

The equations were also used successfully to discover a new planet. Uranus is the seventh planet from the Sun, and was identified as a planet by William Herschel in 1781. Herschel found the planet by chance while making careful observations of the night sky. Further observations of Uranus allowed astronomers to calculate its orbit and to produce tables predicting where it could be observed at future dates. These predictions were not always correct, however, leading to the idea that there must be another planet beyond Uranus whose gravity was affecting the orbit of Uranus. By 1845, astronomers had calculated where this eighth planet should be in the sky, and Neptune was discovered in 1846.

Problems with the theory

For a planet with an elliptical orbit, the point of closest approach to the Sun is called the perihelion. If there were only one planet orbiting the

Sun, the perihelion of its orbit would stay in the same place. However all the planets in our solar system affect each other, so the perihelia precess (rotate) around the Sun. Like all the other planets, Mercury's perihelion precesses, but the precession cannot be completely accounted for using Newton's equations. This was recognized as a problem in 1859. More than 50 years later, Einstein's Theory of General Relativity described gravity as an effect of the curvature of space-time, and calculations based on this theory do account for the observed precession of Mercury's orbit, as well as other observations not linked to Newton's laws.

Newton's laws today

Newton's laws form the basis of what is referred to as "classical mechanics"—a set of equations used to calculate the effects of forces and motion. Although these laws have been superseded by equations based on Einstein's theories of relativity, the two sets of laws agree as long as any motion involved is small compared to the speed of light. So for the calculations involved in designing airplanes or cars, or figuring out how strong the components of a skyscraper need to be, the equations of classical mechanics are both accurate enough and much simpler to use. Newtonian mechanics, while it may not strictly be correct, is still widely used. ■

Nature and nature's laws lay hid in night; God said "Let Newton be" and all was light.
Alexander Pope

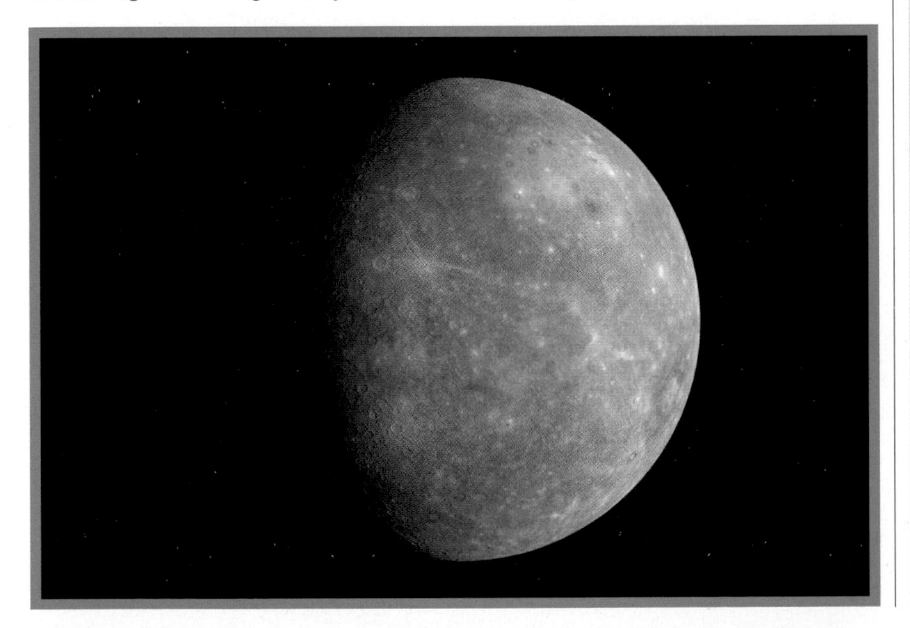

The precession (change in the rotational axis) of the orbit of Mercury was the first phenomenon that could not be explained by Newton's laws.

Isaac Newton

Born on Christmas Day in 1642, Isaac Newton attended school in Grantham, before studying at Trinity College, Cambridge, where he graduated in 1665. During his life, Newton was variously Professor of Mathematics at Cambridge, Master of the Royal Mint, Member of Parliament for Cambridge University, and President of the Royal Society. Besides his dispute with Hooke, Newton became involved in a feud with German mathematician Gottfried Leibnitz over priority in the development of calculus.

In addition to his scientific work, Newton spent much time in alchemical investigations and Biblical interpretation. A devout but unorthodox Christian, he successfully managed to avoid being ordained as a priest, which was normally a requirement for some of the offices he held.

Key works

1684 *On the Motion of Bodies in an Orbit*
1687 *Philosophiae Naturalis Principia Mathematica*
1704 *Opticks*

EXPAND
HORIZO
1700–1800

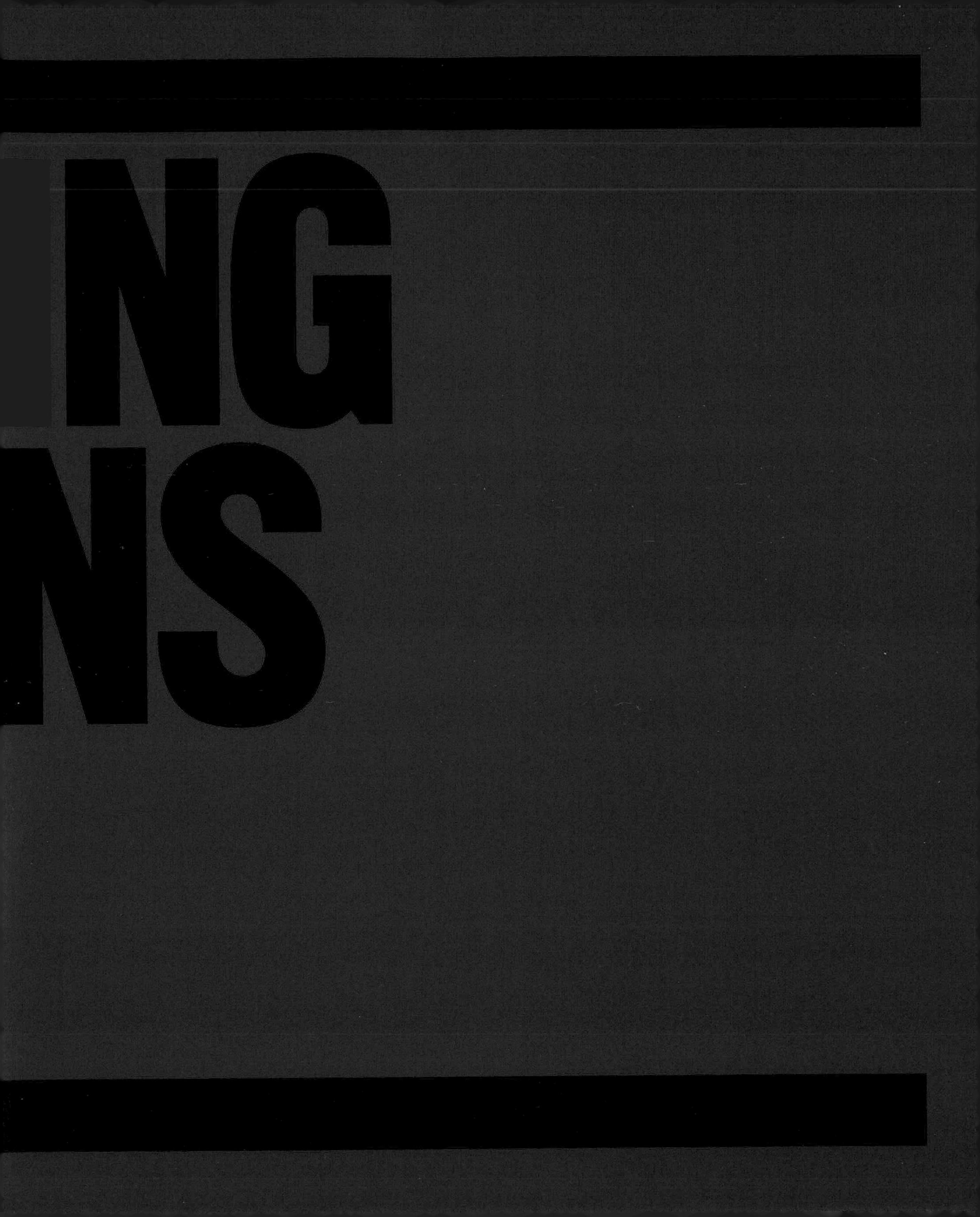

NG
NS

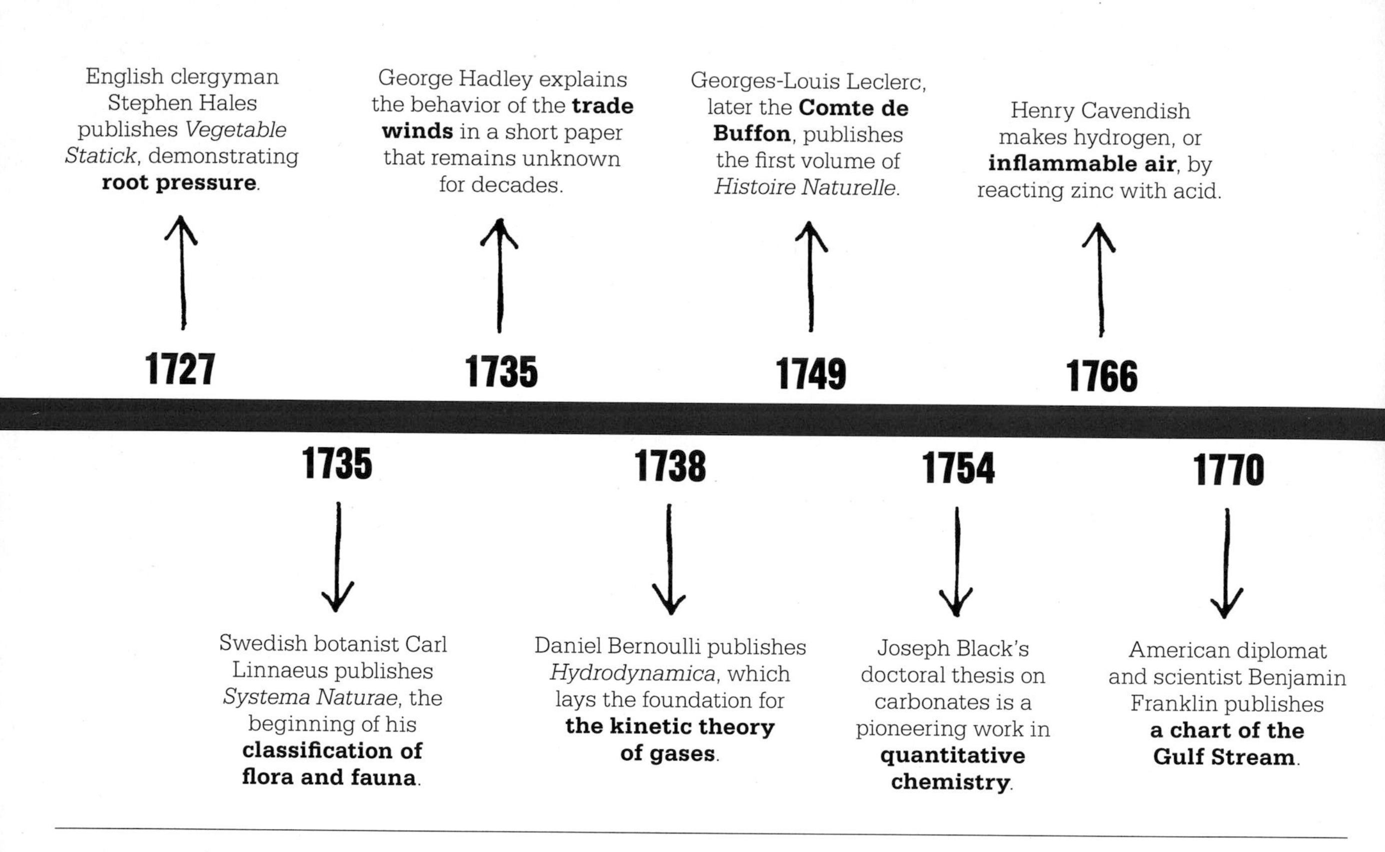

At the end of the 17th century, Isaac Newton set down his laws of motion and gravity, making science more precise and mathematical than it had ever been before. Scientists in various fields identified the underlying principles governing the universe, and the various branches of scientific enquiry became increasingly specialized.

Fluid dynamics

In the 1720s, Stephen Hales, an English curate, performed a series of experiments with plants, discovering root pressure—by which sap rises through plants—and inventing the pneumatic trough, a laboratory apparatus for collecting gases, which was to prove useful for later work identifying the components of air. Daniel Bernoulli, the brightest in a family of Swiss mathematicians, formulated the Bernoulli principle—that the pressure of a fluid falls when it is moving. This allowed him to measure blood pressure. It is also the principle that allows aircraft to fly.

In 1754, Scottish chemist Joseph Black, who would later formulate the theory of latent heat, produced a remarkable doctoral thesis about the decomposition of calcium carbonate and the generation of "fixed air," or carbon dioxide. This sparked a chain reaction of chemical research and discovery. In England, reclusive genius Henry Cavendish isolated hydrogen gas and demonstrated that water is made of two parts of hydrogen to one of oxygen. Dissident minister Joseph Priestley isolated oxygen and several other new gases. Dutchman Jan Ingenhousz picked up where Priestley left off and showed how green plants give off oxygen in sunlight and carbon dioxide in the dark. Meanwhile, in France, Antoine Lavoisier showed that many elements, including carbon, sulfur, and phosphorus, burn by combining with oxygen to form what we now call oxides, thus debunking the theory that combustible materials contain a substance called phlogiston that make them burn. (Unfortunately, French revolutionaries would send Lavoisier to the guillotine.)

In 1793, French chemist Joseph Proust discovered that chemical elements nearly always combine in definite proportions. This was a vital step toward figuring out the formulae of simple compounds.

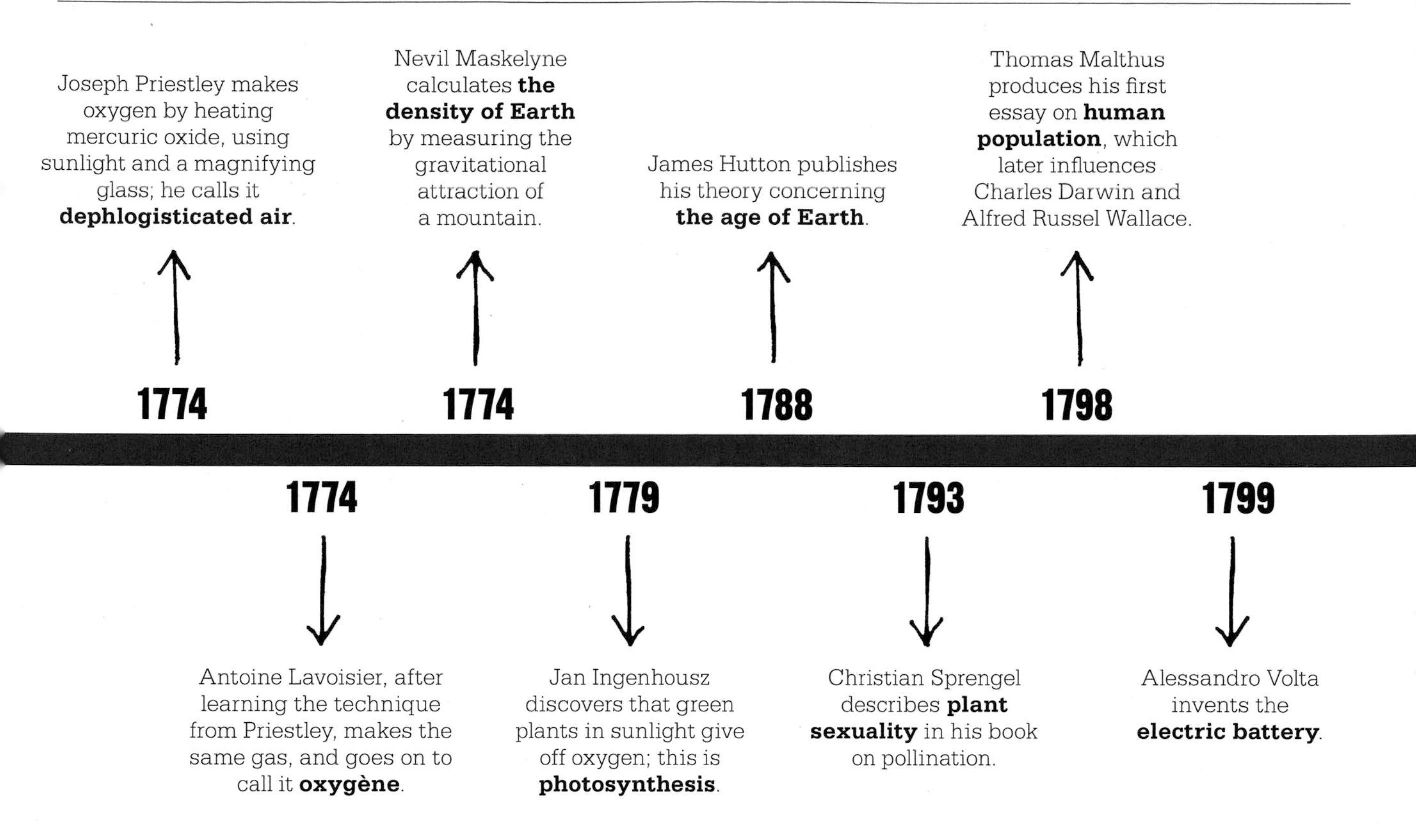

Earth sciences

At the other end of the scale, understanding of Earth processes was making great advances. In the Americas, Benjamin Franklin, in addition to performing a dangerous experiment to prove that lightning is a form of electricity, demonstrated the existence of large-scale ocean currents with his investigations of the Gulf Stream. George Hadley, English lawyer and amateur meteorologist, published a short paper explaining the action of the trade winds in relation to the rotation of Earth, while Nevil Maskelyne seized on an idea from Newton and camped out for several months in terrible weather to measure the gravitational attraction of a Scottish mountain. In doing so, he figured out the density of Earth. James Hutton became interested in geology after inheriting farmland in Scotland, and realized that Earth was a great deal older than anyone had previously thought.

Understanding life

As scientists learned of Earth's extreme age, new ideas about how life originated and evolved began to emerge. Georges-Louis Leclerc, Comte de Buffon, a larger-than-life French author, naturalist, and mathematician, took the first steps toward a theory of evolution. German theologian Christian Sprengel spent much of his life studying the interaction of plants and insects, and noted that bisexual flowers produce male and female flowers at different times, so they cannot fertilize themselves. English parson Thomas Malthus turned his attention to demography and wrote *An Essay on the Principle of Population*, predicting catastrophe as the population grows. Malthus's pessimism has proved unfounded (so far), but his idea that a population will grow to outstrip resources if left unchecked was later to have a profound influence on Charles Darwin.

At the end of the century, Italian physicist Alessandro Volta opened up a new world by inventing the electric battery, which was to accelerate advances in the decades that followed. Such had been the progress through the 18th century that English philosopher William Whewell proposed the creation of a new profession distinct from that of philosopher: "We need very much a name to describe a cultivator of science in general. I should incline to call him a Scientist." ■

NATURE DOES NOT PROCEED BY LEAPS AND BOUNDS

CARL LINNAEUS (1707–1778)

IN CONTEXT

BRANCH
Biology

BEFORE
c.320 BCE Aristotle groups similar organisms on a scale of increasing complexity.

1686 John Ray defines a biological species in his *Historia Plantarum*.

AFTER
1817 French zoologist Georges Cuvier extends the Linnaean hierarchy in his study of fossils as well as living animals.

1859 Charles Darwin's *On the Origin of Species* sets out how species arise and are related in his theory of evolution.

1866 German biologist Ernst Haeckel pioneers the study of evolving lineages, known as phylogenetics.

1950 Willi Hennig bases a new system of classification on cladistics, which looks for evolutionary links.

The classification of the natural world into a clear hierarchy of groups of named and described organisms is a foundation stone of the biological sciences. These groupings help to make sense of life's diversity, allowing scientists to compare and identify millions of individual organisms. Modern taxonomy—the science of identifying, naming, and classifying organisms—began with the Swedish naturalist, Carl Linnaeus. He was the first to devise a systematic hierarchy, based on his wide-ranging and detailed study of physical characteristics of plants and animals. He also pioneered a way of naming different organisms that is still in use today.

The most influential of early classifications was that of the Greek philosopher Aristotle. In his *History of Animals*, he grouped similar animals into broad genera, distinguished the species within each group, and ranked them on a *scala naturae* or "ladder of life" with 11 grades of increasing complexity in form and purpose, from plants at the base to humans at the apex.

Over the ensuing centuries, a chaotic multiplicity of names and descriptions of plants and animals appeared. By the 17th century, scientists were striving to set out a more coherent and consistent system. In 1686, English botanist John Ray introduced the concept of the biological species, defined by the ability of plants or animals to reproduce with one another, and this remains the most widely accepted definition today.

Linnaeus's system groups organisms according to shared characteristics. A tiger belongs to the cat family Felidae, which in turn belongs to the order Carnivora, in the class Mammalia.

See also: Jan Swammerdam 53 ▪ John Ray 60–61 ▪ Jean-Baptiste Lamarck 118 ▪ Charles Darwin 142–49

In 1735, Linnaeus produced a classification in a 12-page booklet that grew into a multivolume 12th edition by 1778 and developed the idea of the genus into a hierarchy of groupings based on shared physical characteristics. At the top were three kingdoms: animals, plants, and minerals. Kingdoms were divided into phyla, then classes, orders, families, genera, and species. He also stabilized the naming of species by using a two-part Latin name, with one name for the genus and another for a species within that genus, as in *Homo sapiens*—Linnaeus was the first to define humans as animals.

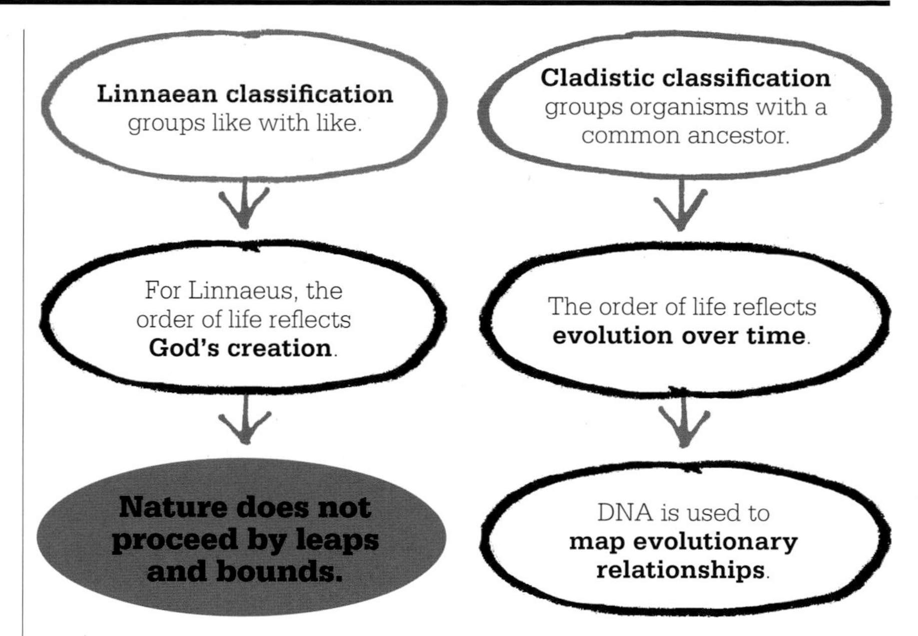

God-given order

For Linnaeus, classification revealed that "nature does not proceed in leaps and bounds" but rather in its God-given order. His work was the fruit of numerous expeditions across Sweden and Europe in search of new species. His classification system paved the way for Charles Darwin, who saw the evolutionary significance of its "natural hierarchy," with all species in a genus or family related by descent and divergence from a common ancestor. A century after Darwin, German biologist Willi Hennig developed a new approach to classification, called cladistics. To reflect their evolutionary links, this groups organisms into "clades" with one or more shared unique characteristics, which they have inherited from their last common ancestor and which are not found in more distant ancestors. The process of classification by clades continues to this day, with species reassigned new positions as fresh, often genetic, evidence is found. ■

Carl Linnaeus

Born in 1707 in rural southern Sweden, Carl Linnaeus studied medicine and botany in the universities of Lund and Uppsala, and earned a degree in medicine in the Netherlands in 1735. Later that year he published a 12-page booklet called *Systema Naturae*, which outlined a system of classification for living organisms. After further travels in Europe, Linnaeus returned to Sweden in 1738 to practice medicine before being appointed professor of medicine and botany at Uppsala University. His students, most famously Daniel Solander, traveled the world collecting plants. With this vast collection, Linnaeus expanded his *Systema Naturae* through 12 editions into a multivolume work, more than 1,000 pages long, encompassing more than 6,000 species of plants and 4,000 animals. By the time he died in 1778, Linnaeus was one of the most acclaimed scientists in Europe.

Key works

1753 *Species Plantarum*
1778 *Systema Naturae, 12th edition*

THE HEAT THAT DISAPPEARS IN THE CONVERSION OF WATER INTO VAPOR IS NOT LOST

JOSEPH BLACK (1728–1799)

IN CONTEXT

BRANCH
Chemistry and physics

BEFORE
1661 Robert Boyle pioneers the isolation of gases.

1750s Joseph Black weighs materials before and after chemical reactions—the first quantitative chemistry—and discovers carbon dioxide.

AFTER
1766 Henry Cavendish isolates hydrogen.

1774 Joseph Priestley isolates oxygen and other gases.

1798 American-born British physicist Benjamin Thompson suggests that heat is produced by the movement of particles.

1845 James Joule studies the conversion of motion into heat and measures the mechanical equivalent of heat, stating that a given quantity of mechanical work generates the same amount of heat.

Heat generally **raises the temperature of water**.

↓

But when water boils, **the temperature stops rising**.

↓

Additional heat is needed to **turn the liquid into vapor**. This latent heat gives steam a **terrible scalding power**.

↓

The heat that disappears in the conversion of water into vapor is not lost.

A professor of medicine at the University of Glasgow and later at Edinburgh, Joseph Black also gave lectures on chemistry. Although he was a notable research scientist, he rarely published his results formally, but instead announced them during his lectures; his students were at the cutting edge of new science.

Some of Black's students were the sons of Scottish whisky distillers, who were concerned about the costs of running their businesses. Why, they asked him, was it so expensive to distill whisky, when all they were doing was boiling the liquid and condensing the vapor.

An idea brought to the boil

In 1761, Black investigated the effects of heat on liquids, and discovered that if a pan of water is heated on a stove, the temperature increases steadily until it reaches

See also: Robert Boyle 46–49 ▪ Joseph Priestley 82–83 ▪ Antoine Lavoisier 84 ▪ John Dalton 112–13 ▪ James Joule 138

212°F (100°C). Then the water begins to boil, but the temperature does not change, even though heat is still going into the water. Black realized that the heat is needed to turn the liquid into vapor—or, in modern terms, to give the molecules enough energy to escape from the bonds that hold them fast in the liquid. This heat does not change the temperature, and seems to disappear—so Black called it latent heat (from the Latin for "hidden"). More precisely, it is the latent heat of evaporation of water. This discovery was the beginning of the science of thermodynamics—the study of heat, its relation to energy, and the conversion of heat energy into motion to do mechanical work.

Water has an unusually high latent heat, meaning that liquid water will boil for a long time before it all turns into gas. This is why steaming is such an effective way of cooking vegetables, why steam has terrible scalding power, and why it is used in heating systems.

Melting ice

Just as heat is needed to turn water into steam, so it is needed to turn ice into water. The latent heat of melting ice means that ice will cool a drink. To melt the ice requires heat, and this heat is extracted from the drink in which it floats, thus cooling down the liquid.

Black explained all this to the distillers, although he was unable to help them save money. He also explained it to a colleague called James Watt, who was trying to figure out why steam engines were so inefficient. Subsequently, Watt came up with the idea of the separate condenser, which condensed the steam without cooling the piston and cylinder. This made the steam engine a far more efficient machine, and made Watt a rich man. ■

Black is shown here visiting the engineer James Watt at his workshop in Glasgow. Watt is demonstrating one of his steam-powered instruments.

Joseph Black

Born in Bordeaux, France, Joseph Black studied medicine at the universities of Glasgow and Edinburgh, conducting chemical experiments in the laboratory of his professor. In his 1754 doctoral thesis, Black showed that when chalk (calcium carbonate) is heated to become quicklime (calcium oxide), it does not gain some fiery principle from the fire, as was commonly believed, but loses weight. Black realized that this loss must be a gas, since no liquid or solid was produced, and called it "fixed air" because it was an air (gas) that had been fixed in the chalk. He also showed that fixed air (which we now know as carbon dioxide) was among the gases that we exhale.

While professor of medicine at Glasgow from 1756, Black conducted his landmark research on heat. Although he did not publish his results, his students circulated his findings. After moving to Edinburgh in 1766, he gave up research to focus on lecturing and—as the Industrial Revolution gathered speed—advising on chemical-based innovations in Scottish industry and agriculture.

INFLAMMABLE AIR

HENRY CAVENDISH (1731–1810)

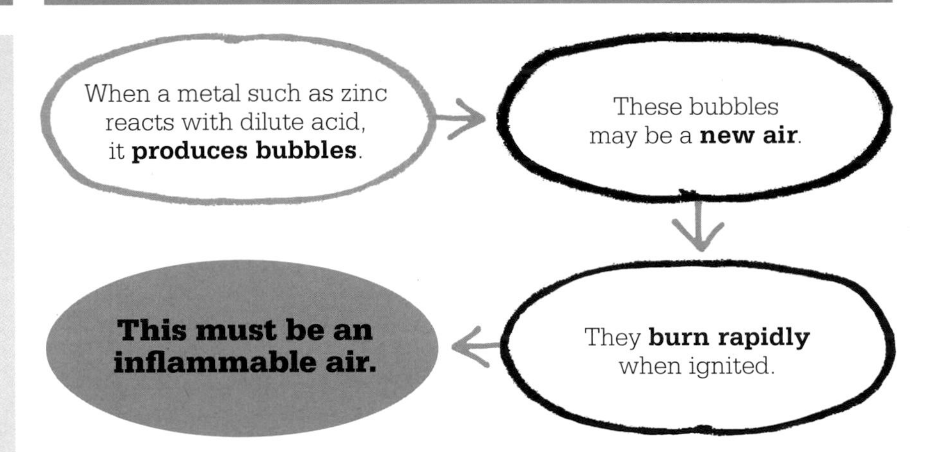

IN CONTEXT

BRANCH
Chemistry

BEFORE
1661 Robert Boyle defines an element, laying the foundations for modern chemistry.

1754 Joseph Black identifies a gas, carbon dioxide, which he calls "fixed air."

AFTER
1772–75 Joseph Priestley and (independently) Sweden's Carl Scheele isolate oxygen, followed by Antoine Lavoisier, who names the gas. Priestley also discovers nitric oxide, nitrous oxide, and hydrogen chloride, and experiments with inhaling oxygen and making soda water.

1799 Humphry Davy suggests nitrous oxide could be useful as an anesthetic in surgery.

1844 Nitrous oxide is first used for anesthesia by American dentist Horace Wells.

In 1754, Joseph Black had described what we now call carbon dioxide (CO_2) as "fixed air." He was not only the first scientist to identify a gas, but also demonstrated that there were various kinds of "air," or gases.

Twelve years later, an English scientist named Henry Cavendish reported to the Royal Society in London that the metals zinc, iron, and tin "generate inflammable air by solution in acids." He called his new gas "inflammable air" because it burned easily, unlike ordinary or "fixed air." Today we call it hydrogen (H_2). This was the second gas to be identified and the first gaseous element to be isolated.

Cavendish set out to measure the weight of a sample of the gas, by measuring the loss of weight of the zinc-acid mixture during the reaction, and by collecting all the gas produced in a bladder and weighing it—first full of the gas, then empty. Knowing the volume, he could calculate its density. He found that inflammable air was 11 times less dense than ordinary air.

The discovery of low-density gas led to aeronautical balloons that were lighter than air. In France in 1783, inventor Jacques Charles launched the first hydrogen balloon, less than two weeks after the Montgolfier brothers launched their first manned hot-air balloon.

See also: Empedocles 21 ▪ Robert Boyle 46–49 ▪ Joseph Black 76–77 ▪ Joseph Priestley 82–83 ▪ Antoine Lavoisier 84 ▪ Humphry Davy 114

It appears from these experiments, that this air, like other inflammable substances, cannot burn without the assistance of common air.
Henry Cavendish

Explosive discoveries

Cavendish also mixed measured samples of his gas with known volumes of air in bottles, and ignited the mixtures by taking the tops off and applying lighted pieces of paper. He found that with nine parts of air to one of hydrogen there was a slow, quiet flame; with increasing proportions of hydrogen the mixture exploded with increasing ferocity; but pure, 100 percent hydrogen did not ignite. Cavendish's thinking was still handicapped by an obsolete notion from alchemy that a firelike element ("phlogiston") was released during combustion. However, he was precise in his experiments and in his reporting: "it appears that 423 measures of inflammable air are nearly sufficient to phlogisticate 1,000 of common air; and that the bulk of the air remaining after the explosion is then very little more than four-fifths of the common air employed. We may conclude that...almost all the inflammable air and about one fifth of the common air...are condensed into the dew which lines the glass."

Defining water

Although Cavendish used the term "phlogisticate," he managed to demonstrate that the only new material produced was water, and deduced that two volumes of inflammable air had combined with one volume of oxygen. In other words, he showed that the composition of water is H_2O. Although he reported his findings to Joseph Priestley, Cavendish was so diffident about publishing the results that his friend the Scottish engineer James Watt was the first to announce the formula, in 1783.

Among his many contributions to science, Cavendish went on to calculate the composition of air as "one part dephlogisticated air [oxygen], mixed with four of phlogisticated [nitrogen]"—the two gases we now know make up 99 percent of Earth's atmosphere. ■

The first hydrogen balloon, inspired by Cavendish, was cheered by a huge crowd of spectators. Since hydrogen is so explosive, modern balloons use helium.

Henry Cavendish

One of the strangest and most brilliant pioneers of 18th century chemistry and physics, Henry Cavendish was born in 1731 in Nice, France. His grandfathers were both dukes, and he was immensely rich. After his studies at the University of Cambridge, he lived and worked alone in his house in London. A man of few words and shy of women, it was said that he ordered his meals by leaving notes for his servants.

Cavendish attended meetings of the Royal Society for about 40 years, and also assisted Humphry Davy at the Royal Institution. He did significant original research into chemistry and electricity, accurately described the nature of heat, and measured Earth's density—or, as people said, "weighed the world." He died in 1810. In 1874, the University of Cambridge named its new physics laboratory in his honor.

Key works

1766 *Three Papers Containing Experiments on Factitious Air*
1784 *Experiments on Air* (*Philosophical Transactions of the Royal Society of London*)

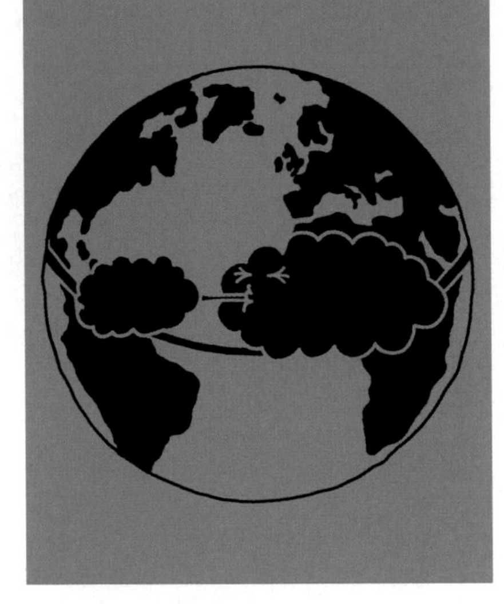

WINDS, AS THEY COME NEARER THE EQUATOR, BECOME MORE EASTERLY

GEORGE HADLEY (1685–1768)

IN CONTEXT

BRANCH
Meteorology

BEFORE
1616 Galileo Galilei points to trade winds as evidence of Earth's rotation.

1686 Edmond Halley proposes that the Sun traveling west through the sky causes air to rise and be replaced by winds from the east.

AFTER
1793 John Dalton publishes *Meteorological Observations and Essays*, which supports Hadley's theory.

1835 De Coriolis builds on Hadley's ideas, describing a "compound centrifugal force" that deflects the wind.

1856 American meteorologist William Ferrel identifies a circulation cell in the mid-latitudes (30–60°) where air pulled into a low-pressure center creates the prevailing westerly winds.

By 1700, it was known that persistent surface winds, or "trade winds," blow from a northeasterly direction between a latitude of 30°N and the equator at 0°. Galileo had suggested that Earth's eastward rotation made it "get ahead" of the air in the tropics, so the winds come from the east. Later, English astronomer Edmond Halley realized that the Sun's heat, at its greatest over the equator, causes air to rise, and that rising air is replaced by winds blowing in from higher latitudes.

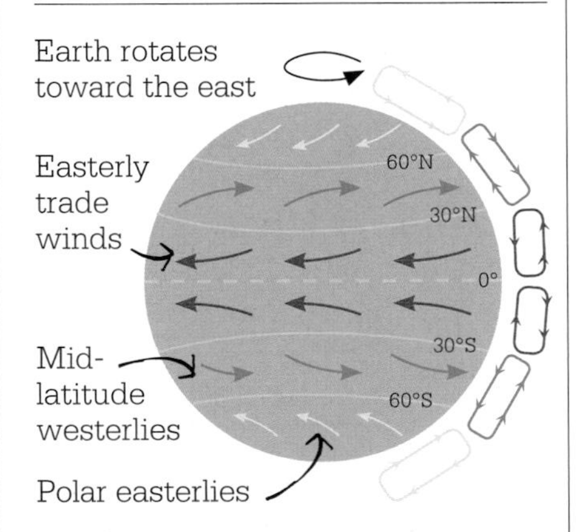

Wind patterns result from Earth's rotation combined with circulation "cells" as hot air rises, cools, and falls in polar cells (shown in gray), Ferrel cells (blue), and Hadley cells (pink).

In 1735, English physicist George Hadley published his theory on trade winds. He agreed that the Sun causes air to rise, but rising air near the equator would only cause winds to flow toward it from the north and south, not from the east. As the air rotates with Earth, air moving from 30° N toward the equator would have its own momentum toward the east. However, Earth's surface moves faster at the equator than at higher latitudes, so the surface speed becomes greater than the air's speed and the winds appear to come from an increasingly easterly direction as they near the equator.

Hadley's idea was a step on the way to understanding wind patterns, but contained errors. The key to the deflection of wind direction is in fact that the wind's angular momentum (causing it to rotate) is conserved, not its linear (straight-line) momentum. ■

See also: Galileo Galilei 42–43 ▪ John Dalton 112–13 ▪ Gaspard-Gustave de Coriolis 126 ▪ Robert FitzRoy 150–55

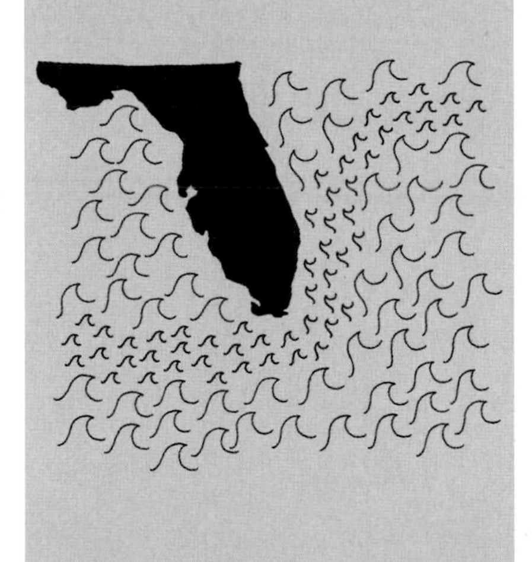

A STRONG CURRENT COMES OUT OF THE GULF OF FLORIDA

BENJAMIN FRANKLIN (1706–1790)

IN CONTEXT

BRANCH
Oceanography

BEFORE
c.2000 BCE Polynesian seafarers use ocean currents to cross between Pacific islands.

1513 Juan Ponce de Léon is the first to describe the strong currents of the Atlantic Ocean's Gulf Stream.

AFTER
1847 US naval officer Matthew Maury publishes his chart of winds and currents, compiled by studying ships' logs and charts in naval archives.

1881 Prince Albert I of Monaco realizes that the Gulf Stream is a gyre (loop) and splits in two—one branch flowing north toward the British Isles, and the other south to Spain and Africa.

1942 Norwegian oceanographer Harald Sverdrup develops a theory of general ocean circulation.

The warm Gulf Stream current that flows eastward across the North Atlantic Ocean is one of the greatest movements of water on Earth. It is driven east by prevailing westerly winds, and is part of a great loop that then recrosses the Atlantic to the Caribbean. The current had been known since 1513, when Spanish explorer Juan Ponce de León found his ship moving back north off Florida despite winds blowing him south. But it was only properly charted in 1770, by US statesman and scientist Benjamin Franklin.

Local advantage

As deputy postmaster of the British American colonies, Franklin was fascinated by why it took British packet ships delivering mail two weeks longer to cross the Atlantic than American merchant ships. Already famous for his invention of the lightning conductor, he asked Nantucket whaling captain Timothy Folger why this might be. Folger explained that American captains knew of the west–east current. They could spot it by whale migrations, differences in temperature and color, and the speed of surface bubbles, and so they crossed over the current to escape it, while the westbound British packet ships battled against it all the way.

With Folger's aid, Franklin charted the current's course as it flowed along the east coast of North America from the Gulf of Mexico to Newfoundland and then streamed east across the Atlantic. He also gave the Gulf Stream its name. ■

Franklin's chart was published in 1770 in Britain, but it would be years before British captains learned to use the Gulf Stream to cut sailing times.

See also: George Hadley 80 ■ Gaspard-Gustave de Coriolis 126 ■ Robert FitzRoy 150–55

DEPHLOGISTICATED AIR

JOSEPH PRIESTLEY (1733–1804)

IN CONTEXT

BRANCH
Chemistry

BEFORE
1754 Joseph Black isolates the first gas, carbon dioxide.

1766 Henry Cavendish prepares hydrogen.

1772 Carl Scheele isolates a third gas, oxygen, two years before Priestley, but does not publish his findings until 1777.

AFTER
1774 In Paris, Priestley demonstrates his method to Antoine Lavoisier, who makes the new gas and publishes his results in May 1775.

1779 Lavoisier gives the gas the name *"oxygène."*

1783 Geneva's Schweppes Company starts making the soda water Priestley invented.

1877 Swiss chemist Raoul Pictet produces liquid oxygen, which will be used in rocket fuel, industry, and medicine.

Following Joseph Black's pioneering discovery of "fixed air," or carbon dioxide (CO_2), an English clergyman named Joseph Priestley became interested in investigating various other "airs," or gases, and identified several more—most notably oxygen.

While a minister in Leeds, Priestley visited the brewery close to his lodgings. The layer of air above the brewing vat was already known to be fixed air. He found that when he lowered a candle over the vat, the candle went out about 12 in (30 cm) above the froth, where the flame entered the layer of fixed air floating there. The smoke drifted across the top of the fixed air, making it visible and revealing the boundary between the two airs. He also noticed that the fixed air flowed over the side of the vat and sank to the floor, because it was denser than "ordinary" air. When Priestley experimented with dissolving fixed air in cold water, sloshing it from one vessel to

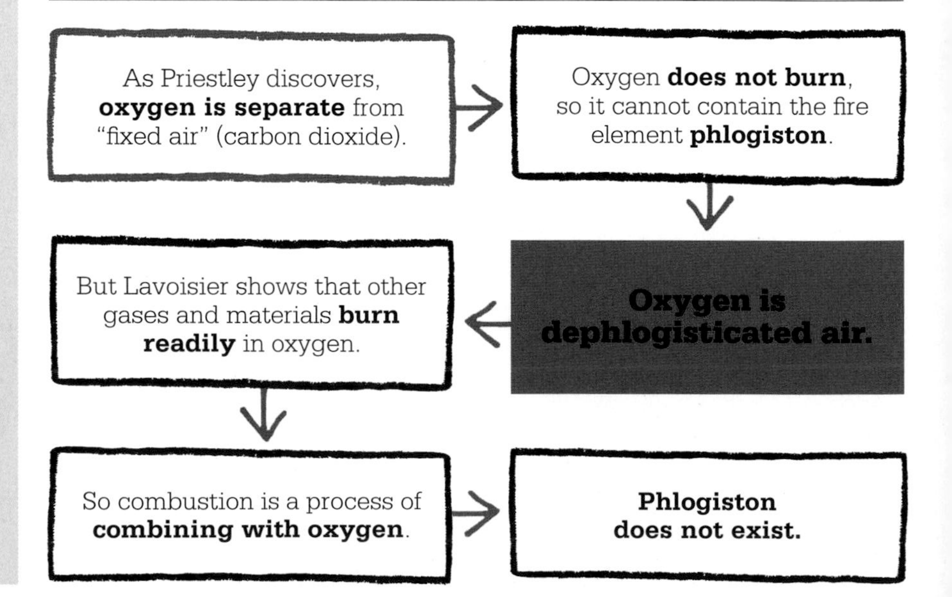

See also: Joseph Black 76–77 ▪ Henry Cavendish 78–79 ▪ Antoine Lavoisier 84 ▪ John Dalton 112–13 ▪ Humphry Davy 114

another, he found that it made a refreshing sparkling drink, which later led to the craze for soda water.

Releasing oxygen

On August 1, 1774, Priestley first isolated his new gas—which we now know as oxygen (O_2)—from mercuric oxide in a sealed glass flask by heating it with sunlight and a magnifying glass. He later discovered that this new gas kept mice alive much longer than ordinary air, was pleasant to breathe and more energizing than ordinary air, and supported the combustion of various substances he burned as fuel. He also showed that plants produce the gas in sunlight—a first hint of the process we call photosynthesis. At the time, however, combustion was thought to involve the release from a fuel of a mysterious material called phlogiston. Because this new gas did not burn, and therefore must contain no phlogiston, he called it "dephlogisticated air."

Priestley isolated several other gases at about this time, but then went on a European tour, and did not publish his results until late the following year. Swedish chemist Carl Scheele had prepared oxygen two years before Priestley, but did not publish his results until 1777. Meanwhile in Paris, Antoine Lavoisier heard of Scheele's work, was given a demonstration by Priestley, and promptly made his own oxygen. His experiments on combustion and respiration proved that combustion is a process of combining with oxygen, not liberating phlogiston. In respiration, oxygen absorbed from the air reacts with glucose and releases carbon dioxide, water, and energy. He named the new gas *oxygène*, or "acid-maker," when he discovered that it reacts with some materials—such as sulfur, phosphorus, and nitrogen—to make acids.

This led many scientists to abandon phlogiston, but Priestley, though a great experimenter, clung to the old theory to explain his discoveries and made little further contribution to chemistry. ▪

> The most remarkable of all the kinds of air I have produced…is, one that is five or six times better than common air, for the purpose of respiration.
>
> **Joseph Priestley**

Priestley's apparatus for his gas experiments appear in his book about his discoveries. At the front, a mouse is kept in oxygen under a jar; on the right, a plant releases oxygen in a tube.

Joseph Priestley

Born on a farm in Yorkshire, Joseph Priestley was brought up as a dissenting Christian, and was intensely religious and political all his life.

Priestley became interested in gases while living in Leeds in the early 1770s, but his best work was done after he moved to Wiltshire as librarian to the Earl of Shelburne. His duties were light and left him time to conduct research. He later fell out with the earl—his political views may have been too radical—and in 1780, he moved to Birmingham. Here he joined the Lunar Society, an informal but influential group of freethinkers, engineers, and industrialists.

Priestley's support for the French Revolution made him unpopular. In 1791, his house and laboratory were burned down, forcing him to move to London and then to America. He settled in Pennsylvania, and died there in 1804.

Key works

1767 *The History and Present State of Electricity*
1774–77 *Experiments and Observations on Different Kinds of Air*

IN NATURE, NOTHING IS CREATED, NOTHING IS LOST, EVERYTHING CHANGES

ANTOINE LAVOISIER (1743–1794)

IN CONTEXT

BRANCH
Chemistry

BEFORE
1667 German alchemist Johann Joachim Becher proposes that things are made to burn by a fire element.

1703 German chemist Georg Stahl renames it phlogiston.

1772 Swedish chemist Carl-Wilhelm Scheele discovers "fire air" (later called oxygen) but does not publish his findings until 1777.

1774 Joseph Priestley isolates "dephlogisticated air" (later called oxygen) and tells Lavoisier about his findings.

AFTER
1783 Lavoisier confirms his ideas on combustion with experiments on hydrogen, oxygen, and water.

1789 Lavoisier's *Elementary Treatise on Chemistry* names 33 elements.

French chemist Antoine Lavoisier brought a new level of precision to science, not least by naming oxygen and quantifying its role in combustion. By taking careful measurements of mass in the chemical reactions that occur during combustion, he demonstrated the conservation of mass—the principle that, in a reaction, the total mass of all the substances taking part is the same as the total mass of all its products.

Lavoisier heated various substances in sealed containers and found that the mass a metal gained when it was heated was exactly equal to the mass of air lost. He also found that burning stopped when the "pure" part of the air (oxygen) had all gone. The air that remained (mostly nitrogen) did not support combustion. He realized that combustion therefore involved a combination of heat, fuel (the burning material), and oxygen.

Published in 1778, Lavoisier's results not only demonstrated the conservation of mass, but also, by identifying oxygen's role in combustion, demolished the theory of a fire element called phlogiston. For the past century, scientists had thought inflammable substances contained phlogiston and released it when they burned. The theory explained why substances such as wood lost mass on burning, but not why others, such as magnesium, gained mass on burning. Lavoisier's careful measurements showed that oxygen was the key, in a process during which nothing was added or lost, but all was transformed. ■

"I consider nature a vast chemical laboratory in which all kinds of composition and decompositions are formed."
Antoine Lavoisier

See also: Joseph Black 76–77 ▪ Henry Cavendish 78–79 ▪ Joseph Priestley 82–83 ▪ Jan Ingenhousz 85 ▪ John Dalton 112–13

THE MASS OF A PLANT COMES FROM THE AIR

JAN INGENHOUSZ (1730–1799)

IN CONTEXT

BRANCH
Biology

BEFORE
1640s Flemish chemist Jan Baptista van Helmont deduces that a potted tree gains weight by absorbing water from soil.

1699 English naturalist John Woodward shows that water is both taken in and given off by plants, so their growth needs another source of matter.

1754 Swiss naturalist Charles Bonnet notices that plant leaves produce bubbles of air under water when illuminated.

AFTER
1796 Swiss botanist Jean Sénébier shows that it is the green parts in plants that release oxygen and absorb carbon dioxide.

1882 German scientist Théodore Engelman pinpoints chloroplasts as the oxygen-making parts in plant cells.

In the 1770s, Dutch scientist Jan Ingenhousz set out to discover why plants, as earlier scientists had noticed, put on weight. He went to England and did his research at Bowood House—where Joseph Priestley discovered oxygen in 1774—and was about to find the keys to photosynthesis: sunlight and oxygen.

Bubbling weeds

Ingenhousz had read how plants in water produce bubbles of gas, but the bubbles' precise composition and origin were unclear. In a series of experiments, he saw that sunlit leaves gave off more bubbles than leaves in the dark. He collected the gas produced only in sunlight, and found that it re-lit a glowing splint—this was oxygen. The gas given off by plants in the dark put out a flame—this was carbon dioxide.

Ingenhousz knew that plants put on weight with little change in the weight of the soil they grew from. In 1779, he correctly reasoned that gas exchange with the atmosphere, especially the absorption of the gas carbon dioxide, was at least partly the source of a plant's increased organic matter—that is, its extra mass came from air.

Pondweed bubbles at night show respiration as plants convert glucose into energy, absorbing oxygen and releasing carbon dioxide.

As we now know, plants make their food by photosynthesis—converting energy from sunlight into glucose by reacting the water and carbon dioxide that plants absorb, and releasing oxygen as waste. As a result, plants supply both the oxygen that is vital to life, and—as food for others—the energy. In a reverse process called respiration, plants use the glucose as food and release carbon dioxide, day and night. ■

See also: Joseph Black 76–77 ▪ Henry Cavendish 78–79 ▪ Joseph Priestley 82–83 ▪ Joseph Fourier 122–23

DISCOVERING NEW PLANETS

WILLIAM HERSCHEL (1738–1822)

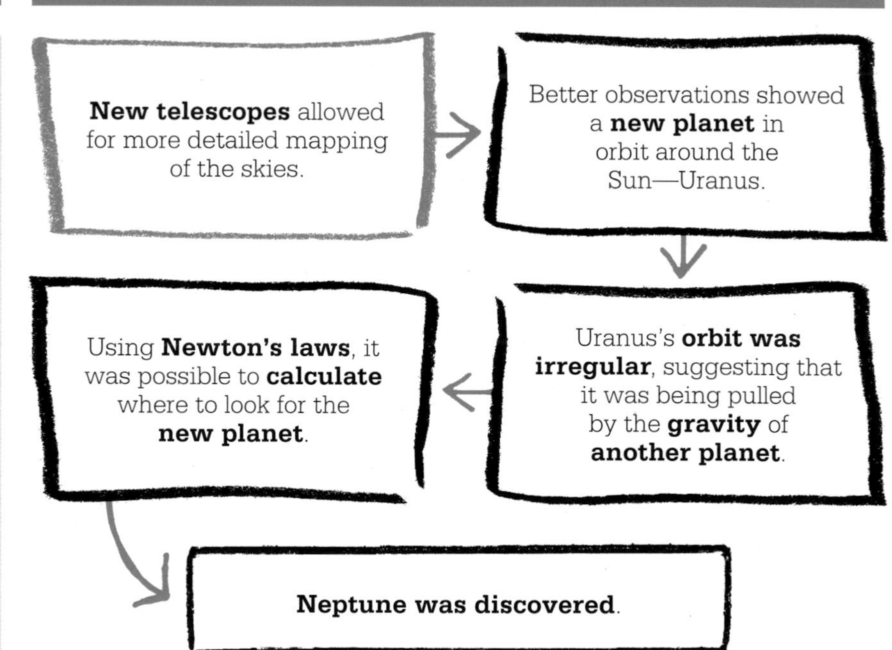

IN CONTEXT

BRANCH
Astronomy

BEFORE
Early 1600s The lens-based refracting telescope is invented, but mirror-based telescopes are not developed until the 1660s, by Isaac Newton and others.

1774 French observer Charles Messier publishes his astronomical survey, inspiring Herschel to begin work on a survey of his own.

AFTER
1846 Unexplained changes to the orbit of Uranus lead French mathematician Urbain Le Verrier to predict the existence and position of an eighth planet—Neptune.

1930 US astronomer Clyde Tombaugh discovers Pluto, which is initially recognized as a ninth planet, but now seen as the brightest member of the Kuiper Belt of small icy worlds.

In 1781, German scientist William Herschel identified the first new planet to be seen since ancient times, although Herschel himself initially thought it was a comet. His discovery would also lead to the discovery of another planet as a result of predictions based on Newton's laws.

By the late 18th century, astronomical instruments had advanced significantly—not least through the construction of reflecting telescopes that used mirrors rather than lenses to gather light, avoiding many of the problems associated with lenses at the time. This was the age of the first great astronomical surveys, as astronomers scoured the sky and identified a wide variety of "nonstellar" objects—star clusters and nebulae that looked like amorphous clouds of gas or dense balls of light.

See also: Ole Rømer 58–59 ▪ Isaac Newton 62–69 ▪ Nevil Maskelyne 102–03 ▪ Geoffrey Marcy 327

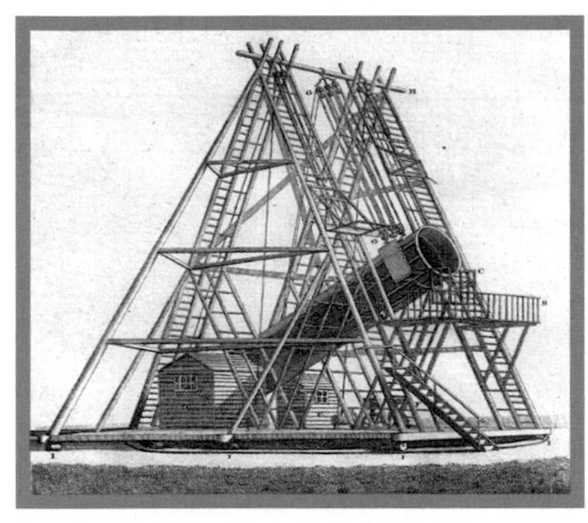

In the 1780s, Herschel built his "40-foot" telescope with a 47 in (1.2 m) wide primary mirror and a 40 ft (12 m) focal length. It remained the largest telescope in the world for 50 years.

Assisted by his sister Caroline, Herschel systematically quartered the sky, recording curiosities such as the unexpectedly large number of double and multiple stars. He even attempted to compile a map of the Milky Way galaxy based on the number of stars he counted in different directions.

On March 13, 1781, Herschel was scanning the constellation Gemini when he spotted a faint green disk that he suspected might be a comet. He returned to it a few nights later, and found that it had moved, confirming that it was not a star. Upon looking at Herschel's discovery, Nevil Maskelyne realized that the new object was moving far too slowly to be a comet, and might in fact be a planet in a distant orbit. Swedish-Russian Anders Johan Lexell and German Johann Elert Bode independently computed the orbit for Herschel's discovery, confirming that it was indeed a planet, roughly twice as far away as Saturn. Bode suggested naming it after Saturn's mythological father, the ancient Greek sky god Uranus.

Irregular orbit

In 1821, French astronomer Alexis Bouvard published a detailed table describing the orbit of Uranus as it should be according to Newton's laws. However, his observations of the planet soon showed substantial discrepancies with his table's predictions. The irregularities of its orbit suggested a gravitational pull from an eighth, more distant planet. By 1845, two astronomers—Frenchman Urbaine Le Verrier and Briton John Couch Adams—were independently using Bouvard's data to calculate where in the sky to look for the eighth planet. Telescopes were trained on the predicted area, and on September 23, 1846, Neptune was discovered within just one degree of where Le Verrier had predicted it would be. Its existence confirmed Bouvard's theory and provided powerful evidence of the universality of Newton's laws. ■

I looked for the Comet or Nebulous Star and found that it is a Comet, for it has changed its place.
William Herschel

William Herschel

Born in Hanover, Germany, Frederick William Herschel emigrated to Britain at 19 to make a career in music. His studies of harmonics and mathematics led to an interest in optics and astronomy, and he set out to make his own telescopes.

Following his discovery of Uranus, Herschel discovered two new moons of Saturn and the largest two moons of Uranus. He also proved that the solar system is in motion relative to the rest of the galaxy. While studying the Sun in 1800, Herschel discovered a new form of radiation. He performed an experiment using a prism and a thermometer to measure the temperatures of different colors of sunlight, and found that the temperature continued to rise in the region beyond visible red light. He concluded that the Sun emitted an invisible form of light, which he termed "calorific rays" and which today we call infrared.

Key works

1781 *Account of a Comet*
1786 *Catalogue of 1,000 New Nebulae and Clusters of Stars*

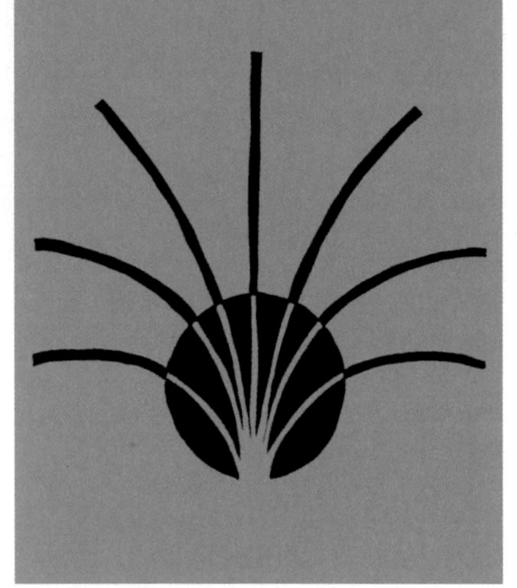

THE DIMINUTION OF THE VELOCITY OF LIGHT

JOHN MICHELL (1724–1793)

IN CONTEXT

BRANCH
Cosmology

BEFORE
1686 Isaac Newton formulates his law of universal gravitation, in which the strength of the gravitational attraction between objects is proportional to their masses.

AFTER
1796 Pierre-Simon Laplace independently theorizes about the possibility of black holes.

1915 Albert Einstein shows that gravity is a warping of the space-time continuum, which is why massless light photons are affected by gravity.

1916 Karl Schwarzschild proposes the event horizon, beyond which no data can be received about a black hole.

1974 Stephen Hawking predicts that quantum effects at the event horizon will emit infrared radiation.

Newton shows that the **gravitational attraction** of an object is **proportional to its mass**.

If light is affected by gravity, a **massive enough object** will have such a strong gravitational field that **no light will be able to escape it**.

The velocity of light will appear to diminish.

Einstein explains gravity as a **distortion of space-time**, meaning that massless **light is affected by gravity**.

In a 1783 letter to Henry Cavendish at the Royal Society, British polymath John Michell set out his thoughts on the effect of gravity. The letter was rediscovered in the 1970s and found to contain a remarkable description of black holes. Newton's law of gravity states that an object's gravitational pull increases with its mass. Michell considered what might happen to light if it is affected by gravity. He wrote: "If the semidiameter of a sphere of the same density with the sun were to exceed the sun in the proportion of 500 to 1, a body falling from an infinite height toward it would have acquired at its surface a greater velocity than that of light, & consequently, supposing light to be attracted by the same force… all light emitted from such a body would be made to return towards it." In 1796, French mathematician Pierre-Simon Laplace came up with a similar idea in his *Exposition du Système du Monde*.

However, the idea of a black hole would lie dormant until Albert Einstein's 1915 paper on general

See also: Henry Cavendish 78–79 ▪ Isaac Newton 62–69 ▪ Albert Einstein 214–21 ▪ Subrahmanyan Chandrasekhar 248 ▪ Stephen Hawking 314

Matter swirls around a black hole in a doughnut-shaped "accretion disk" before being sucked in. Heat in the swirling disk causes the hole to emit energy—as narrow beams of X-rays.

relativity, which described gravity as a result of the curving of space-time. Einstein showed how matter can wrap space-time around itself, making a black hole within a region called the Schwarzschild radius, or event horizon. Matter—and also light—can enter it, but cannot leave. In this picture, the speed of light is unchanged. Rather, it is the space the light travels through that changes, but Michell's intuition now had a mechanism by which the velocity of light would at least appear to diminish.

From theory to reality

Einstein himself doubted whether black holes existed in reality. It was not until the 1960s that they began to acquire general acceptance as indirect evidence of their existence grew. Today, most cosmologists think that black holes form when massive stars collapse under their own gravity, and grow as they assimilate ever more matter, and that a giant black hole lurks at the center of every galaxy. Black holes pull matter in, but nothing escapes, other than faint infrared radiation, known as Hawking radiation after Stephen Hawking, the physicist who proposed it. An astronaut falling into a black hole would feel nothing and notice nothing unusual on the approach to the event horizon, but if he or she dropped a clock toward the black hole, the clock would appear to slow down, and approach but never quite reach the event horizon, gradually fading from sight.

Problems with the theory still exist, however. In 2012, physicist Joseph Polchinski suggested that effects at the quantum scale would create a "firewall" at the event horizon that would burn any astronaut falling through it to a crisp. In 2014, Hawking changed his mind and concluded that black holes cannot exist after all. ■

John Michell

John Michell was a true polymath. He became professor of geology at the University of Cambridge in 1760, but also taught arithmetic, geometry, theology, philosophy, Hebrew, and Greek. In 1767, he retired to become a clergyman, and focused on his science.

Michell speculated on the properties of stars, investigated earthquakes and magnetism, and invented a new method for measuring the density of Earth. He built the apparatus for "weighing the world"—a delicate torsion balance—but died in 1793 before he could use it. He left it to his friend Henry Cavendish, who performed the experiment in 1798, and obtained a value close to the currently accepted figure. Ever since, this has somewhat unfairly been known as "the Cavendish experiment."

Key work

1767 *An Inquiry into the Probable Parallax and Magnitude of the Fixed Stars*

SETTING THE ELECTRIC FLUID IN MOTION

ALESSANDRO VOLTA (1745–1827)

IN CONTEXT

BRANCH
Physics

BEFORE
1754 Benjamin Franklin proves that lightning is natural electricity with his famous kite experiment.

1767 Joseph Priestley publishes a comprehensive account of static electricity.

1780 Luigi Galvani conducts his frog's legs experiments with "animal electricity."

AFTER
1800 English chemists William Nicholson and Anthony Carlisle use a Voltaic pile to split water into its two elements, oxygen and hydrogen.

1807 Humphry Davy isolates the elements potassium and sodium using electricity.

1820 Hans Christian Ørsted reveals the link between magnetism and electricity.

For centuries, philosophers had wondered at the terrifying power of lightning, and also at the way in which sparks can be drawn from solids such as amber when rubbed with a silk cloth. The Greek word for amber was "electron," and the sparking phenomenon became known as static electricity.

In an experiment of 1754, Benjamin Franklin flew a kite into a thunderstorm and showed that these two phenomena were closely related. When he saw sparks flying from a brass key tied to the kite's line, he proved that the clouds were electrified, and that lightning is also a form of electricity. Franklin's work inspired Joseph Priestley to publish a comprehensive work on *The History and Present State of Electricity* in 1767. But it was the Italian Luigi Galvani, a lecturer in anatomy at the University of Bologna, who, in 1780, took the first major steps toward understanding electricity when he noticed a frog's leg twitch.

Galvani was investigating a theory that animals are driven by "animal electricity," whatever that was, and was dissecting frogs to look for evidence of this. He noticed that if there was a machine nearby generating static electricity, a frog's leg lying on the bench suddenly twitched, even though the frog was long dead. The same thing happened when a frog's leg was hung on a brass hook that came into contact with an iron fence. Galvani believed this evidence supported his belief that electricity was coming from the frog itself.

Luigi Galvani is shown here conducting his famous frog's legs experiment. He believed that animals were driven by an electrical force, which he called "animal electricity."

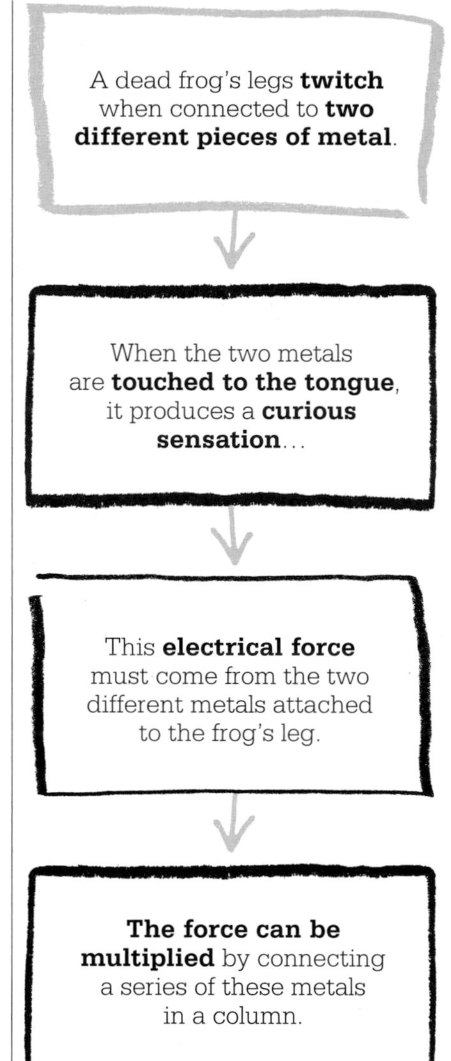

Volta's breakthrough

Galvani's younger colleague Alessandro Volta, a professor of natural philosophy, was intrigued by Galvani's observations and was initially convinced by his theory.

Volta himself had a notable background in electricity experiments. In 1775, he had invented the "electrophorus," a device that provided an instant source of electricity for an experiment (the modern equivalent is a capacitor). It consisted of a

See also: Henry Cavendish 78–79 ▪ Benjamin Franklin 81 ▪ Joseph Priestley 82–83 ▪ Humphry Davy 114 ▪ Hans Christian Ørsted 120 ▪ Michael Faraday 121

resin disk rubbed with cat fur to give it a static electric charge. Each time a metal disk was placed over the resin, the charge was transferred, electrifying the metal disk.

Volta stated that Galvani's animal electricity was "among the demonstrated truths." But he soon began to have his doubts. He came to the conclusion that the electricity causing the frog's legs to twitch on the hook came from the touching of the two different metals (the brass and the iron). He published his ideas in 1792 and 1793, and began investigating the phenomenon.

Volta found that a single junction of two different metals did not produce much electricity, although there was enough for him to feel a curious sensation with his tongue. But then he had the brilliant idea of multiplying the effect by making a series of such junctions connected by salt water. He took a small disk of copper, then placed a disk of zinc on top, then a piece of cardboard soaked in salt water, then another disk of copper, zinc, salty wet cardboard, copper, zinc, and so on, until he had a column, or stack. In other words, he created a pile, or "battery." The point of the salty wet cardboard was to carry the electricity without letting the metals on either side of it come into contact with each other.

The result was, literally, electrifying. Volta's crude battery probably produced only a few volts (the electrical unit named after him), but that was enough to make a tiny spark when the two ends were connected by a piece of wire, and enough to give him a mild electric shock.

The news spreads

Volta made his discovery in 1799, and news spread rapidly. He demonstrated the effect to Napoleon Bonaparte in 1801, but more importantly, in March 1800, he had reported his results in a long letter to Sir Joseph Banks, president of the Royal Society in »

> Each metal has a certain power, which is different from metal to metal, of setting the electric fluid in motion.
> **Alessandro Volta**

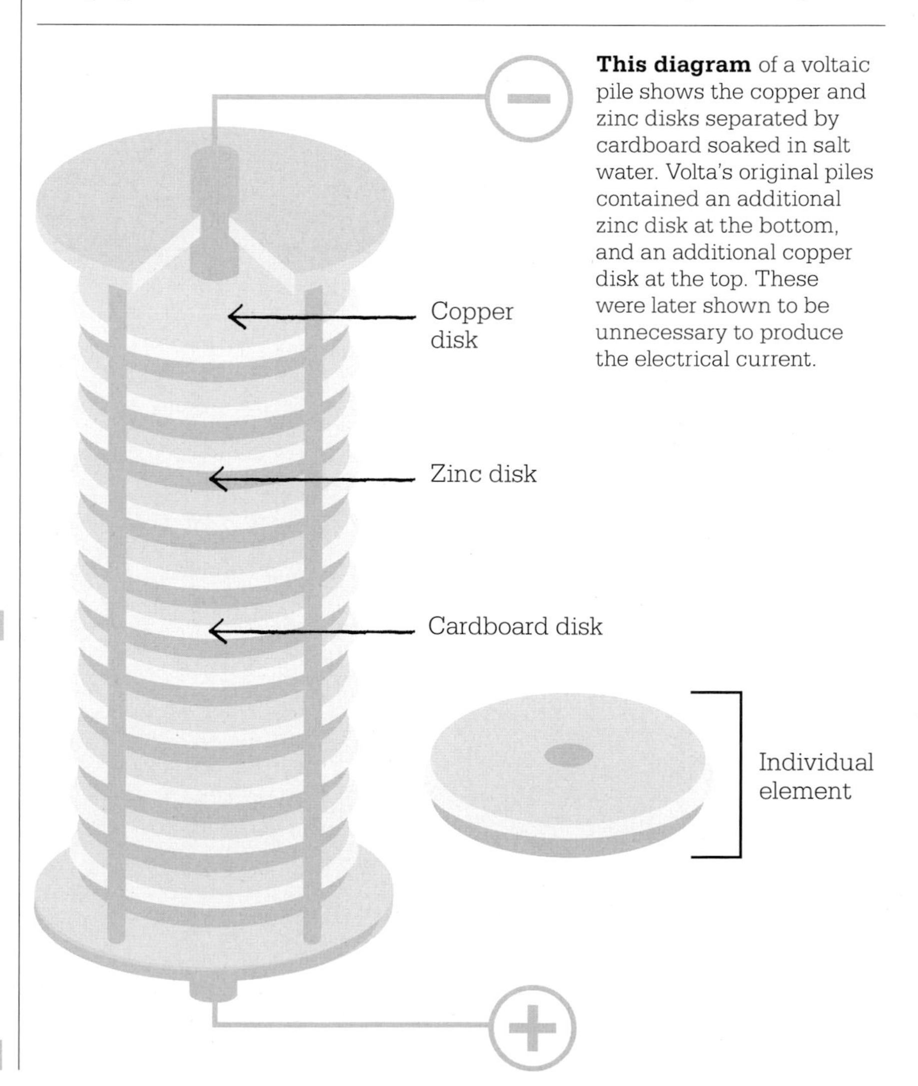

This diagram of a voltaic pile shows the copper and zinc disks separated by cardboard soaked in salt water. Volta's original piles contained an additional zinc disk at the bottom, and an additional copper disk at the top. These were later shown to be unnecessary to produce the electrical current.

Britain. The letter was titled "On the electricity excited by the mere Contact of conducting Substances of different Kinds," and in it Volta describes his apparatus: "I place then horizontally, on a table or any other stand, one of the metallic pieces, for example one of silver, and over the first I adapt one of zinc; on the second I place one of the moistened discs, then another plate of silver followed immediately by another of zinc…I continue to form…a column as high as possible without any danger of its falling."

Without a buzzer or a semiconductor to detect voltage, Volta used his body as a detector, and did not seem to mind getting electric shocks: "I receive from a column formed of twenty pairs of pieces (not more) shocks which affect the whole finger with considerable pain." He then describes a more elaborate apparatus, consisting of a series of cups or drinking glasses, each containing salt water, arranged in a line or a circle. Each pair is connected by a piece of metal that dips into the liquid in each cup. One end of this metal is silver, the other zinc, and these metals may be soldered together or connected by a wire of any metal, provided that only the silver dips into the liquid in one cup, and only the zinc into the next. He explains that this is in some ways more convenient than the solid pile, albeit more cumbersome.

Volta describes in detail the various unpleasant sensations that result from putting one hand in the bowl at one end of the chain and touching a wire attached to the other end to the forehead, eyelid, or tip of the nose: "I feel nothing for some moments; afterward, however, there begins at the part applied to the end of the wire, another sensation, which is a sharp pain (without shock), limited precisely to the point of contact, a quivering, not only continued, but which always goes on increasing to such a degree, that in a little time it becomes insupportable, and does not cease till the circle is interrupted."

Battery mania

That his letter reached Banks at all is surprising, since the Napoleonic wars were in progress,

Volta demonstrated his electric pile to Napoleon Bonaparte at the French National Institute in Paris in 1801. Napoleon was sufficiently impressed to make Volta a count the same year.

The language of experiment is more authoritative than any reasoning: facts can destroy our ratiocination [logical argument]—not vice versa.
Alessandro Volta

but Banks immediately spread the word to anyone who might be interested. Within weeks, people all over Britain were making electric batteries and investigating the properties of current electricity. Before 1800, scientists had had to work with static electricity, which is difficult and unrewarding. Volta's invention allowed them to find out how a range of materials—liquids, solids, and gases—react to a live electrical current.

Among the first to work with Volta's discovery were William Nicholson, Anthony Carlisle, and William Cruickshank, who, in May 1800, made their own "pile of thirty-six half crowns with the correspondent pieces of zinc and pasteboard" and passed the current through platinum wires into a tube filled with water. The bubbles of gas that appeared were identified as two parts of hydrogen and one part of oxygen. Henry Cavendish had shown that the formula of water is H_2O, but this was the first time anyone had split water into its separate elements.

Volta's pile was the ancestor of all modern batteries, used in everything from hearing aids to trucks and aircraft. Without batteries, many of our everyday devices would not work.

Reclassifying metals

In addition to kick-starting the study of current electricity, and thereby not only creating a new branch of physics but rapidly advancing the development of modern technology, Volta's pile led to a whole new chemical classification of metals, for he tried a variety of pairs of metals in his pile, and found that some worked much better than others. Silver with zinc made an excellent combination, as did copper with tin, but if he tried silver and silver, or tin and tin, he got no electricity at all; the metals had to be different. He showed that metals could be arranged in a sequence such that each became positive when placed in contact with the next one below it in the series. This electrochemical series has been invaluable to chemists ever since.

Who was right?

An ironic aspect of this story is that Volta started investigating the touching of different metals only because he doubted Galvani's hypothesis. Yet Galvani was not entirely wrong—our nerves do indeed work by sending electrical impulses around the body—while Volta himself did not get his theory entirely right. He believed that the electricity arose from just the touching together of two different metals, whereas Humphry Davy later showed that something could not come from nothing. When electricity is being generated, something else must be consumed. Davy suggested that there was a chemical reaction going on, and this led him to further important discoveries about electricity. ■

Alessandro Volta

Born in 1745 in Como, northern Italy, Alessandro Giuseppe Antonio Anastasio Volta was brought up in an aristocratic, religious family who hoped that he would become a priest. Instead he became interested in static electricity, and, in 1775, he made an improved device for generating it, called the "electrophorus." He discovered methane in the atmosphere at Lake Maggiore in 1776, and investigated its combustion by the novel method of igniting it with an electrical spark inside a sealed glass vessel.

In 1779, Volta was appointed professor of physics at the University of Pavia, a post he held for 40 years. Toward the end of his life, he pioneered the remotely operated pistol, whereby an electric current traveled 30 miles (50 km) from Como to Milan and fired a pistol. This was the forerunner of the telegraph, which uses electricity to communicate. The unit of electrical potential, the volt, is named after him.

Key work

1769 *On the Attractive Force of Electrical Fire*

NO VESTIGE OF A BEGINNING AND NO PROSPECT OF AN END

JAMES HUTTON (1726–1797)

IN CONTEXT

BRANCH
Geology

BEFORE
10th century Al-Biruni uses fossil evidence to argue that land must once have been under the sea.

1687 Isaac Newton argues that Earth's age can be calculated scientifically.

1779 The Comte de Buffon's experiments suggest an age of 74,832 years for Earth.

AFTER
1860 John Phillips calculates Earth's age at 96 million years.

1862 Lord Kelvin calculates Earth's cooling to produce an age of 20–400 million years, later settling on 20–40 million.

1905 Ernest Rutherford uses radiation to date a mineral.

1953 Clair Patterson puts Earth's age at 4.55 billion years.

For millennia, human cultures have pondered the age of Earth. Before the advent of modern science, estimates were based on beliefs rather than evidence. It was not until the 17th century that a growing understanding of Earth's geology provided the means to determine the planet's age.

Biblical estimates

In the Judaeo-Christian world, ideas about Earth's age were based on descriptions in the Old Testament. However, since these texts only presented the creation story in brief outline, they were subject to much interpretation, especially over the complex genealogical chronologies that followed the appearance of Adam and Eve.

Best known of these Biblical calculations is that by James Ussher, the protestant Primate of all Ireland. In 1654, Ussher pinpointed the date of Earth's creation to the night preceding Sunday October 23, 4004 BCE. This date became virtually enshrined in Christian culture when it was printed in many Bibles as part of the Old Testament chronology.

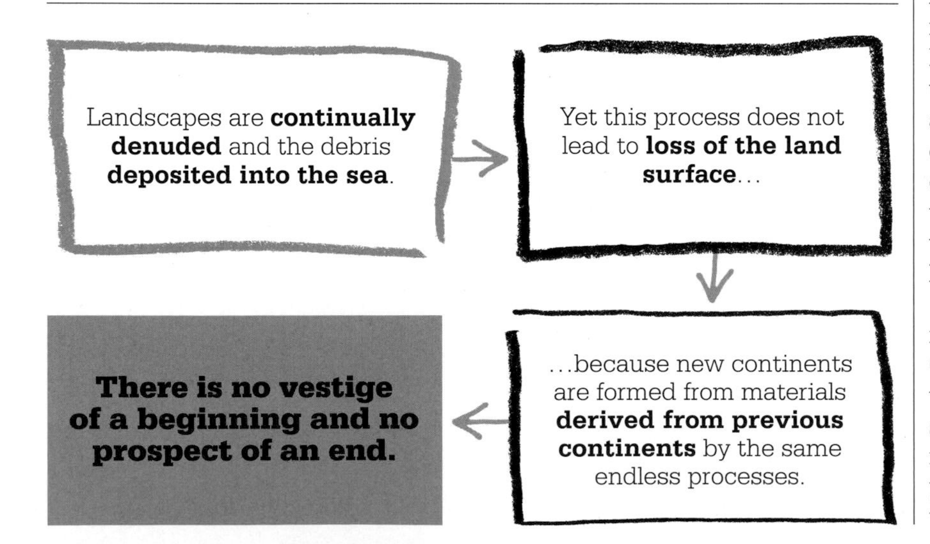

All the years from the creation of the world amount to a total of 5,698 years.
Theophilus of Antioch

A scientific approach

During the 10th century CE, scholars in Persia began to consider the question of Earth's age more empirically. Al-Biruni, a pioneer of experimental science, reasoned that if marine fossils were found on dry land, then that land must once have been under the sea. Earth, he concluded, must be evolving over long periods of time. Another Persian scholar, Avicenna, suggested that layers of rock had been laid down one upon another.

In 1687, a scientific approach to the problem was suggested by Isaac Newton. He argued that it would take a large body like Earth about 50,000 years to cool if it were made of molten iron. He derived this figure by scaling up the cooling time taken for a "globe of iron of an inch in diameter, exposed red hot to open air." Newton had opened the door to a scientific challenge to previous understandings of Earth's formation.

Following Newton's lead, French naturalist Georges-Louis Leclerc, Comte de Buffon, experimented with a large ball of red-hot iron, and showed that if Earth were made of molten iron, it would take 74,832 years to cool. In private, Buffon thought that Earth must be far

See also: Isaac Newton 62–69 ▪ Louis Agassiz 128–29 ▪ Charles Darwin 142–49 ▪ Marie Curie 190–95 ▪ Ernest Rutherford 206–13

older, since eons of time would be needed for chalk mountains to build up from the remains of marine fossils, but he did not want to publish this view without evidence.

Secrets of the rocks

In Scotland, quite a different approach to the problem of Earth's age was being taken by James Hutton, one of the preeminent natural philosophers of the Scottish Enlightenment. Hutton was a pioneer of geological fieldwork, and used field evidence to demonstrate his arguments to the Royal Society of Edinburgh in 1785.

Hutton was impressed by the apparent continuity of the processes by which landscape was denuded and its debris deposited into the sea. And yet all these processes did not lead to loss of the land surface, as might be expected. Perhaps thinking of the famous steam engine built by his friend James Watt, Hutton saw Earth as "a material machine moving in all its parts," with a new world constantly reshaped and recycled from the ruins of the old.

Hutton formulated his Earth-machine theory before he had found the supporting evidence, but, in 1787, he found the "unconformities" he was looking for—breaks in the continuity of sedimentary rocks. He saw that much of the land had once been seabed, where layers of sediment had been laid down and compressed. In many places these layers had been pushed upward, so that they were above sea level, and often distorted, so that they were not horizontal. He repeatedly found that rock material from the truncated upper boundary of older strata was incorporated into the base of the younger rocks above.

Such unconformities showed that there had been many episodes in Earth's history when the sequence of erosion, transportation, and deposition of rock debris had been repeated, and when rock strata had been moved by volcanic activity. Today, this is known as the geological cycle. From this evidence, Hutton declared that all continents are formed from materials derived from previous continents by the same processes, and that these processes still operate today. Famously, he wrote that "the result, therefore, of this present enquiry is, that we find no vestige of a beginning—no prospect of an end."

The popularization of Hutton's ideas about "deep time" was primarily due to John Playfair, a Scottish scientist who published Hutton's observations in an illustrated book, and to British geologist Charles Lyell, who transformed Hutton's ideas into a system called uniformitarianism. This held that the laws of nature »

In 1770, Hutton built a house overlooking Salisbury Crags in Edinburgh, Scotland. Among the crags he found evidence of volcanic penetration through sedimentary rock.

have always been the same, and therefore the clues to the past lie in the present. However, while Hutton's insights concerning the antiquity of the planet rang true to geologists, there was still no satisfactory method of determining just how old the planet was.

An experimental approach

Since the end of the 18th century, scientists had recognized that Earth's crust comprises successive layers of sedimentary strata. Geological mapping of these strata revealed that cumulatively they are very thick and many contain the fossil remains of the organisms that lived in their respective depositional environments. By the 1850s, the geological column of strata (also known as the stratigraphic column) had been more or less carved up into some eight named systems of strata and fossils, each of which represented a period of geological time.

Geologists were impressed by the overall thickness of the strata, estimated to be 16–70 miles (25–112 km) thick. They had

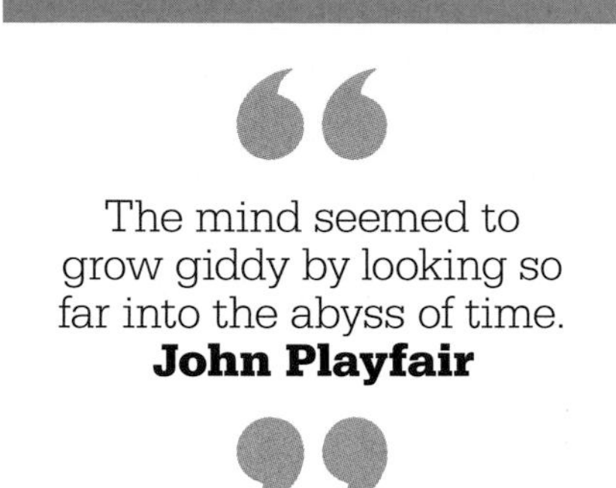

> The mind seemed to grow giddy by looking so far into the abyss of time.
> **John Playfair**

observed that the processes of erosion and deposition of the rock materials that make up such strata were very slow—estimated to be a few inches (centimeters) every 100 years. In 1858, Charles Darwin made a somewhat ill-judged foray into the debate when he estimated that it had taken some 300 million years for erosion to cut through the Tertiary and Cretaceous period rocks of the Weald in southern England. In 1860, John Phillips, a geologist at Oxford University, estimated that Earth is about 96 million years old.

But in 1862, such geological calculations were scorned by the eminent Scottish physicist William Thomson (Lord Kelvin) for being unscientific. Kelvin was a strict empiricist and argued that he could use physics to determine an accurate age for Earth, which he thought was constrained by the age of the Sun. Understanding of Earth's rocks, their melting points and conductivity, had vastly improved since Buffon's day. Kelvin took Earth's initial temperature at 7,000°F (3,900°C) and applied the observation that temperature increases as you go downward from the surface—by about 1°F (0.5°C) over every 50 ft (15 m) or so. From this, Kelvin calculated that it had taken 98 million years for Earth to cool to its present state, which he later reduced to 40 million years.

Lord Kelvin pronounced the world to be 40 million years old in 1897, the year in which radioactivity was discovered. He did not know that radioactive decay in Earth's crust provides heat that greatly slows the rate of cooling.

A radioactive "clock"

Such was Kelvin's prestige that his measure was accepted by most scientists. Geologists, however, were left feeling that 40 million years was simply not long enough for the observed rates of geological processes, accumulated deposits, and history. However, they had no scientific method with which to contradict Kelvin.

In the 1890s, the discovery of naturally occurring radioactive elements in some of Earth's minerals and rocks provided the key that would resolve the impasse between Kelvin and the geologists, since the rate at which atoms decay makes a reliable timer. In 1903, Ernest Rutherford predicted rates of radioactive decay and suggested that radioactivity might be used as a "clock" to date minerals and the rocks that contain them.

In 1905, Rutherford obtained the very first radiometric dates of formation for a mineral from Glastonbury, Connecticut: 497–500 million years. He warned that these were minimum dates. In 1907, American radiochemist Bertram Boltwood improved on Rutherford's technique to produce the first radiometric dates of minerals in rocks with a known geological context. These included a 2.2-billion-year-old rock from Sri Lanka, whose age increased previous estimates by an order

An uncomformity is a buried surface separating two rock strata of different ages. This diagram shows an angular unconformity, similar to those discovered by James Hutton on the east coast of Scotland. Here, layers of rock strata have been tilted by volcanic activity or movements in Earth's crust, producing an angular discordance with overlying, younger layers.

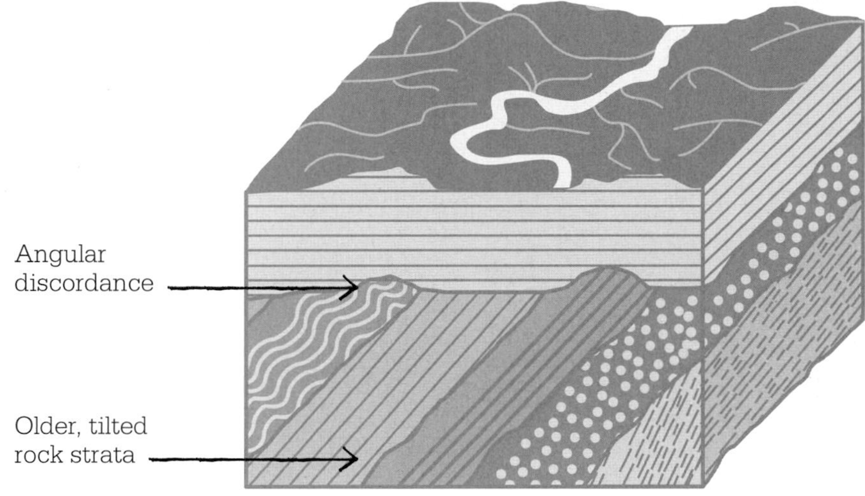

of magnitude. By 1946, British geologist Arthur Holmes had made some isotope measurements from lead-bearing rocks from Greenland, which gave an age of 3.015 billion years. This was one of the first reliable minimum ages for Earth. Holmes went on to estimate the age of the uranium from which the lead was derived, obtaining a date of 4.46 billion years, but he thought that must be the age of the gas cloud from which Earth formed.

Finally, in 1953, American geochemist Clair Patterson obtained the first generally accepted radiometric age of 4.55 billion years for Earth's formation. There are no known minerals or rocks dating from Earth's origin, but many meteorites are thought to originate from the same event in the solar system. Patterson calculated the radiometric date for lead minerals in the Canyon Diablo meteorite at 4.51 billion years. Comparing it with the average radiometric age of 4.56 billion years for granite and basalt igneous rocks in Earth's crust, he concluded that the similarity of dates was indicative of the age of Earth's formation. By 1956, he had made further measurements, which increased his confidence in the accuracy of the date of 4.55 billion years. This remains the figure accepted by scientists today. ■

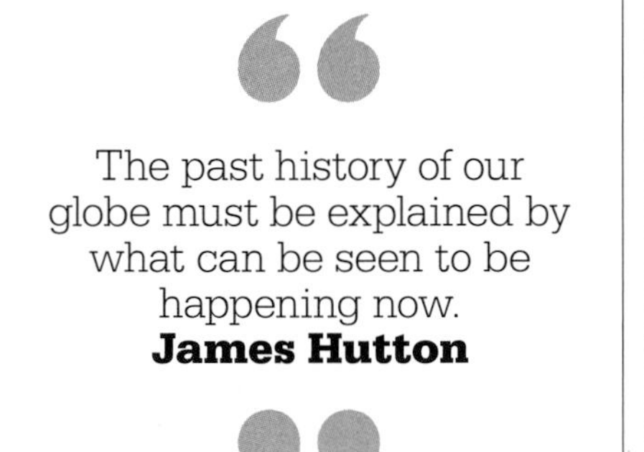

James Hutton

Born in 1726 to a respected merchant in Edinburgh, Scotland, James Hutton studied humanities at Edinburgh University. He became interested in chemistry and then medicine, but did not practice as a doctor. Instead, he studied the new agrarian techniques being used in East Anglia, England, where his exposure to soils and the rocks they were derived from led to an interest in geology. This took him on field expeditions all over England and Scotland.

Returning to Edinburgh in 1768, Hutton became acquainted with some of the major figures of the Scottish Enlightenment, including the engineer James Watt and the moral philosopher Adam Smith. Over the next 20 years, Hutton developed his famous theory of Earth's age and discussed it with his friends before finally publishing a long outline in 1788 and a much longer book in 1795. He died in 1797.

Key work

1795 *Theory of the Earth with Proofs and Illustrations*

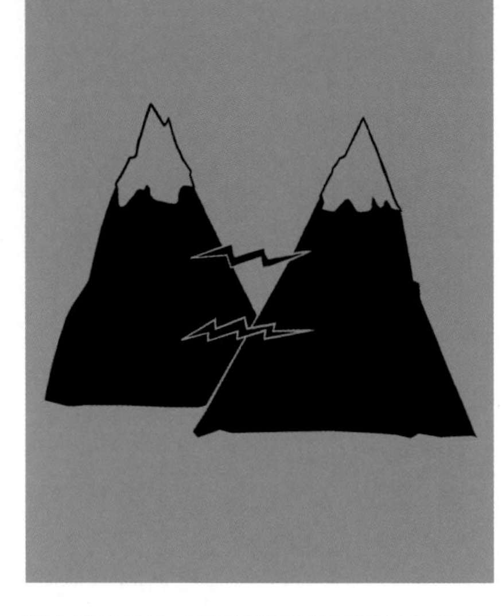

THE ATTRACTION OF MOUNTAINS

NEVIL MASKELYNE (1732–1811)

IN CONTEXT

BRANCH
Earth science and physics

BEFORE
1687 Isaac Newton publishes the *Principia*, in which he suggests experiments for calculating Earth's density.

1692 In an effort to explain Earth's magnetic field, Edmond Halley suggests that the planet consists of three concentric hollow spheres.

1738 Pierre Bouguer attempts Newton's experiment, without success, on Chimborazo, a volcano in Ecuador.

AFTER
1798 Henry Cavendish uses a different method to calculate the density of Earth, and finds it to be 340 lb/ft^3 (5,448 kg/m^3).

1854 George Airy figures out Earth's density using pendulums in a mine.

The **gravitational mass** of a mountain should **attract a plumb bob**.

The plumb line will **hang at an angle** that depends on the **relative density** of the mountain and Earth.

Measuring the deviation should allow calculation of Earth's mass.

In the 17th century, Isaac Newton had suggested methods for "weighing the Earth"—or calculating Earth's density. One of these involved measuring the angle of a plumb line on each side of a mountain to find out how far the gravitational attraction of the mountain pulled it from the vertical. This deviation could be measured by comparing the plumb line to a vertical calculated using astronomical methods. If the density and volume of the mountain could be ascertained, then, by extension, so could the density of Earth. However, Newton himself dismissed the idea because he thought the deviation would be too small to be measured with the instruments of the day.

In 1738, Pierre Bouguer, a French astronomer, attempted the experiment on the slopes of Chimborazo in Ecuador. Weather and altitude caused problems, however, and Bouguer did not think his measurements were accurate.

In 1772, Nevil Maskelyne proposed to the Royal Society in London that the experiment could be conducted in Britain. The Society agreed, and sent a surveyor

See also: Isaac Newton 62–69 ▪ Henry Cavendish 78–79 ▪ John Michell 88–89

Schiehallion was chosen as the site for the experiment because it was symmetrically shaped and isolated (and therefore less affected by the gravitational pull of other mountains).

to select an appropriate mountain. He chose Schiehallion in Scotland, and Maskelyne spent nearly four months making observations from both sides of the mountain.

The density of rocks

The orientation of the plumb line compared to the stars should have been different at the two stations even without any gravitational effects, because of the difference in latitude. However, even when this was accounted for, there was still a difference of 11.6 seconds of arc (just over 0.003 degrees). Maskelyne used a survey of the shape of the mountain and a measurement of the density of its rocks to figure out the mass of Schiehallion. He was assuming that the whole Earth had the same density as Schiehallion, but the deviation of the plumb lines showed a measured value of less than half of what he was expecting. Maskelyne realized that the density assumption was not correct—the density of Earth was clearly much greater than that of its surface rocks, probably, he reasoned, due to the planet having a metallic core. The actual observed angle was used to figure out that the overall density of Earth is about double that of Schiehallion's rocks.

This result disproved one theory of the time, advocated by English astronomer Edmond Halley, that said Earth was hollow. It also allowed the mass of Earth to be extrapolated from its volume and average density. Maskelyne's value for the overall density of Earth was 280 lb/ft^3 (4,500 kg/m^3). Compared with today's accepted value of 344 lb/ft^3 (5,515 kg/m^3), he had figured out the density of Earth with an error of less than 20 percent, and in the process had proved Newton's law of gravitation. ■

> ...the mean density of the earth is at least double of that at the surface...the density of the internal parts of the earth is much greater than near the surface.
> **Nevil Maskelyne**

Nevil Maskelyne

Born in London in 1732, Nevil Maskelyne became interested in astronomy at school. After graduating from Cambridge University and being ordained a priest, he became a member of the Royal Society in 1758, and was the Astronomer Royal from 1765 until his death.

In 1761, the Royal Society sent Maskelyne to the Atlantic island of St. Helena to observe the transit of Venus. Measurements taken as the planet passed across the Sun's disk allowed astronomers to calculate the distance between Earth and the Sun. He also spent much time trying to solve the problem of measuring longitude while at sea—a major issue of the day. His method involved carefully measuring the distance between the Moon and a given star, and consulting published tables.

Key works

1764 *Astronomical Observations Made at the Island of St Helena*
1775 *An Account of Observations Made on the Mountain Schehallien for Finding its Attraction*

THE MYSTERY OF NATURE IN THE STRUCTURE AND FERTILIZATION OF FLOWERS

CHRISTIAN SPRENGEL (1750–1816)

IN CONTEXT

BRANCH
Biology

BEFORE
1694 German botanist Rudolph Camerarius shows that flowers carry a plant's reproductive parts.

1753 Carl Linnaeus publishes *Species Plantarum*, devising a classification system guided by flower structure.

1760s Josef Gottlieb Kölreuter, a German botanist, proves that pollen grains are needed to fertilize a flower.

AFTER
1831 Scottish botanist Robert Brown describes how pollen grains germinate on a flower's stigma (female part).

1862 Charles Darwin publishes *Fertilisation of Orchids*, a detailed study of the relationship between flowers and pollinating insects.

In the mid-18th century, Swedish botanist Carl Linnaeus realized that flower parts parallel the reproductive organs of animals. Forty years later, a German botanist called Christian Sprengel figured out how insects played a major role in the pollination, and so fertilization, of flowering plants.

Mutual benefit

In the summer of 1787, Sprengel noticed insects visiting open flowers to feed on the nectar inside. He began to wonder whether the nectar was being "advertised" by the petals' special color and pattern, and deduced that the insects were being enticed onto the flowers so that pollen from the stamen (male part) of one flower stuck to the insect and was carried to the pistil (female part) of another flower. The insect's reward was a drink of energy-rich nectar.

Sprengel discovered that some flowering plants, if they lack color and scent, rely on wind to disperse their pollen. He also observed that many flowers contain both male and female parts, and that in these, the parts mature at different times, preventing self-fertilization.

Published in 1793, Sprengel's work was largely underappreciated during his lifetime. However, it was finally given due credit when Charles Darwin used it as a springboard for his own studies on the coevolution of flowering plants and the particular species of insects that pollinate them and ensure cross-fertilization—to their mutual benefit. ■

A honeybee lands on the sexual parts displayed at the center of these brightly colored petals. Honeybees account for 80 percent of all insect pollination and pollinate a third of all food crops.

See also: Carl Linnaeus 74–75 ■ Charles Darwin 142–49 ■ Gregor Mendel 166–71 ■ Thomas Hunt Morgan 224–25

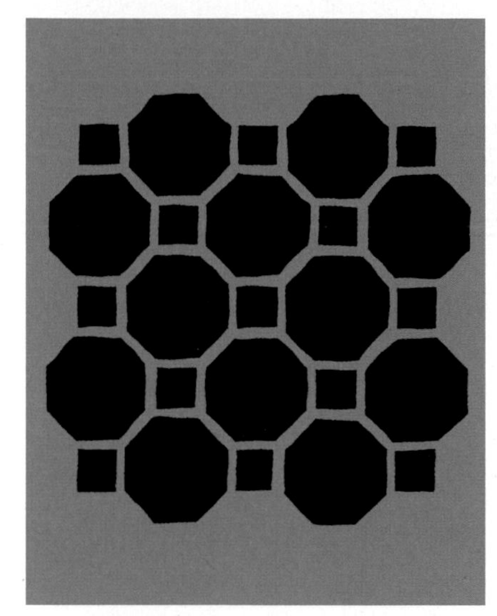

ELEMENTS ALWAYS COMBINE THE SAME WAY

JOSEPH PROUST (1754–1836)

IN CONTEXT

BRANCH
Chemistry

BEFORE
c.400 BCE The Greek thinker Democritus proposes that the world is ultimately made of tiny indivisible particles—atoms.

1759 English chemist Robert Dossie argues that substances combine when they are in the right proportion, which he calls the "saturation proportion."

1787 Antoine Lavoisier and Claude Louis Berthollet devise the modern system of naming chemical compounds.

AFTER
1805 John Dalton shows that elements are made up of atoms of a particular mass, which combine to make compounds.

1811 Italian chemist Amedeo Avogadro makes a distinction between atoms and the molecules that are formed by atoms to make compounds.

The Law of Definite Proportions, published by French chemist Joseph Proust in 1794, shows that no matter how elements combine, the proportions of each element in a compound are always precisely the same. This theory was one of the fundamental ideas about elements that emerged at this period to form the basis of modern chemistry.

In making his discovery, Proust was following a trend in French chemistry, pioneered by Antoine Lavoisier, which advocated careful measurement of weights, ratios, and percentages. Proust studied the percentages in which metals combined with oxygen in metal oxides. He concluded that when metal oxides formed, the proportion of metal and oxygen was constant. If the same metal combined with oxygen in a different proportion, it formed a different compound with different properties.

Not everyone agreed with Proust, but in 1811, the Swedish chemist Jöns Jakob Berzelius realized that Proust's theory fit John Dalton's new atomic theory of elements—that elements are each made of their own unique atoms. If a compound is always made from the same combination of atoms, Proust's argument that elements always combine in fixed proportions must be true. This is now accepted as one of the key laws of chemistry. ■

See also: Henry Cavendish 78–79 ▪ Antoine Lavoisier 84 ▪ John Dalton 112–13 ▪ Jöns Jakob Berzelius 119 ▪ Dmitri Mendeleev 174–79

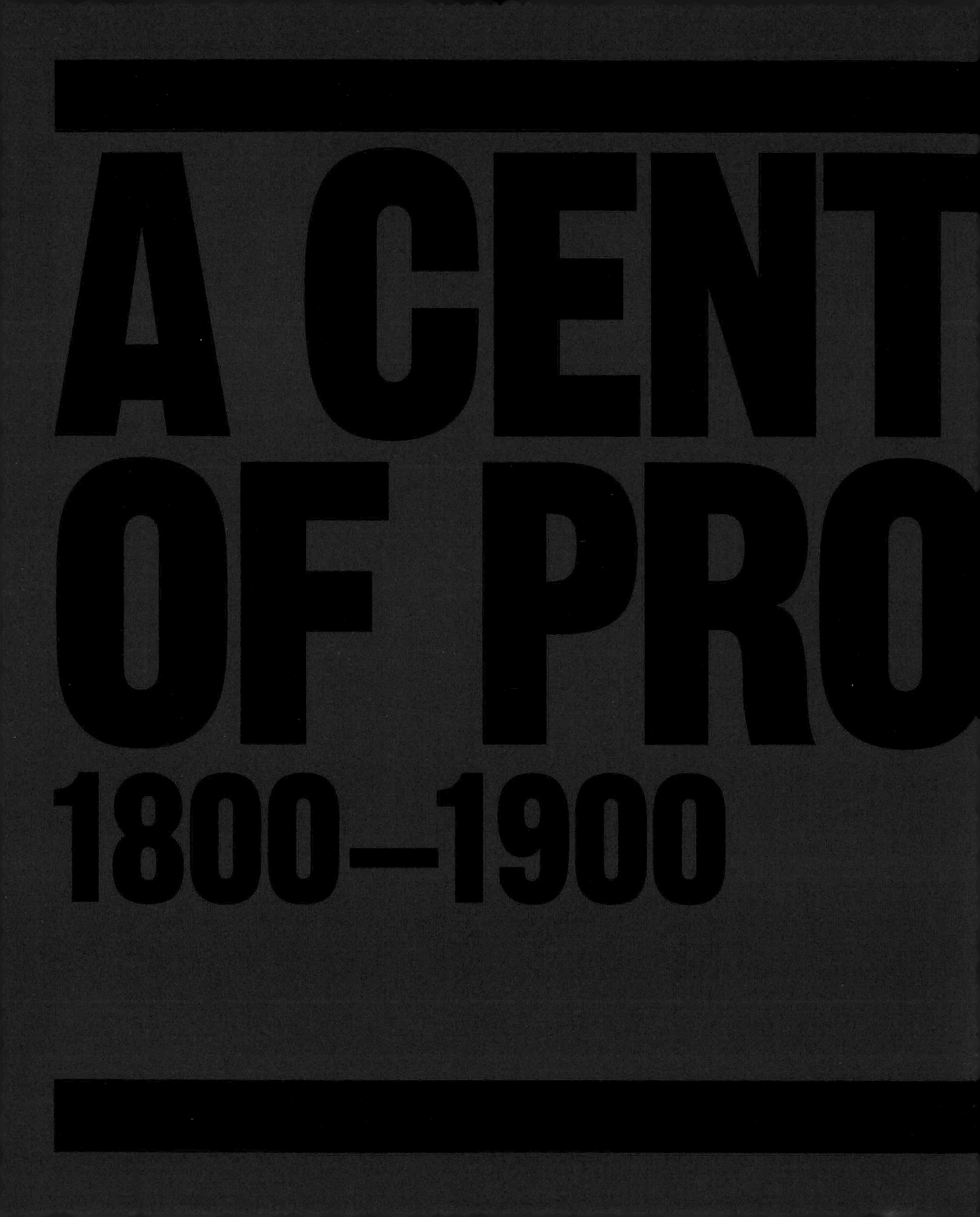
A CENT
OF PRO
1800–1900

URY
GRESS

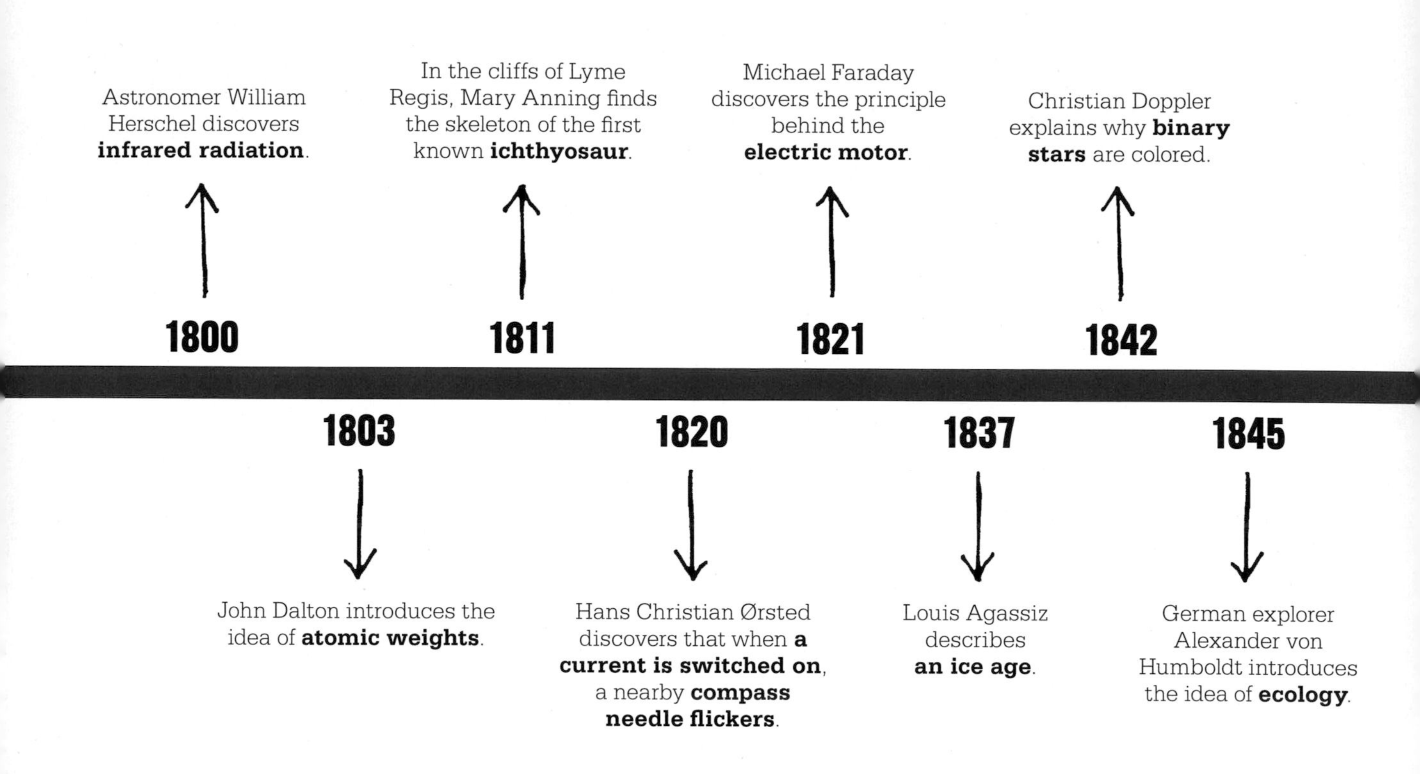

The invention of the electric battery in 1799 opened up whole new fields of scientific research. In Denmark, Hans Christian Ørsted accidentally discovered a connection between electricity and magnetism. At London's Royal Institution, Michael Faraday imagined the shapes of magnetic fields, and invented the world's first electric motor. In Scotland, James Clerk Maxwell picked up Faraday's ideas and figured out the complex mathematics of electromagnetism.

Seeing the invisible

Invisible forms of electromagnetic waves were discovered before they were understood or the laws governing their behavior were figured out. Working in Bath, Britain, German astronomer William Herschel used a prism to separate the various colors of sunlight to study their temperatures; he found that his thermometer showed a higher temperature beyond the red end of the visible spectrum. Herschel had stumbled upon infrared radiation, and ultraviolet radiation was discovered the following year—proving that there was more to the spectrum than visible light. In a similar accidental way, Wilhelm Röntgen later discovered X-rays in his laboratory in Germany. British physician Thomas Young devised a clever double-slit experiment to determine whether light is really a wave or a particle. His discovery of wavelike interference appeared to settle the argument. In Prague, Austrian physicist Christian Doppler explained the color of binary stars using the idea that light is a wave with a spectrum of various frequencies, laying out the phenomenon now known as the Doppler effect. Meanwhile, in Paris, French physicists Hippolyte Fizeau and Léon Foucault measured the speed of light, and showed that it travels more slowly through water than through air.

Chemical changes

British meteorologist John Dalton tentatively suggested that atomic weights might be a useful concept for chemists and ventured to estimate a few of them. Fifteen years later, Swedish chemist Jöns Jakob Berzelius drew up a much more complete list of atomic weights. His student, the German chemist Friedrich Wöhler, turned an inorganic salt into an organic

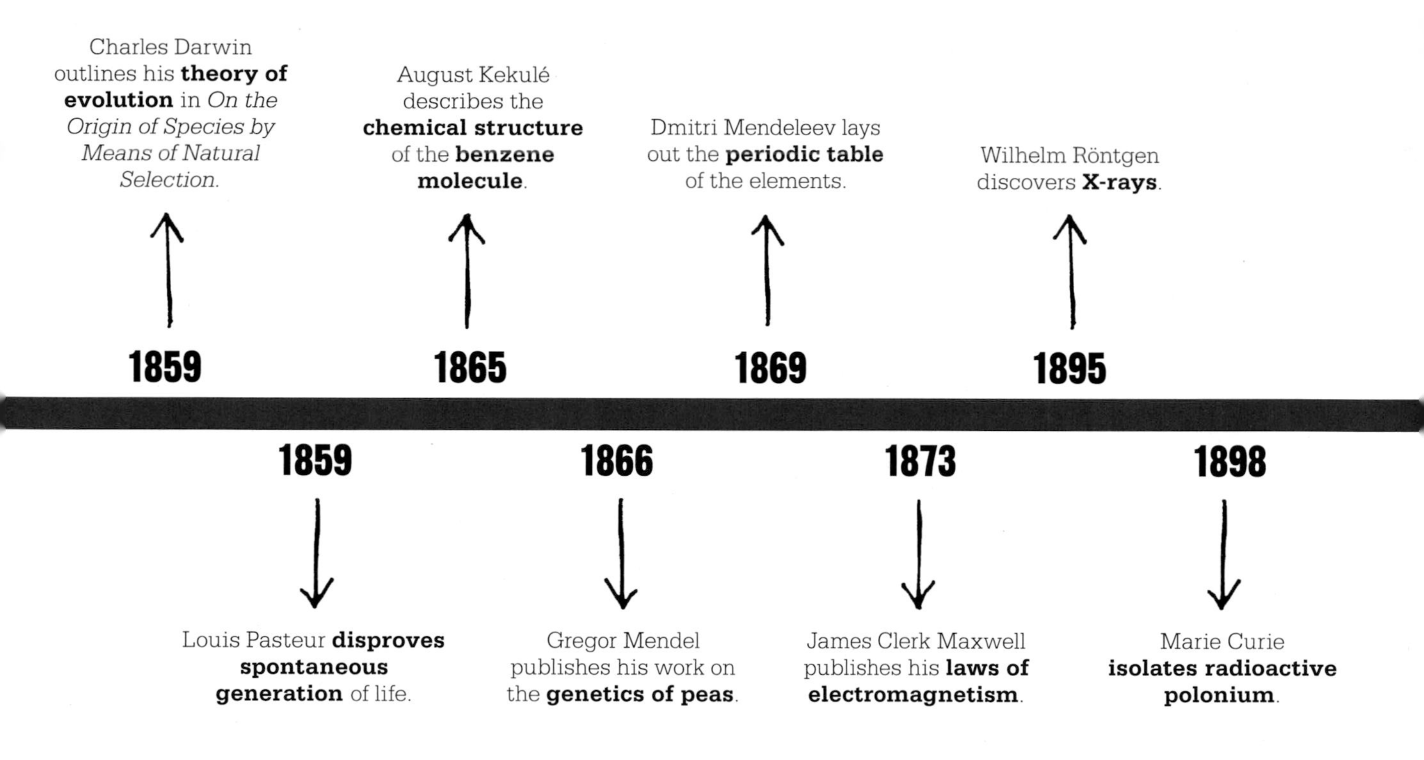

compound, and so disproved the idea that life chemistry operated according to separate rules. In Paris, Louis Pasteur further showed that life cannot be generated spontaneously. Inspiration for new ideas came from various quarters. The structure of the benzene molecule came to German chemist August Kekulé as he drifted off to sleep, while Russian chemist Dmitri Mendeleev used a pack of cards to crack the problem of the periodic table of the elements. Marie (Skłodowska) Curie isolated polonium and radium, and became the only person to win Nobel prizes in both Chemistry and Physics.

Clues from the past

The century saw nothing short of a revolution in the understanding of life. On the south coast of England, Mary Anning documented a series of fossils of extinct creatures she had dug out of the cliffs. Soon afterward, Richard Owen coined the word "dinosaurs" to describe the "terrible lizards" that had once roamed the planet. Swiss geologist Louis Agassiz suggested that large parts of Earth had once been covered with ice, further expanding the idea that Earth has experienced very different conditions through its history. Alexander von Humboldt used cross-disciplinary insights to uncover the connections in nature and established the study of ecology. In France, Jean-Baptiste Lamarck outlined a theory of evolution, mistakenly believing that the passing on of acquired characteristics was its driving force. Then, in the 1850s, British naturalists Alfred Russel Wallace and Charles Darwin both hit on the idea of evolution by means of natural selection. T. H. Huxley demonstrated that birds may well have evolved from dinosaurs, and the evidence to support evolution mounted. Meanwhile, a German-speaking Silesian friar named Gregor Mendel sorted out the basic laws of genetics by studying thousands of pea plants. Mendel's work would be neglected for some decades, but its rediscovery would provide the genetic mechanism for natural selection.

In 1900, British physicist Lord Kelvin is alleged to have said "There is nothing new to be discovered in physics now. All that remains is more and more precise measurement." Little can he have suspected what shocks were just around the corner. ■

THE EXPERIMENTS MAY BE REPEATED WITH GREAT EASE WHEN THE SUN SHINES

THOMAS YOUNG (1773–1829)

IN CONTEXT

BRANCH
Physics

BEFORE
1678 Christiaan Huygens first proposes that light travels as waves. He publishes his *Treatise on Light* in 1690.

1704 In his book *Opticks*, Isaac Newton suggests that light comprises streams of particles, or "corpuscles."

AFTER
1905 Albert Einstein argues that light must be thought of as particles, later called photons, as well as waves.

1916 US physicist Robert Andrews Millikan proves Einstein correct through experiment.

1961 Claus Jönsson repeats Young's double-slit experiment with electrons, and shows that, like light, they can behave as waves as well as particles.

If **light** is made of **particles** that **travel in straight lines**, then this can be proved in a simple experiment...

↓

Shine a light through two adjacent slits onto a screen. **Two pools of light** should be seen on the screen.

↓

But instead, it creates **interfering patterns of light and dark**, just as water waves would if water flowed through two slits.

↓

Light must travel as waves.

At the turn of the 19th century, scientific opinion was divided over the question of the nature of light. Isaac Newton had argued that a beam of light is made of countless, tiny, fast-moving "corpuscles" (particles). If light consists of these bulletlike corpuscles, he said, this would explain why light travels in straight lines and casts shadows.

But Newton's corpuscles did not explain why light refracts (bends when it enters glass) or splits into the colors of the rainbow—also an effect of refraction. Christiaan Huygens had argued that light comprises not particles, but waves. If light travels as waves, Huygens said, it is easy to explain these effects. However, Newton's stature was such that most scientists backed the particle theory.

Then, in 1801, British physician and physicist Thomas Young hit on a design for a simple yet ingenious experiment that would, he believed, settle the question one way or the other. The idea began when Young was looking at the patterns of light made by a candle shining through a mist of fine water droplets. The pattern showed colored rings around a bright

See also: Christiaan Huygens 50–51 ▪ Isaac Newton 62–69 ▪ Léon Foucault 136–37 ▪ Albert Einstein 214–21

center, and Young wondered if the rings might be caused by interacting waves of light.

The double-slit experiment

Young made two slits in a piece of cardboard and shone a beam of light onto them. On a paper screen placed behind the slits, the light created a pattern that convinced Young that it was waves. If light were streams of particles, as Newton said, there should simply have been a strip of light directly beyond each slit. Instead, Young saw alternating bright and dark bands, like a fuzzy bar code. He argued that as light waves spread out beyond the slits, they interact. If two waves ripple up (peak) or down (trough) at the same time, they make a wave twice as big (constructive interference)—creating the bright bands. If one wave ripples up as the other ripples down, they cancel each other out (destructive interference)—creating the dark bands. Young also showed that different colors of light create different interference patterns. This demonstrated that the color of light depends on its wavelength. For a century, Young's double-slit experiment convinced scientists that light is a wave, not a particle. Then in 1905, Albert Einstein showed that light also behaves as if it were a stream of particles—it can behave like a wave and a particle. Such was the simplicity of Young's experiment that, in 1961, German physicist Claus Jönsson used it to show that the subatomic particles electrons produce similar interference, so that they, too, must also be waves. ■

> Scientific investigations are a sort of warfare carried on against all one's contemporaries and predecessors.
> **Thomas Young**

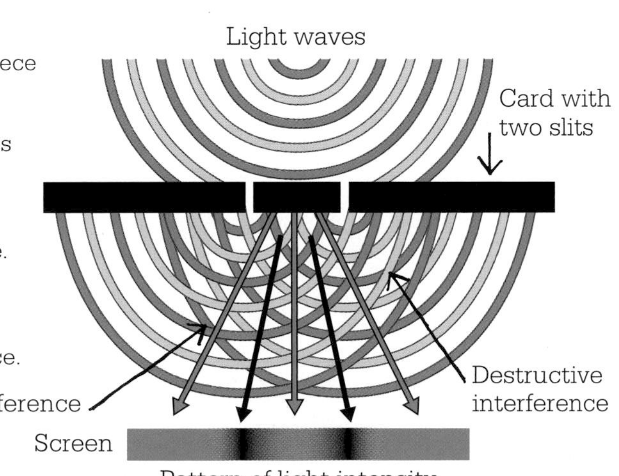

Here, light travels through two slits in a piece of card, and reaches a screen. The light waves passing through the slits interfere. Where peaks (yellow) intersect with troughs (blue), there is destructive interference. Where peaks intersect with peaks and troughs with troughs, there is constructive interference.

Thomas Young

The eldest of 10 children raised by Quaker parents in Somerset, England, Thomas Young's brilliant mind made him a child prodigy, and he was nicknamed the "Young Phenomenon." At 13, he could read five languages fluently—as an adult, he made the first modern translation of Egyptian hieroglyphics.

After medical training in Scotland, Young set up as a physician in London in 1799, but he was a true polymath who, in his spare time, conducted inquiries into everything from a theory of musical tuning to linguistics. He is most famous, however, for his work on light. In addition to establishing the principle of interference of light, he devised the first modern scientific theory of color vision, arguing that we see colors as varying proportions of the three main colors: blue, red, and green.

Key works

1804 *Experiments and Calculations Relative to Physical Optics*
1807 *Course of Lectures on Natural Philosophy and the Mechanical Arts*

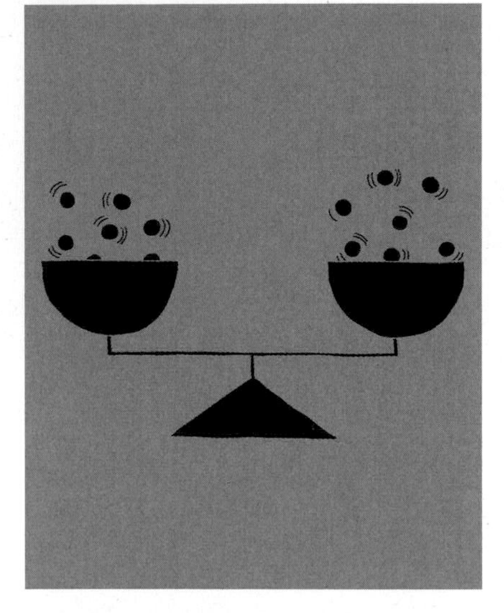

ASCERTAINING THE RELATIVE WEIGHTS OF ULTIMATE PARTICLES

JOHN DALTON (1766–1844)

IN CONTEXT

BRANCH
Chemistry

BEFORE
c.400 BCE Democritus proposes that the world is made of indivisible particles.

8th century CE Persian polymath Jabir ibn Hayyan (or Geber) classifies elements into metals and non-metals.

1794 Joseph Proust shows that compounds are always made of elements combined in the same proportions.

AFTER
1811 Amedeo Avogadro shows that equal volumes of different gases contain equal numbers of molecules.

1869 Dmitri Mendeleev draws up a periodic table, displaying elements by atomic weight.

1897 Through his discovery of the electron, J. J. Thomson shows that atoms are not the smallest possible particle.

Elements combine with each other to make compounds in simple **fixed ratios**.

These fixed ratios must depend on the **relative weight of the atoms** of each element.

Therefore, the **atomic weight** of an element **can be calculated** from the **weight of each element** involved in a **compound**.

Tables of elements should be based on the weights of their ultimate particles.

Toward the end of the 18th century, scientists had begun to realize that the world is made up of a range of basic substances, or chemical elements. But no one was certain what an element was. It was John Dalton, an English meteorologist, who, through his study of weather, saw that each element is made wholly of its own unique, identical atoms, and it is this special atom that distinguishes and defines an element. In developing the atomic theory of elements, Dalton established the basis of chemistry.

The idea of atoms dates back to ancient Greece, but it had always been assumed that all atoms were identical. Dalton agreed with Isaac Newton who, a century earlier, had described the atoms that made up the elements as "solid, massy, hard, impenetrable, moveable particles." Dalton's breakthrough was to understand that each element is made from different atoms.

Dalton's ideas originated in his study of the way in which air pressure affected how much water could be absorbed by air. He became convinced that air is

See also: Joseph Proust 105 ▪ Dmitri Mendeleev 174–79

An inquiry into the relative weight of the ultimate particles of bodies is a subject, as far as I know, entirely new.
John Dalton

a mixture of different gases. As he experimented, he observed that a given quantity of pure oxygen will take up less water vapor than the same amount of pure nitrogen, and he jumped to the remarkable conclusion that this is because oxygen atoms are bigger and heavier than nitrogen atoms.

Weighty matters

In a flash of insight, Dalton realized that atoms of different elements could be distinguished by differences in their weights. He saw that the atoms, or "ultimate particles," of two or more elements combined to make compounds in very simple ratios, so he could figure out the weight of each atom by the weight of each element involved in a compound. Very quickly, he figured out the atomic weight of each element then known.

Hydrogen, Dalton realized, was the lightest gas, so he assigned it an atomic weight of 1. Because of the weight of oxygen that combined with hydrogen in water, he assigned oxygen an atomic weight of 7. However, there was a flaw in Dalton's method, because he did not realize that atoms of the same element can combine. He always assumed that a compound of atoms—a molecule—had only one atom of each element. But Dalton's work had put scientists on the right track, and within a decade Italian physicist Amedeo Avogadro had devised a system of molecular proportions to calculate atomic weights correctly. Yet the basic idea of Dalton's theory—that each element has its own unique-sized atoms—has proved to be true. ■

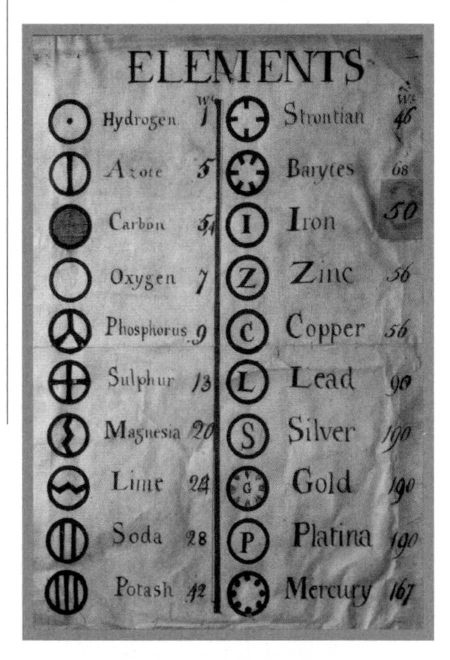

ELEMENTS

Element	Wt	Element	Wt
Hydrogen	1	Strontian	46
Azote	5	Barytes	68
Carbon	5.4	Iron	50
Oxygen	7	Zinc	56
Phosphorus	9	Copper	56
Sulphur	13	Lead	90
Magnesia	20	Silver	190
Lime	24	Gold	190
Soda	28	Platina	190
Potash	42	Mercury	167

Dalton's table shows symbols and atomic weights of different elements. Dalton was drawn to atomic theory through meteorology, when he asked himself why air and water particles could mix.

John Dalton

Born into a Quaker family in England's Lake District in 1766, John Dalton made regular observations of the weather from the age of 15. These provided many key insights, such as that atmospheric moisture turns to rain when the air cools. In addition to his meteorological studies, Dalton became fascinated by a condition he and his brother shared: color blindness. His scientific paper on the subject gained him admission to the Manchester Literary and Philosophical Society, of which he was elected president in 1817. He wrote hundreds of scientific papers for the Society, including those about his atomic theory. The atomic theory was quickly accepted, and Dalton became a celebrity in his own lifetime—more than 40,000 people attended his funeral in Manchester in 1844.

Key works

1805 *Experimental Enquiry into the Proportion of the Several Gases or Elastic Fluids, Constituting the Atmosphere*
1808–27 *New System of Chemical Philosophy*

THE CHEMICAL EFFECTS PRODUCED BY ELECTRICITY

HUMPHRY DAVY (1778–1829)

IN CONTEXT

BRANCH
Chemistry

BEFORE
1735 Swedish chemist Georges Brandt discovers cobalt, the first of many new metallic elements to be found over the next 100 years.

1772 Italian physician Luigi Galvani notices the effect of electricity on a frog and believes electricity is biological.

1799 Alessandro Volta shows that touching metals produce electricity, and creates the first battery.

AFTER
1834 Davy's former assistant Michael Faraday publishes the laws of electrolysis.

1869 Dmitri Mendeleev arranges the known elements into a periodic table, creating a group for the soft alkali metals that Davy had been the first to identify in 1807.

In 1800, Alessandro Volta invented the "voltaic pile"—the world's first battery, and soon many other scientists began to experiment with batteries.

English chemist Humphry Davy realized that the battery's electricity is produced by a chemical reaction. Electric charge flows as the pile's two different metals (the electrodes) react via the brine-soaked paper between them. In 1807, Davy found that he could use the electric charge from a pile to split chemical compounds, discovering new elements, and pioneering a process that was later called electrolysis.

New metals

Davy inserted two electrodes into dry potassium hydroxide (potash), which he moistened by exposing it to the damp air in his laboratory so that it would conduct electricity. To his delight, metallic globules began to form on the negatively charged electrode. The globules were a new element: the metal potassium. A few weeks later, he electrolyzed sodium hydroxide (caustic soda) in the same way and produced the metal sodium. In 1808, he used electrolysis to discover four more metallic elements—calcium, barium, strontium, and magnesium—and the metalloid boron. Like electrolysis, their commercial use would prove highly valuable. ■

Davy used apparatus similar to this in his lectures at London's Royal Institution to show how electrolysis splits water into its two elements, hydrogen and oxygen.

See also: Alessandro Volta 90–95 ▪ Jöns Jakob Berzelius 119 ▪ Hans Christian Ørsted 120 ▪ Michael Faraday 121 ▪ Dmitri Mendeleev 174–79

MAPPING THE ROCKS OF A NATION

WILLIAM SMITH (1769–1839)

IN CONTEXT

BRANCH
Geology

BEFORE
1669 Nicolas Steno publishes the principles of stratigraphy that will guide geologists' understanding of rock strata.

1760s In Germany, geologists Johann Lehmann and Georg Füchsel make some of the first measured sections and maps of geological strata.

1813 English geologist Robert Bakewell makes the first geognostic map of rock types in England and Wales.

AFTER
1835 The Geological Survey of Great Britain is founded to conduct systematic geological mapping of the country.

1878 The first International Geological Congress is held in Paris. Congresses have been held every three to five years ever since.

In the mid to late 18th century, the need to find fuels and ores to power Europe's Industrial Revolution spurred a growing interest in producing geological maps. German mineralogists Johann Lehmann and Georg Füchsel produced detailed aerial views showing topography and rock strata. Many subsequent geological maps did little more than show the surface distribution of different rock types—until the pioneering work of Georges Cuvier and Alexandre Brongniart in France, who mapped the geology of the Paris Basin in 1811, and William Smith in Britain.

Organized fossils are to the naturalist as coins to the antiquary.
William Smith

First national map

Smith was a self-taught engineer and surveyor who produced the first nationwide geological map in 1815, showing England, Wales, and part of Scotland. By amassing samples from mines, quarries, cliffs, canals, and road and railroad cuttings, Smith established the succession of rock strata, using Steno's principles of stratigraphy and identifying each stratum by its characteristic fossils. He also drew vertical sections of the succession of strata and the geological structures into which they had been formed by earth movements.

Over the next few decades, the first national geological surveys were established, and they set about methodically mapping their entire countries. The correlation of strata of similar age across national boundaries was achieved by international agreement in the latter part of the 19th century. ■

See also: Nicolas Steno 55 ■ James Hutton 96–101 ■ Mary Anning 116–17 ■ Louis Agassiz 128–29

SHE KNOWS TO WHAT TRIBE THE BONES BELONG

MARY ANNING (1799–1847)

IN CONTEXT

BRANCH
Paleontology

BEFORE
11th century Persian scholar Avicenna (Ibn Sina) suggests that rocks can be formed from petrified fluids, leading to the formation of fossils.

1753 Carl Linnaeus includes fossils in his system of biological classification.

AFTER
1830 British artist Henry De la Beche paints one of the first paleo-reconstructions of a scene from "deep time."

1854 Richard Owen and Benjamin Waterhouse Hawkins make the first life-size reconstructions of extinct plants and animals.

Early 20th century The development of radiometric dating techniques allows scientists to date fossils according to the rock strata in which they are found.

Fossils are the **preserved remains** of plants and animals.

Fossils have been found of **large animals no longer around** today.

In the past, **very different animals** lived on Earth.

By the end of the 18th century, it was generally accepted that fossils were the remains of once living organisms that had been petrified as the sediment around them hardened into rock. Both fossils and living organisms had been classified for the first time into a hierarchy of species, genera, and families by naturalists such as the Swedish taxonomist Carl Linnaeus. However, fossil remains were still seen in isolation from their environmental and biological context.

In the early 19th century, the discovery of large fossilized bones unlike those of any living animal raised many new questions. Where did they fit into the classification systems, and when had they become extinct? Within the Judeo-Christian culture of the Western world, it was generally thought that a benevolent God would not have allowed any of his creations to die out.

Monsters from the abyss

Some of the first of these large and distinctive fossil remains were found by the Anning family of fossil collectors around Lyme Regis on the coast of southern England. Here, Jurassic-period limestone and shale strata outcrop in the cliffs, where they are eroded by the sea to reveal abundant remains of ancient marine organisms. In 1811, Joseph

See also: Carl Linnaeus 74–75 ▪ Charles Darwin 142–49 ▪ Thomas Henry Huxley 172–73

Anning found a 4 ft- (1.2 m-) long skull with a curiously elongated toothed beak. His sister Mary found the rest of the skeleton, which they sold for about $37 (£23). Exhibited in London, this was the first entire skeleton of an extinct "monster of the abyss" and attracted a great deal of popular attention. It was identified as an extinct marine reptile and named an ichthyosaur, meaning "fish-lizard."

The Anning family went on to find more ichthyosaurs and the first complete specimen of another marine reptile, the plesiosaur, in addition to the first British specimen of a flying reptile, new fossil fish, and shellfish. Among the fish they found were cephalopods known as belemnites, some with the ink-sac preserved. The family, and especially Mary, had a talent for fossil hunting. Although poor, Mary was literate and taught herself geology and anatomy, which made her a far more effective fossil hunter. As Lady Harriet Sylvester observed in 1824, Mary Anning was "so thoroughly acquainted with the science that the moment she finds any bones she knows to what tribe they belong." She became an authority on many kinds of fossils, especially coprolites—fossilized dung.

The picture of life in ancient Dorset revealed by Anning's fossils was one of a tropical coast where a wide variety of now-extinct animals thrived. In 1854, Anning's fossils provided models for the first life-size reconstruction of an ichthyosaur, made for London's Crystal Palace park by the sculptor Benjamin Waterhouse Hawkins and the paleontologist Richard Owen. It was Owen who coined the word "dinosaur," but Anning who had provided the first glimpse of the richness of Jurassic life. ■

In 1830, Henry De la Beche painted this reconstruction of life in the Jurassic seas around Dorset based on Anning's fossil discoveries.

Mary Anning

Several biographies and novels have been written about the life of Mary Anning, a self-taught fossil collector. She was one of two surviving children out of 10 born into an impoverished Dorset family of religious dissenters who lived in the coastal village of Lyme Regis. The family eked out a precarious living collecting fossils for sale to the growing numbers of tourists. However, it was Mary who found and sold their most significant finds—fossils of Jurassic reptiles that lived 201–145 million years ago.

Due to a combination of her gender, humble social standing, and religious unorthodoxy, Anning received little formal recognition of her work in her lifetime, and she noted in a letter, "The world has used me unkindly, I fear it has made me suspicious of everyone." However, she was widely known in geological circles and various scientists sought out her expertise. When her health failed, Anning was provided with a small annual pension of about $40 (£25) in recognition of her contribution to science. She died of breast cancer at 47.

THE INHERITANCE OF ACQUIRED CHARACTERISTICS

JEAN-BAPTISTE LAMARCK (1744–1829)

IN CONTEXT

BRANCH
Biology

BEFORE
c.1495 Leonardo da Vinci suggests in his notebook that fossils are relics of ancient life.

1796 Georges Cuvier proves that fossil bones belong to extinct mastodons.

1799 William Smith shows the succession of fossils in rock strata of different ages.

AFTER
1858 Charles Darwin introduces his theory of evolution by natural selection.

1942 The "modern synthesis" reconciles Gregor Mendel's genetics with Darwin's natural selection, paleontology, and ecology in trying to explain how new species arise.

2005 Eva Jablonka and Marion Lamb claim that nongenetic, environmental, and behavioral changes can affect evolution.

In 1809, French naturalist Jean-Baptiste Lamarck introduced the first major theory that life on Earth has evolved over time. The impetus to his theory was the discovery of fossils of creatures unlike any alive today. In 1796, French naturalist Georges Cuvier had shown that fossilized elephant-like bones were markedly different in anatomy from the bones of modern elephants, and must come from extinct creatures now called mammoths and mastodons.

Cuvier explained the vanished creatures of the past as victims of catastrophes. Lamarck challenged this idea, and argued that life had "transmutated," or evolved, gradually and continuously through time, developing from the simplest life forms to the most complex. A change in the environment, he suggested, could spur a change in the characteristics of an organism. Those changes could then be inherited through reproduction. Characteristics that were useful developed further; those that were not useful might disappear.

> What nature does in the course of long periods we do every day when we suddenly change the environment in which some species of living plant is situated.
> **Jean-Baptiste Lamarck**

Lamarck believed characteristics were acquired during a creature's life and passed on. Later, Darwin showed that changes occur because mutations at conception survive to be passed on through natural selection, and the idea of "acquired characteristics" was ridiculed. But recently, scientists have argued that the environment—chemicals, light, temperature, and food—can in fact alter genes and their expression. ■

See also: William Smith 115 ▪ Mary Anning 116–17 ▪ Charles Darwin 142–49 ▪ Gregor Mendel 166–71 ▪ Thomas Hunt Morgan 224–25 ▪ Michael Syvanen 318–19

EVERY CHEMICAL COMPOUND HAS TWO PARTS

JÖNS JAKOB BERZELIUS (1779–1848)

IN CONTEXT

BRANCH
Chemistry

BEFORE
1704 Isaac Newton suggests that atoms are bonded by some force.

1800 Alessandro Volta shows that placing two different metals next to each other can produce electricity, and so creates the first battery.

1807 Humphry Davy discovers sodium and other metal elements by splitting salts with electrolysis.

AFTER
1857–58 August Kekulé and others develop the idea of valency—the number of bonds an atom can form.

1916 US chemist Gilbert Lewis proposes the idea of the covalent bond in which electrons are shared, while German physicist Walther Kossel suggests the idea of ionic bonds.

The leading light of a generation of chemists inspired by Alessandro Volta's creation of the battery, Sweden's Jöns Jakob Berzelius conducted a series of experiments looking at the effect of electricity on chemicals. He developed a theory called electrochemical dualism, published in 1819, which proposed that compounds are created by the coming together of elements with opposite electrical charges.

The habit of an opinion often leads to the complete conviction of its truth, and makes us incapable of accepting the proofs against it.
Jöns Jakob Berzelius

In 1803, Berzelius had teamed up with a mine owner to make a voltaic pile and see how electricity splits salts. Alkali metals and alkaline earths migrated to the pile's negative pole, while oxygen, acids, and oxidized substances migrated to the positive pole. He concluded that salt compounds combine a basic oxide, which is positively charged, and an acidic oxide, which is negatively charged.

Berzelius developed his dualistic theory to suggest that compounds are bonded by the attraction of opposite electrical charges between their constituent parts. Though later shown to be incorrect, the theory triggered further research into chemical bonds. In 1916, it was found that electrical bonding occurs as "ionic" bonding, in which atoms lose or gain electrons to become mutually attractive charged atoms, or ions. In fact, this is just one of several ways in which the atoms in a compound bind—another is the "covalent" bond, in which electrons are shared between atoms. ■

See also: Isaac Newton 62–69 ▪ Alessandro Volta 90–95 ▪ Joseph Proust 105 ▪ Humphry Davy 114 ▪ August Kekulé 160–65 ▪ Linus Pauling 254–59

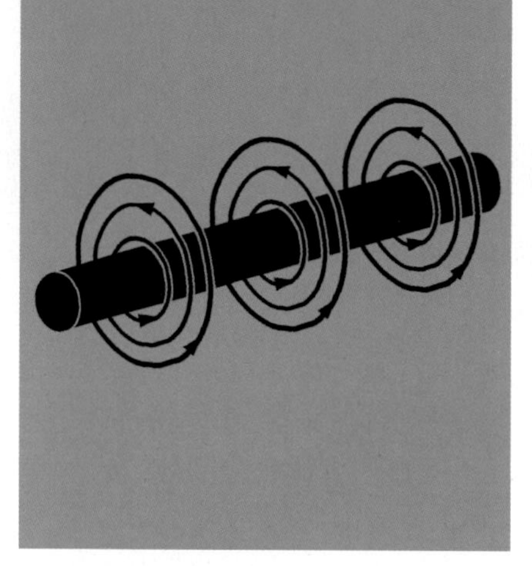

THE ELECTRIC CONFLICT IS NOT RESTRICTED TO THE CONDUCTING WIRE

HANS CHRISTIAN ØRSTED (1777–1851)

IN CONTEXT

BRANCH
Physics

BEFORE
1600 William Gilbert conducts the first scientific experiments on electricity and magnetism.

1800 Alessandro Volta creates the first electric battery.

AFTER
1820 André-Marie Ampère develops a mathematical theory of electromagnetism.

1821 Michael Faraday is able to show electromagnetic rotation in action, by creating the first electric motor.

1831 Faraday and US scientist Joseph Henry independently discover electromagnetic induction; Faraday uses it in the first generator to convert motion into electricity.

1864 James Clerk Maxwell formulates a set of equations to describe electromagnetic waves—including light waves.

The quest to discover an underlying unity to all forces and matter is as old as science itself, but the first big break came in 1820, when the Danish philosopher Hans Christian Ørsted found a link between magnetism and electricity. The link had been suggested to him by the German chemist and physicist Johann Wilhelm Ritter, whom he had met in 1801. Already influenced by the philosopher Immanuel Kant's idea that there is a unity to nature, Ørsted now investigated the possibility in earnest.

It appears that the electric conflict is not restricted to the conducting wire, but that it has a rather extended sphere of activity around it.
Hans Christian Ørsted

Chance discovery

Lecturing at the University of Copenhagen, Ørsted wanted to show his students how the electric current from a voltaic pile (the battery invented by Alessandro Volta in 1800) can heat up a wire and make it glow. He noticed that a compass needle standing near the wire moved every time the current was switched on. This was the first proof of a link between electricity and magnetism. Further study convinced him that the current produced a circular magnetic field as it flowed through the wire.

Ørsted's discovery rapidly prompted scientists across Europe to investigate electromagnetism. Later that year, French physicist André-Marie Ampère formulated a mathematical theory for the new phenomenon and, in 1821, Michael Faraday demonstrated that electromagnetic force could convert electrical into mechanical energy. ■

See also: William Gilbert 44 ▪ Alessandro Volta 90–95 ▪ Michael Faraday 121 ▪ James Clerk Maxwell 180–85

ONE DAY, SIR, YOU MAY TAX IT

MICHAEL FARADAY (1791–1867)

IN CONTEXT

BRANCH
Physics

BEFORE
1800 Alessandro Volta invents the first electric battery.

1820 Hans Christian Ørsted discovers that electricity creates a magnetic field.

1820 André-Marie Ampère formulates a mathematical theory of electromagnetism.

AFTER
1830 Joseph Henry creates the first powerful electromagnet.

1845 Faraday demonstrates the link between light and electromagnetism.

1878 Designed by Sigmund Schuckert, the first steam-driven power station generates electricity for the Linderhof Palace in Bavaria, Germany.

1882 US inventor Thomas Edison builds a power station to power electric lighting in Manhattan, New York City.

British scientist Michael Faraday's discovery of the principles of both the electric motor and the electric generator paved the way for the electrical revolution that would transform the modern world, bringing everything from lightbulbs to telecommunications. Faraday himself foresaw the value of his discoveries—and the tax revenues they could generate for government.

In 1821, a few months after hearing of Hans Christian Ørsted's discovery of the link between electricity and magnetism, Faraday demonstrated how a magnet will move around an electric wire, and an electric wire will move around a magnet. The electric wire produces a circular magnetic field around it, which generates a tangential force on the magnet, producing circular motion. This is the principle behind the electric motor. A spinning motion is set up by alternating the direction of the current, which alternates the direction of the magnetic field in the wire.

In Faraday's apparatus for showing electromagnetic induction, a current flows through the small magnetic coil, which is moved in and out of the large coil, inducing a current in it.

Generating electricity

Ten years later, Faraday made an even more important discovery—that a moving magnetic field can create or "induce" a current of electricity. This discovery—which was also made independently by the US physicist Joseph Henry around the same time—is the basis for generating all electricity. Electromagnetic induction converts the kinetic energy in a spinning turbine into electrical current. ■

See also: Alessandro Volta 90–95 ▪ Hans Christian Ørsted 120 ▪ James Clerk Maxwell 180–85

HEAT PENETRATES EVERY SUBSTANCE IN THE UNIVERSE

JOSEPH FOURIER (1777–1831)

IN CONTEXT

BRANCH
Physics

BEFORE
1761 Joseph Black discovers latent heat—the heat taken up by ice to melt and water to boil without changing temperature. He also studies specific heat—required by substances to raise their temperature by a certain amount.

1783 Antoine Lavoisier and Pierre-Simon Laplace measure latent heat and specific heat.

AFTER
1824 By developing the first theory of heat engines, which turn heat energy into mechanical energy, Nicolas Sadi Carnot provides the foundations for the theory of thermodynamics.

1834 Émile Clapeyron shows that energy must always become more diffuse, formulating the second law of thermodynamics.

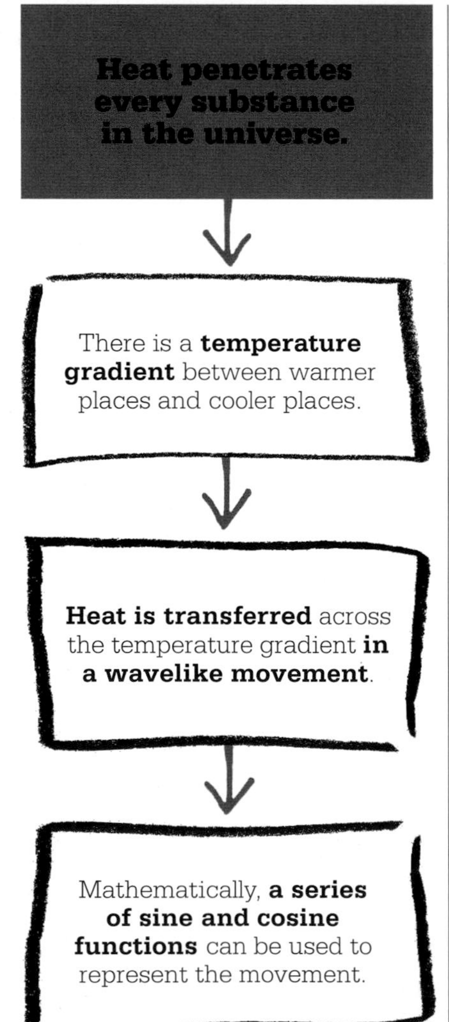

Today, one of the most fundamental laws of physics is that energy is neither created nor destroyed, but only changes from one form to another or moves from one place to another. French mathematician Joseph Fourier was a pioneer in the study of heat and how heat moves from warm places to cool places.

Fourier was interested in both how heat diffused through solids by conduction and how things cooled down by losing heat. His compatriot Jean-Baptiste Biot had imagined the spread of heat as "action at a distance," in which it spreads by jumping from warm places to cool. Biot represented the heat flow in a solid as a series of slices, which allowed it to be studied with conventional equations showing the heat jumping from one slice to the next.

Temperature gradients

Fourier looked at heat flow in an entirely different way. He focused on temperature gradients—continuous gradations between warm and cool places. These could not be quantified with conventional equations, so he devised new mathematical techniques.

See also: Isaac Newton 62–69 ■ Joseph Black 76–77 ■ Antoine Lavoisier 84 ■ Charles Keeling 294–95

Mathematics compares the most diverse phenomena and discovers the secret analogies that unite them.
Joseph Fourier

Fourier focused on the idea of waves, and finding a way to represent them mathematically. He saw that every wavelike movement, which is what a temperature gradient is, can be approximated mathematically by adding together simpler waves, whatever the shape of the wave to be represented. The simpler waves that are to be added together are sines and cosines, derived from trigonometry, and can be written out mathematically as a series.

These individual waves each move uniformly from a peak to a trough. Adding more and more of these simple waves together produces increasing complexity that can approximate any other type of wave. These infinite series are now called Fourier series.

Fourier published his idea in 1807, but it attracted criticism, and it was not until 1822 that his work was finally accepted. Continuing his study of heat, in 1824, Fourier examined the difference between the heat that Earth gains from the Sun and the heat it loses to space. He realized that the reason Earth is pleasantly warm, considering how far it is from the Sun, is because gases in its atmosphere trap heat and stop it from being radiated back into space—the phenomenon now called the greenhouse effect.

Today, Fourier analysis is applied not only to heat transfer but also to a host of problems at the cutting edge of science, ranging from acoustics, electrical engineering, and optics to quantum mechanics. ■

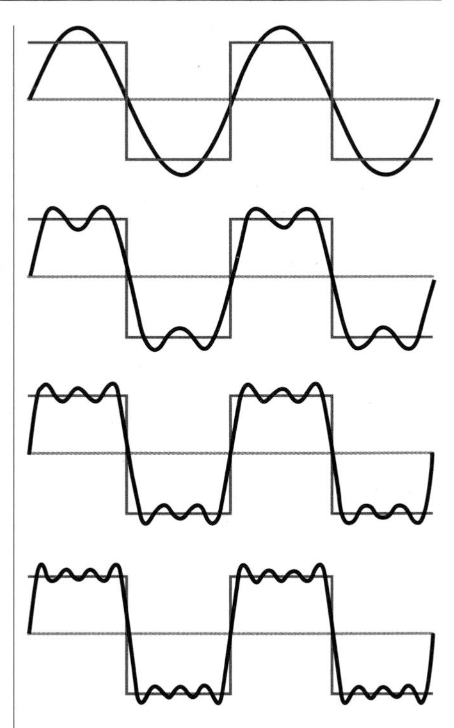

A Fourier series can approximate a wave of any shape—even a square one (shown here in pink). Adding more sine waves to the series gives a closer and closer approximation of the square wave. The first four approximations in the series (shown here in black) each incorporate an extra sine wave.

Joseph Fourier

The son of a tailor, Joseph Fourier was born in Auxerre, France. Orphaned at 10, he was taken into a local convent before going on to a military school, where he excelled at mathematics. France was in the throes of revolution, and during the Terror of 1794, he was briefly imprisoned after falling out with fellow revolutionaries.

After the Revolution, Fourier accompanied Napoleon on an expedition to Egypt in 1798. He was made governor of Egypt and put in charge of the study of ancient Egyptian relics. Returning to France in 1801, Fourier was made governor of Isère in the Alps. In between administrative duties overseeing road building and drainage planning, he published a groundbreaking study of ancient Egypt and started his studies of heat. He died in 1831 after tripping and falling down a flight of stairs.

Key works

1807 *On the Propagation of Heat in Solid Bodies*
1822 *The Analytic Theory of Heat*

THE ARTIFICIAL PRODUCTION OF ORGANIC SUBSTANCES FROM INORGANIC SUBSTANCES

FRIEDRICH WÖHLER (1800–1882)

IN CONTEXT

BRANCH
Chemistry

BEFORE
1770s Antoine Lavoisier and others show that water and salt can return to their former state after heating, but sugar or wood cannot.

1807 Jöns Jakob Berzelius suggests a fundamental difference between organic and inorganic chemicals.

AFTER
1852 British chemist Edward Franklin suggests the idea of valency, the ability of atoms to combine with other atoms.

1858 British chemist Archibald Couper suggests the idea of bonds between atoms, explaining how valency works.

1858 Couper and August Kekulé propose that organic chemicals are made by chains of bonded carbon atoms with side branches of other atoms.

In 1807, the Swedish chemist Jöns Jakob Berzelius suggested that a fundamental difference existed between the chemicals involved in living things and all other chemicals. These unique, "organic" chemicals, Berzelius argued, could only be assembled by living things themselves and, once broken down, could not be remade artificially. His idea conformed with the prevailing theory known as "vitalism," which held that life was special and that living things were endowed with a "life force" beyond the understanding of chemists. So it came as a surprise when the pioneering experiments of a German chemist named Friedrich Wöhler showed that organic chemicals are not unique at all, but behave according to the same basic rules as all chemicals.

We now know that organic chemicals comprise a multitude of molecules based on the element carbon. These carbon-based molecules are indeed essential components of life, but many can be synthesized from inorganic chemicals—as Wöhler discovered.

Widely used in fertilizers, urea is rich in nitrogen, which is essential to the growth of plants. Synthetic urea, first made by Wöhler, is now a key raw material in the chemical industry.

Chemistry rivals

Wöhler's breakthrough came about because of a scientific rivalry. In the early 1820s, Wöhler and fellow chemist Justus von Liebig both came up with identical chemical analyses for what seemed to be two very different substances—silver fulminate, which is explosive, and silver cyanate, which is not. Both men assumed that the other had the wrong results, but after corresponding, they found they were both right. This group of compounds led chemists to realize that substances are defined not just

See also: Antoine Lavoisier 84 ▪ John Dalton 112–13 ▪ Jöns Jakob Berzelius 119 ▪ Leo Baekeland 140–41 ▪ August Kekulé 160–65

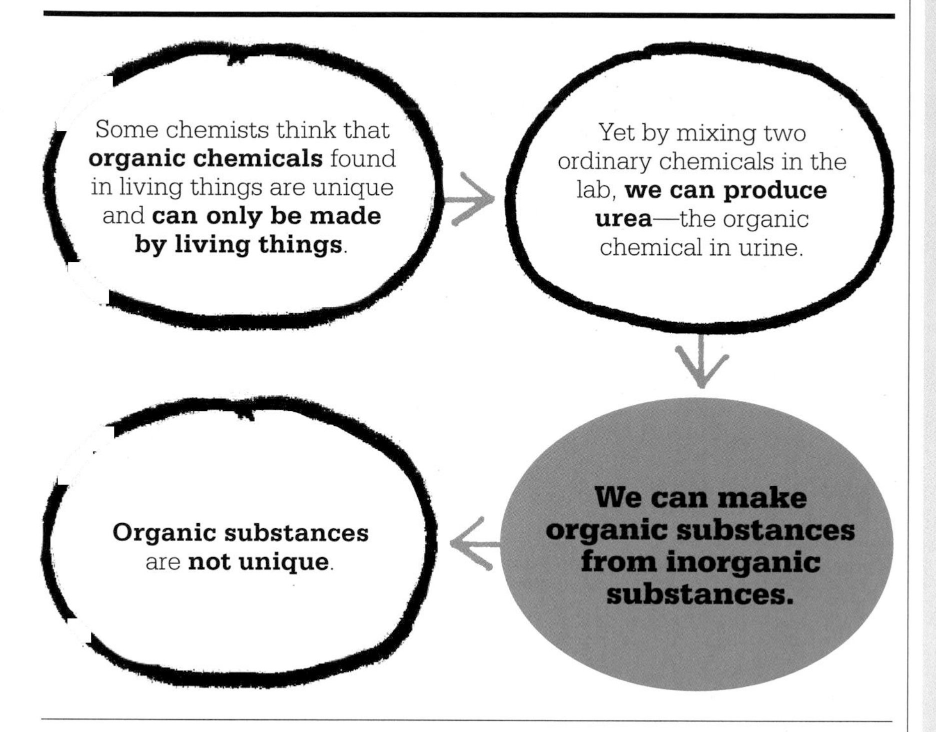

by the number and kinds of atoms in the molecule but also by the atoms' arrangement. The same formula may apply to different structures with different properties—these different structures were later named isomers by Berzelius.

Wöhler and Liebig went on to forge a brilliant partnership, but it was Wöhler alone who, in 1828, stumbled upon the truth about organic chemicals.

The Wöhler synthesis

Wöhler was mixing silver cyanate with ammonium chloride, expecting to get ammonium cyanate. Instead, he got a white substance that had different properties from ammonium cyanate. The same powder appeared when he mixed lead cyanate with ammonium hydroxide. Analysis showed the white powder to be urea—an organic substance that is a key component of urine, and has the same chemical formula as ammonium cyanate. According to Berzelius's theory, it could be made only by living things—yet Wöhler had synthesized it from inorganic chemicals. Wöhler wrote to Berzelius: "I must tell you that I can make urea without the use of kidneys," explaining that urea was in fact an isomer of ammonium cyanate.

The significance of Wöhler's discovery took many years to sink in. Even so, it paved the way for the development of modern organic chemistry, which not only reveals how all living things depend on chemical processes, but enables the artificial synthesis of valuable organic chemicals on a commercial scale. In 1907, a synthetic polymer called Bakelite was produced from two such chemicals and ushered in the "Age of Plastics" that shaped the modern world. ■

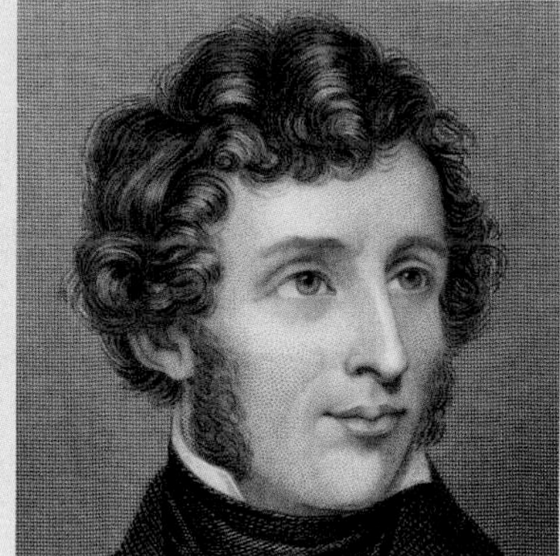

Friedrich Wöhler

Born in Eschersheim, near Frankfurt in Germany, Friedrich Wöhler trained in obstetrics at the University of Heidelberg. But chemistry was his passion and, in 1823, he went to study with Jöns Jakob Berzelius in Stockholm. On his return to Germany, he embarked on a remarkable and varied career in chemical research and innovation.

Besides the first artificial synthesis of an organic substance, Wöhler's many discoveries—often made with Justus von Liebig—included aluminum, beryllium, yttrium, titanium, and silicon. He also helped to develop the idea of "radicals"—basic molecular groups from which other substances are built. Although later disproved, this theory paved the way for today's understanding of how molecules assemble. In later years, Wöhler became an authority on the chemistry of meteorites and helped set up a factory for purifying nickel.

Key works

1830 *Summary of Inorganic Chemistry*
1840 *Summary of Organic Chemistry*

WINDS NEVER BLOW IN A STRAIGHT LINE

GASPARD-GUSTAVE DE CORIOLIS (1792–1843)

IN CONTEXT

BRANCH
Meteorology

BEFORE
1684 Isaac Newton introduces the idea of centripetal force, stating that any motion in a curved path must be the result of a force acting on it.

1735 George Hadley suggests that trade winds blow toward the equator because Earth's rotation deflects air currents.

AFTER
1851 Léon Foucault shows how the swing of a pendulum is deflected by Earth's rotation.

1856 US meteorologist William Ferrel shows that winds blow parallel to isobars—lines that connect points of equal atmospheric pressure.

1857 Dutch meteorologist Christophorus Buys Ballot formulates a rule stating that if the wind is blowing on your back, an area of low pressure is to your left.

Air and ocean currents do not flow in straight lines. As the currents move, they are deflected to the right in the northern hemisphere, and to the left in the southern. In the 1830s, French scientist Gaspard-Gustave de Coriolis discovered the principle behind this effect, now known as the Coriolis effect.

Deflected by rotation

Coriolis got his ideas from studying turning waterwheels, but meteorologists later realized that the ideas apply to the way winds and ocean currents move.

Coriolis showed how, when an object is moving across a rotating surface, its momentum seems to carry it on a curved path. Imagine throwing a ball out from the center of a spinning merry-go-round. The ball appears to curve around—even though to anyone watching from outside the merry-go-round it is actually moving in a straight line.

Winds on the rotating Earth are deflected in the same way. Without the Coriolis effect, winds would simply blow straight from high pressure areas to low pressure areas. The wind direction is in fact a balance between the pull of low pressure and the Coriolis deflection. This is mostly why winds circle counterclockwise into low pressure zones in the northern hemisphere, and clockwise in the southern hemisphere. Similarly, ocean surface currents circulate in giant loops or gyres, clockwise in the northern hemisphere and counterclockwise in the south. ■

Earth's rotation causes winds to be deflected to the right in the northern hemisphere and left in the southern.

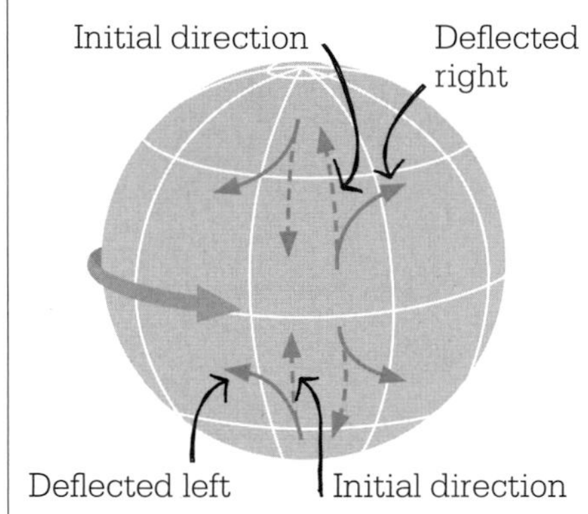

See also: George Hadley 80 ■ Robert FitzRoy 150–55

ON THE COLORED LIGHT OF THE BINARY STARS

CHRISTIAN DOPPLER (1803–1853)

IN CONTEXT

BRANCH
Physics

BEFORE
1677 Ole Rømer estimates the speed of light by studying Jupiter's moons.

AFTER
1840s Dutch meteorologist Christophorus Buys Ballot applies the Doppler shift to sound waves, as does French physicist Hippolyte Fizeau to electromagnetic waves.

1868 British astronomer William Huggins uses redshift to find the velocity of a star.

1929 Edwin Hubble relates the redshift of galaxies to their distance from Earth, showing the expansion of the universe.

1988 The first extrasolar planet is detected, using the Doppler shift of light from the star that it orbits—the star appears to "wobble" as the planet's gravitational pull disrupts its rotation.

The color of light depends on its frequency, which is the number of waves per second. If something moving toward us is emitting waves, the second wave will have a shorter distance to travel than the first wave, so it will arrive sooner than it would if the source were stationary. Thus the frequency of waves increases if the source and receiver are getting closer to each other, and decreases if they are moving apart. This effect applies to all types of wave, including sound, and is responsible for the changing pitch of a siren as an ambulance passes.

To the naked eye, most stars appear to be white, but through a telescope many can be seen to be red, yellow, or blue. In 1842, an Austrian physicist named Christian Doppler suggested that the red color of some stars is due to the fact that they are moving away from the Earth, which would shift their light to longer wavelengths. Since the longest wavelength of visible light is red, this became known as redshift (as illustrated on p.241). The colors of stars are now known to be mainly due to their temperature (the hotter the star, the more blue it appears), but the movement of some stars can be detected through Doppler shifts. Binary stars are pairs of stars orbiting each other. Their rotation causes an alternating redshift and blueshift in the light they emit. ■

The heavens presented an extraordinary appearance, for all the stars directly behind me were now deep red, while those directly ahead were violet. Rubies lay behind me, amethysts ahead of me.

Olaf Stapledon
From his novel, *Star Maker* (1937)

See also: Ole Rømer 58–59 ■ Edwin Hubble 236–41 ■ Geoffrey Marcy 327

THE GLACIER WAS GOD'S GREAT PLOUGH

LOUIS AGASSIZ (1807–1873)

IN CONTEXT

BRANCH
Earth science

BEFORE
1824 Norwegian Jens Esmark suggests that glaciers are responsible for the creation of fjords, erratics, and moraines.

1830 Charles Lyell argues that the laws of nature have always been the same, so the clues to the past lie in the present.

1835 Swiss geologist Jean de Charpentier argues that erratics near Lake Geneva were transported by ice from the Mont Blanc area in an "Alpine glaciation."

AFTER
1875 Scottish scientist James Croll argues that variations in Earth's orbit could explain the temperature changes that cause an ice age.

1938 Serbian physicist Milutin Milankovic relates changes in climate to periodic changes in Earth's orbit.

Retreating glaciers leave **particular features** behind them in the landscape.

↓

These features are found in areas where **there are no glaciers**.

↓

There must have been **glaciers** in these places some time **in the past**.

When glaciers sweep across a landscape, they leave signature features behind them. Glaciers can scour rocks flat or leave them smoothly rounded, often with striations (scratch marks) showing the direction in which the ice moved. They also leave behind erratics—boulders that have been carried long distances by the ice. These can usually be identified because their composition is different from the rocks on which they lie. Many erratics are too large to have been moved by rivers, which is the usual way that rocks are carried across a landscape. A rock of a different kind from rocks around it, therefore, is a telltale sign that a glacier once passed by. Another is the presence of moraines in valleys. These are piles of boulders that were pushed aside when the glacier was growing, and left behind when it retreated.

Riddle of the rocks

Geologists in the 19th century recognized such features as striations, erratics, and moraines as evidence of glaciers. What they

See also: WIlliam Smith 115 ▪ Alfred Wegener 222–23

could not explain was why such features were found in areas on Earth that had no glaciers. One theory argued that rocks were moved by repeated flooding. Floods could explain the "boulder drift" (the sands, clays, and gravels that included erratic boulders) that overlay much of the bedrock of Europe. The material might have been deposited when the last flood retreated. The largest erratics could have been caught up in icebergs, which deposited the rocks when they melted. But the theory could not explain all of the features.

The ice age revealed

During the 1830s, Swiss geologist Louis Agassiz spent several vacations in the European Alps studying glaciers and their valleys. He realized that glacial features everywhere, not just in the Alps, could be explained if Earth had once been covered in far more ice than at present. The glaciers of today must be the remnants of ice sheets that had at one time covered most of the globe. But before he published his theory Agassiz wanted to convince others. He had met William Buckland, a prominent English geologist, while excavating fossil fishes in the Old Red Sandstone rocks in the Alps. When Agassiz showed him the evidence for his theory of an ice age, Buckland was convinced, and in 1840 the two men toured Scotland to look for evidence of glaciation there. After the tour, Agassiz presented his ideas to the Geological Society of London. Although he had convinced Buckland and Charles Lyell—two of the leading geologists of the day—the other members of the society were unimpressed. A nearly global glaciation seemed no more probable than a global flood. However, the idea of ice ages gradually gained acceptance, and today there is evidence from many different fields of geology that ice has covered much of Earth's surface many times in the past. ■

Agassiz was the first to suggest that large erratics, such as these in the Caher Valley of Ireland, were deposited by ancient glaciers.

Louis Agassiz

Born in a small Swiss village in 1807, Louis Agassiz studied to be a physician, but became a professor of natural history at the University of Neuchâtel. His first scientific work, under the French naturalist Georges Cuvier, involved classifying freshwater fish from Brazil, and Agassiz went on to undertake extensive work on fossilized fish. In the late 1830s, his interests spread to glaciers and zoological classification. In 1847, he took a post at Harvard University in the US.

Agassiz never accepted Darwin's theory of evolution, believing that species were "ideas in the mind of God" and that all species had been created for the regions they inhabited. He advocated "polygenism," a belief that different human races did not share a common ancestor, but were created separately by God. In recent years, his reputation has been tarnished by his apparent advocacy of racist ideas.

Key works

1840 *Study on Glaciers*
1842–46 *Nomenclator Zoologicus*

NATURE CAN BE REPRESENTED AS ONE GREAT WHOLE

ALEXANDER VON HUMBOLDT (1769–1859)

IN CONTEXT

BRANCH
Biology

BEFORE
5th–4th century BCE Ancient Greek writers observe the web of interrelationships between plants, animals, and their environment.

AFTER
1866 Ernst Haeckel coins the word "ecology."

1895 Eugenius Warming publishes the first university course book on ecology.

1935 Alfred Tansley coins the word "ecosystem."

1962 Rachel Carson warns of the dangers of pesticides in *Silent Spring*.

1969 Friends of the Earth and Greenpeace are established.

1972 James Lovelock's Gaia hypothesis presents Earth as a single organism.

The study of the interrelationship between the animate and inanimate world, known as ecology, only became a subject of rigorous and methodical scientific investigation over the last 150 years. The term "ecology" was coined in 1866 by the German evolutionary biologist, Ernst Haeckel, and is derived from the Greek words *oikos*, meaning house or dwelling place, and *logos*, meaning study or discourse. But it is an earlier German polymath named Alexander von Humboldt who is regarded as the pioneer of modern ecological thinking.

Through extensive expeditions and writings, Humboldt promoted a new approach to science. He sought to understand nature as a unified whole, by interrelating all of the physical sciences and employing the latest scientific equipment, exhaustive observation, and meticulous analysis of data on an unprecedented scale.

The principal impulse by which I was directed was the earnest endeavour to comprehend the phenomena of physical objects in their general connection, and to represent nature as one great whole, moved and animated by internal forces.
Alexander von Humboldt

The crocodile's teeth

Although Humboldt's holistic approach was new, the concept of ecology developed from early investigations of natural history by ancient Greek writers, such as Herodotus in the 5th century BCE. In one of the first accounts of interdependence, technically known as mutualism, he describes crocodiles on the Nile River in Egypt opening their mouths to allow birds to pick their teeth clean.

A century later, observations by the Greek philosopher Aristotle and his pupil Theophrastus on species' migration, distribution, and behavior provided an early version of the concept of the ecological niche—the particular place in nature that shapes and is shaped by a species' way of life. Theophrastus studied and wrote extensively on plants, realizing the importance of climate and soils to their growth and distribution. Their ideas influenced natural philosophy for the next 2,000 years.

Humboldt's team climbed Mexico's Jorullo volcano in 1803, just 44 years after it first appeared. Humboldt linked geology to meteorology and biology by studying where different plants lived.

See also: Jean-Baptiste Lamarck 118 ▪ Charles Darwin 142–49 ▪ James Lovelock 315

Nature's unifying forces

Humboldt's approach to nature followed in the late 18th-century Romantic tradition that reacted to rationalism by insisting on the value of senses, observation, and experience in understanding the world as a whole. Like his contemporaries, the poets Johann Wolfgang von Goethe and Friedrich Schiller, Humboldt promoted the idea of the unity (or *Gestalt* in German) of nature—and of natural philosophy and the humanities. His studies ranged from anatomy and astronomy to mineralogy and botany, commerce, and linguistics, and provided him with the breadth of knowledge necessary for his exploration of the natural world beyond the confines of Europe.

As Humboldt explained, "The sight of exotic plants, even of dried specimens in a herbarium, fired my imagination and I longed to see the tropical vegetation in southern countries with my own eyes." His five-year exploration of Latin America with the French botanist Aimé Bonpland was his most important expedition. Setting out in June 1799, he declared, "I shall collect plants and fossils, and make astronomical observations with the best of instruments. Yet this is not the main purpose of my journey. I shall endeavor to discover how nature's forces act upon one another and in what manner the geographic environment exerts its influence on animals and plants. In short, I must find out about the harmony in nature." And he did just that.

Among many other projects, Humboldt measured ocean water temperature and suggested the use of "isolines," or isothermal lines, to connect points of equal temperature as a means of characterizing and mapping the global environment, especially the climate, and then comparing the climatic conditions in various countries.

Humboldt was also one of the first scientists to study how physical conditions—such as climate, altitude, latitude, and soils—affected the distribution of life. With Bonpland's assistance, he mapped the changes in flora and fauna between sea level and high altitude in the Andes. In 1805, the year after his return from the Americas, he published a now-celebrated work on the geography of the area, summarizing the interconnectedness of nature and illustrating the altitudinal zones of vegetation. Years later, in 1851, he showed the global application of these zones by comparing the Andean zones with those of the European Alps, Pyrenees, Lapland, Tenerife, and the Asian Himalayas.

Defining ecology

When Haeckel coined the word "ecology," he too was following in the tradition of viewing a *Gestalt* (unity) of the living and inanimate world. An enthusiastic evolutionist, he was inspired by Charles Darwin, whose publication of *On the Origin of Species* in 1859 banished the notion of Earth as an immutable world. Haeckel questioned the role of natural selection, but believed that the environment played an important role in both evolution and ecology. »

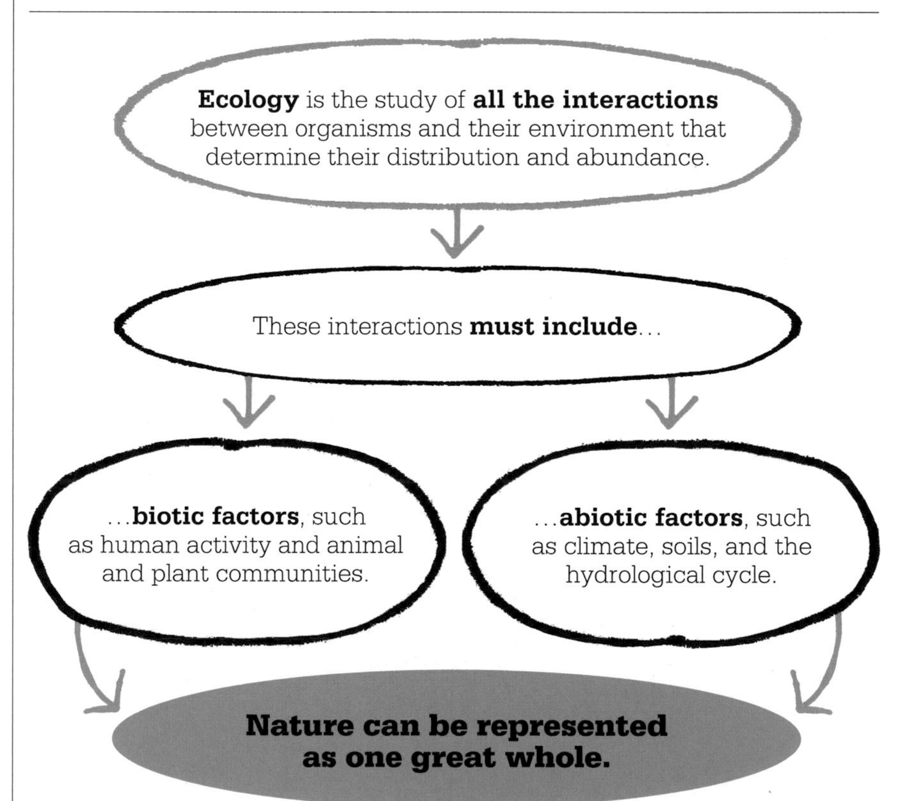

By the end of the 19th century, the first university course in ecology was being taught by the Danish botanist Eugenius Warming, who also wrote the first ecology textbook *Plantesamfund* (*Plant Ecology*) in 1895. From Humboldt's pioneering work, Warming developed the global geographical subdivision of plant distribution known as biomes, such as the tropical rain-forest biome, which are largely based on the interaction of plants with the environment, especially climate.

Individuals and community

Early in the 20th century, the modern definition of ecology developed as the scientific study of the interactions that determine the distribution and abundance of organisms. These interactions include an organism's environment, encompassing all those factors that influence it—both biotic (living organisms) and abiotic (nonliving factors such as soil, water, air, temperature, and sunlight). The scope of modern ecology ranges from the individual organism to populations of individuals of the same species, and the community, made up of populations that share a particular environment.

Many of the basic terms and concepts of ecology came from the work of several pioneer ecologists in the first few decades of the 20th century. The formal concept of the biological community was first developed in 1916 by the American botanist Frederic Clements. He believed that the plants of a given area develop a succession of communities over time, from an initial pioneer community to an optimal climax community within which successive communities of different species adjust to one another to form a tightly integrated and interdependent unit, similar to the organs of a body. Clements' metaphor of the community as a "complex organism" was criticized at first but influenced later thinking.

This whole chain of poisoning, then, seems to rest on a base of minute plants which must have been the original concentrators.
Rachel Carson

The idea of further ecological integration at a higher level than the community was introduced in 1935 with the concept of the ecosystem, developed by the English botanist Arthur Tansley. An ecosystem consists of both living and nonliving elements. Their interaction forms a stable unit with a sustaining flow of energy from the environmental to the living part (through the food chain) and can operate on all scales, from a puddle to an ocean or the whole planet.

Studies of animal communities by the English zoologist Charles Elton led him to develop in 1927 the concept of the food chain and food cycle, subsequently known as the "food web." A food chain is formed by the transfer of energy through an ecosystem from primary producers (such as green plants on land) through a series of consuming

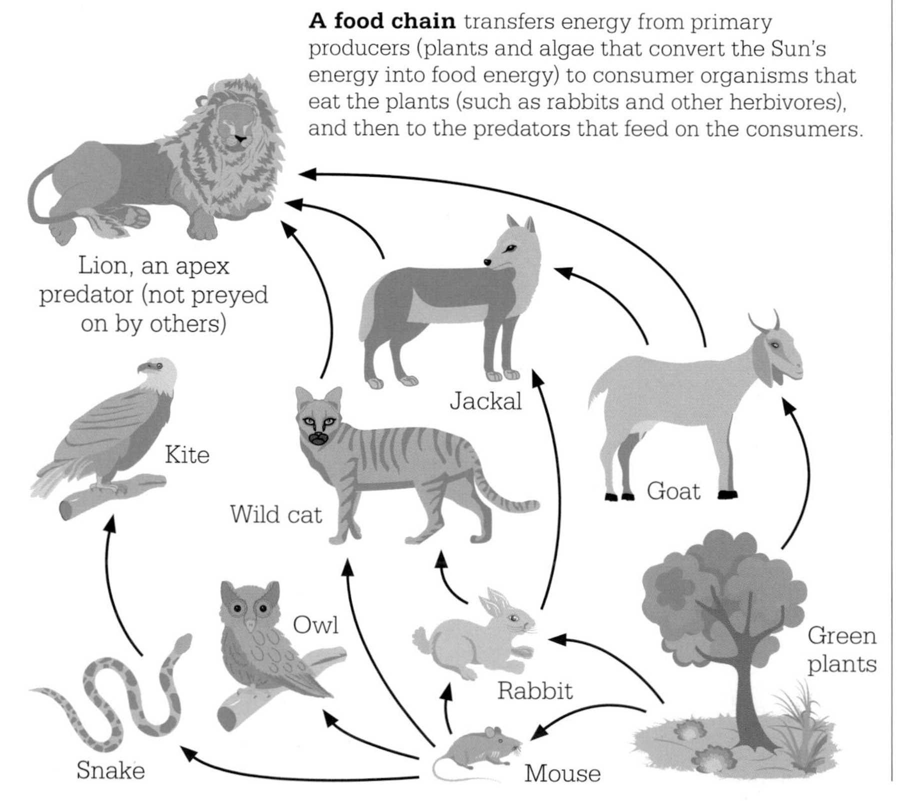

A food chain transfers energy from primary producers (plants and algae that convert the Sun's energy into food energy) to consumer organisms that eat the plants (such as rabbits and other herbivores), and then to the predators that feed on the consumers.

Rachel Carson (far right) made a significant contribution to the science and public understanding of ecology by drawing attention to the destructive impact of pollution on the environment.

organisms. Elton also recognized that particular groups of organisms occupied certain niches in the food chain for periods of time. Elton's niches include not only the habitats but also the resources upon which the occupying organisms rely for sustenance. The dynamics of energy transfer through trophic (feeding) levels were studied by the American ecologists Raymond Lindeman and Robert MacArthur, whose mathematical models helped change ecology from primarily a descriptive science into an experimental one.

The green movement

A boom in popular and scientific interest in ecology in the 1960s and 1970s led to the development of the environmental movement with a whole range of concerns, stimulated by powerful advocates such as the American marine biologist Rachel Carson. Her 1962 book *Silent Spring* documented the harmful effects on the environment of man-made chemicals such as the pesticide DDT. The first image of Earth seen from space, taken by *Apollo 8* astronauts in 1968, awakened public awareness of the planet's fragility. In 1969, the organizations Friends of the Earth and Greenpeace were established, with the mission to "ensure the ability of the Earth to nurture life in all its diversity." Environmental protection, along with clean and renewable energy, organic foods, recycling, and sustainability, were all on the political agenda in both North America and Europe, and national conservation agencies were established based on the science of ecology. Recent decades have seen growing concern over global climate change and its impact on the environment and present ecosystems, many of which are already threatened from human activity. ■

Alexander von Humboldt

Born in Berlin to a wealthy and well-connected family, Humboldt studied finance at the University of Frankfurt, natural history and linguistics in Göttingen, language and commerce in Hamburg, geology in Freiburg, and anatomy in Jena. The death of his mother in 1796 provided Humboldt with the means to fund an expedition to the Americas from 1799 to 1804, accompanied by botanist Aimé Bonpland. Using the latest scientific equipment, Humboldt measured everything from plants to population statistics and minerals to meteorology.

On his return, Humboldt was honored across Europe. Based in Paris, he took 21 years to process and publish his data in over 30 volumes, and then synthesized his ideas in four volumes titled *Kosmos*. A fifth volume was completed after his death in Berlin at 89. Darwin called him "the greatest scientific traveller who ever lived."

Key works

1825 *Journey to the Equinoctial Regions of the New Continent*
1845–1862 *Kosmos*

LIGHT TRAVELS MORE SLOWLY IN WATER THAN IN AIR

LÉON FOUCAULT (1819–1868)

IN CONTEXT

BRANCH
Physics

BEFORE
1676 Ole Rømer makes the first successful estimate of the speed of light, using eclipses of Io, one of Jupiter's moons.

1690 Christiaan Huygens publishes his *Treatise on Light*, in which he proposes that light is a type of wave.

1704 Isaac Newton's *Opticks* suggests that light is a stream of "corpuscles."

AFTER
1864 James Clerk Maxwell realizes that the speed of electromagnetic waves is so nearly the same as the speed of light that light must be a form of electromagnetic wave.

1879–83 German-born US physicist Albert Michelson refines Foucault's method and obtains a measurement for the speed of light (through air) that is very close to today's value.

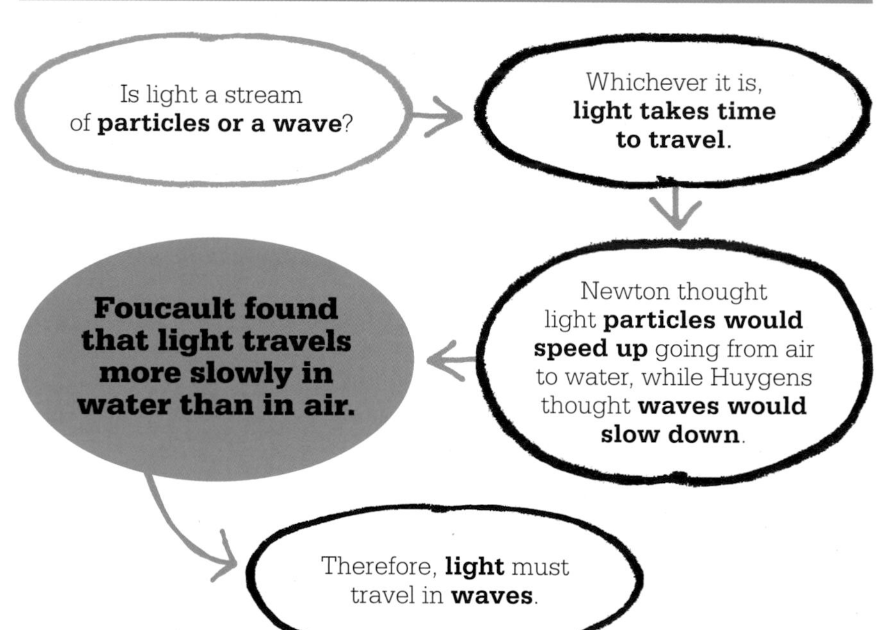

In the 17th century, scientists began to investigate light, and whether it had a finite, measurable speed. In 1690, Christiaan Huygens published his theory that light is a pressure wave, moving in a mysterious fluid called ether. Huygens thought of light as a longitudinal wave, and predicted that the wave would travel more slowly through glass or water than through air. In 1704, Isaac Newton published his theory of light as a stream of "corpuscles," or particles. Newton's explanation for refraction—the bending of a beam of light as it passes from one transparent material to another—assumed that light travels faster after it passes from air into water.

Estimates for the speed of light relied on astronomical phenomena, showing how fast light travels through space. The first terrestrial

See also: Christiaan Huygens 50–51 ▪ Ole Rømer 58–59 ▪ Isaac Newton 62–69 ▪ Thomas Young 110–11 ▪ James Clerk Maxwell 180–85 ▪ Albert Einstein 214–21 ▪ Richard Feynman 272–73

Above all we must be accurate, and it is an obligation which we intend to fulfill scrupulously.
Léon Foucault

measurement was carried out by French physicist Hippolyte Fizeau in 1849. A beam of light was shone through a gap between the teeth of a rotating cogwheel. That light was then reflected by a mirror that was positioned 5 miles (8 km) away, and passed back through the next gap between the wheel's teeth. Taking the precise speed of rotation that allowed this to happen, together with time and distance, Fizeau calculated the speed of light as 194,489 miles/s (313,000 km/s).

Contradicting Newton

In 1850, Fizeau collaborated with fellow physicist León Foucault, who adapted his apparatus—and made it much smaller—by reflecting the beam of light off a rotating mirror instead of passing it through the cogwheel. Light shining at the rotating mirror would only be reflected toward the distant mirror when the rotating mirror was at the correct angle. Light returning from the fixed mirror was reflected by the rotating mirror again, but since this mirror had moved while the light was traveling, it was not reflected directly back toward the source. The speed of light could now be calculated from the angle between the light going to and from the rotating mirror and the speed of rotation of the mirror.

The speed of light in water could be measured by putting a tube of water in the apparatus between the rotating and stationary mirrors. Using this apparatus, Foucault established that light traveled more slowly in water than in air. As such, he argued, light could not be a particle, and the experiment was viewed at the time as a refutation of Newton's theory of corpuscles. Foucault refined his apparatus further, and in 1862, measured the speed of light in air as 185,168 miles/s (298,000 km/s)—remarkably close to today's value of 186,282 miles/s (299,792 km/s). ■

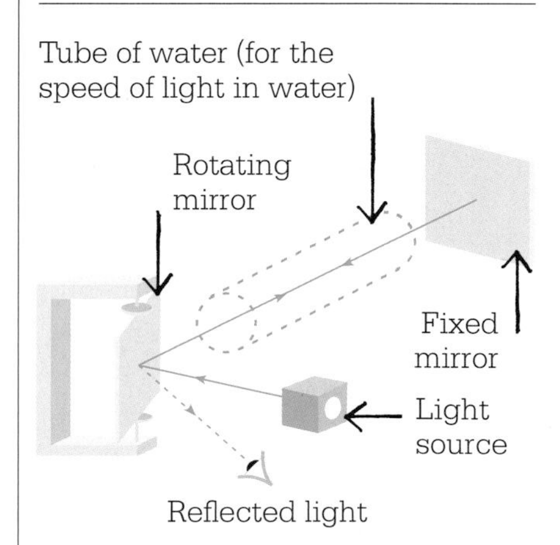

In Foucault's experiment, the speed of light was calculated from the difference in angle as a beam of light reflected back and forth between a rotating mirror and a fixed mirror.

Léon Foucault

Born in Paris, France, Léon Foucault was educated mainly at home before entering medical school, where he studied under the bacteriologist Alfred Donné. Since he could not bear the sight of blood, Foucault soon gave up his studies, became Donné's laboratory assistant, and devised a way of taking photographs through a microscope—he later teamed up with Hippolyte Fizeau to take the first ever photograph of the Sun. In addition to measuring the speed of light, Foucault is best known for providing experimental evidence of Earth's rotation, using a pendulum in 1851 and later a gyroscope. Although he had no formal training in science, a post was created for Foucault at the Imperial Observatory in Paris. He was also made a member of several scientific societies, and is one of 72 French scientists named on the Eiffel Tower.

Key works

1851 *Demonstration of Physical Movement of Rotation of the Earth by Means of the Pendulum*
1853 *On the Relative Velocities of the Light in Air and in Water*

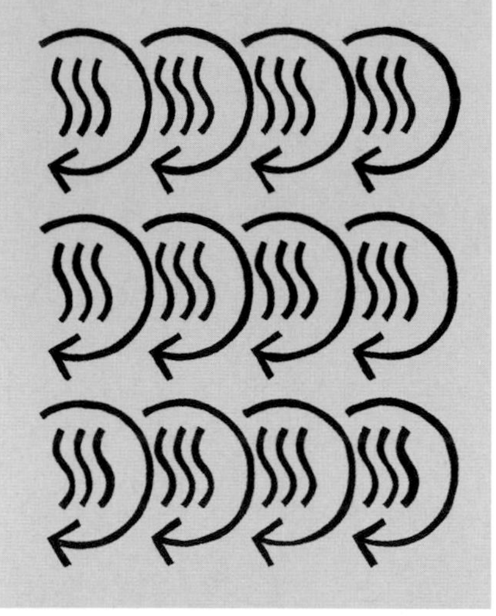

LIVING FORCE MAY BE CONVERTED INTO HEAT

JAMES JOULE (1818–1889)

IN CONTEXT

BRANCH
Physics

BEFORE
1749 French mathematician Émilie du Châtelet derives her law of the conservation of energy from Newton's laws.

1824 French engineer Sadi Carnot states that there are no reversible processes in nature, paving the way for the second law of thermodynamics.

1834 French physicist Émile Clapeyron develops Carnot's work, stating a version of the second law of thermodynamics.

AFTER
1850 German physicist Rudolf Clausius gives the first clear statement of the first and second laws of thermodynamics.

1854 Scottish engineer William Rankine adds the concept that is later named entropy (a measure of disorder) in the transformation of energy.

The principle of the conservation of energy states that energy is never lost but only changed in form. But in the 1840s, scientists had only a vague idea of what energy was. It was a British brewer's son, James Joule, who showed that heat, mechanical movement, and electricity are interchangeable forms of energy, and that when one is changed to another the total energy remains the same.

Converting energy

Joule began his experiments in a laboratory in the family home. In 1841, he figured out how much heat an electric current generates. He experimented with converting mechanical movement into heat, and developed an experiment in which a falling weight turns a paddle wheel in water, heating the water. By measuring the rise in temperature of the water, Joule was able to figure out the exact amount of heat a certain amount of mechanical work would create. He went on to assert that no energy was ever lost in this conversion. His ideas were largely ignored until 1847, when German physicist Hermann Helmholtz published a paper summarizing the theory of the conservation of energy, and Joule then presented his work at the British Association in Oxford. The standard unit of energy, a joule, is named after him. ■

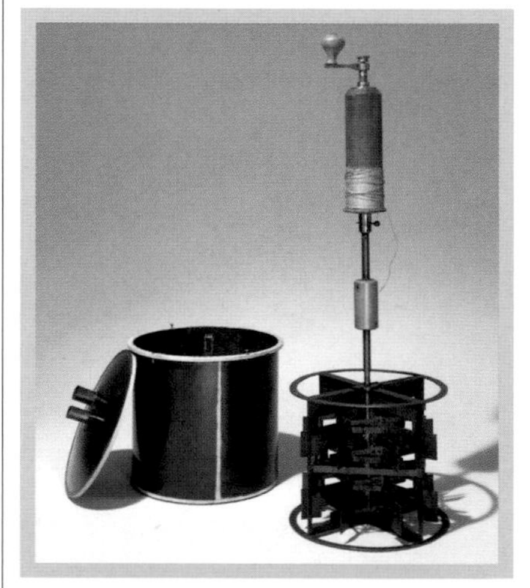

In Joule's experiment, a falling weight drove a paddle that turned inside a bucket of water. The energy of the movement was changed into heat.

See also: Isaac Newton 62–69 ▪ Joseph Black 76–77 ▪ Joseph Fourier 122–23

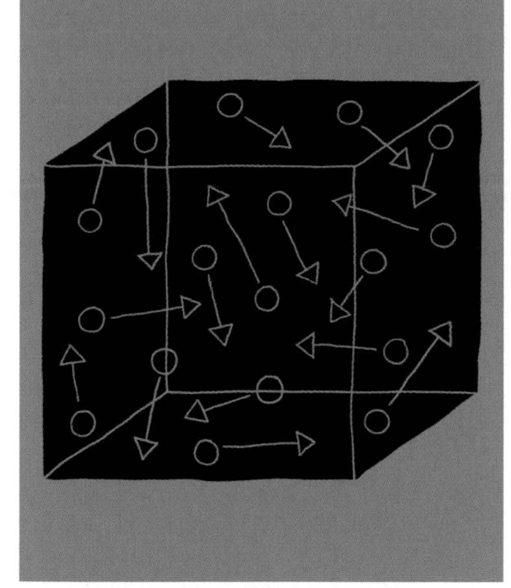

STATISTICAL ANALYSIS OF MOLECULAR MOVEMENT

LUDWIG BOLTZMANN (1844–1906)

IN CONTEXT

BRANCH
Physics

BEFORE
1738 Daniel Bernoulli suggests that gases are made of moving molecules.

1827 Scottish botanist Robert Brown identifies the movement of pollen in water, which becomes known as Brownian motion.

1845 Scottish physicist John Waterston describes how energy among gas molecules is distributed according to statistical rules.

1857 James Clerk Maxwell calculates the mean speed of molecules and the mean distance between collisions.

AFTER
1905 Albert Einstein analyzes Brownian motion mathematically, showing how it is the result of the impact of molecules.

By the middle of the 19th century, atoms and molecules had become central ideas in chemistry, and most scientists understood that they were the key to the identity and behavior of elements and compounds. Few thought they had much relevance to physics, but in the 1880s, Austrian physicist Ludwig Boltzmann developed the kinetic theory of gases, putting atoms and molecules right at the heart of physics, too.

Available energy is the main object at stake in the struggle for existence and the evolution of the world.
Ludwig Boltzmann

In the early 18th century, Swiss physicist Daniel Bernoulli had suggested that gases are made of a multitude of moving molecules. It is their impact that creates pressure and their kinetic energy (the energy of their movement) that creates heat. In the 1840s and 1850s, scientists had begun to realize that the properties of gases reflect the average movement of the countless particles. In 1859, James Clerk Maxwell calculated the speed of molecules and how far they traveled before colliding, showing that temperature is a measure of the average speed of the molecules.

Centrality of statistics

Boltzmann revealed how important the statistics are. He showed that the properties of matter are simply a combination of the basic laws of motion and the statistical rules of probability. Following this principle, he calculated a number now called the Boltzmann constant, providing a formula linking the pressure and volume of a gas to the number and energy of its molecules. ■

See also: John Dalton 112–13 ▪ James Joule 138 ▪ James Clerk Maxwell 180–85 ▪ Albert Einstein 214–21

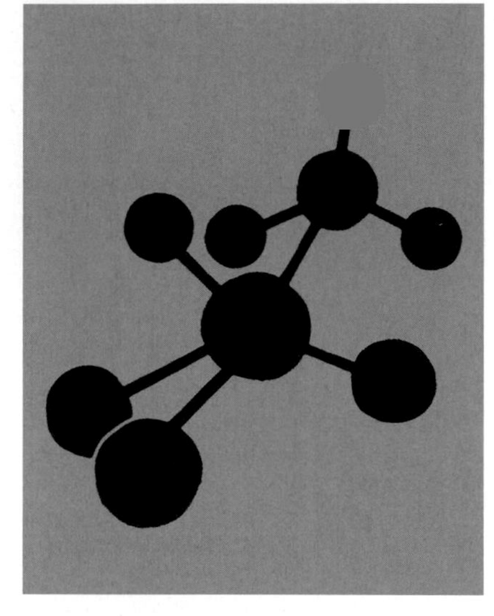

PLASTIC IS NOT WHAT I MEANT TO INVENT

LEO BAEKELAND (1863–1944)

IN CONTEXT

BRANCH
Chemistry

BEFORE
1839 Berlin apothecary Eduard Simon distils styrol resin from the Turkish sweet-gum tree. A century later, this is developed into polystyrene by the German IG Farben company.

1862 Alexander Parkes develops the first synthetic plastic, Parkesine.

1869 American John Hyatt creates celluloid, which is soon used instead of ivory to make billiard balls.

AFTER
1933 British chemists Eric Fawcett and Reginald Gibson of the ICI company create the first practical polythene.

1954 Italian Giulio Natta and German Karl Rehn independently invent polypropylene, now the most widely used plastic.

The discovery of synthetic plastics in the 19th century opened the way to the creation of a huge range of solid materials unlike anything that had ever been known before—light, noncorroding, and capable of being molded into almost any imaginable shape. While plastics can occur naturally, all of the plastics now in widespread use are entirely synthetic. In 1907, Belgian-born American inventor Leo Baekeland created one of the first commercially successful plastics, now known as Bakelite.

What gives plastic its special quality is the shape of its molecules. With only a few exceptions, plastics are made from long organic molecules, known as polymers, strung together from many smaller molecules, or monomers. A few polymers occur naturally, such as cellulose, the main woody substance in plants.

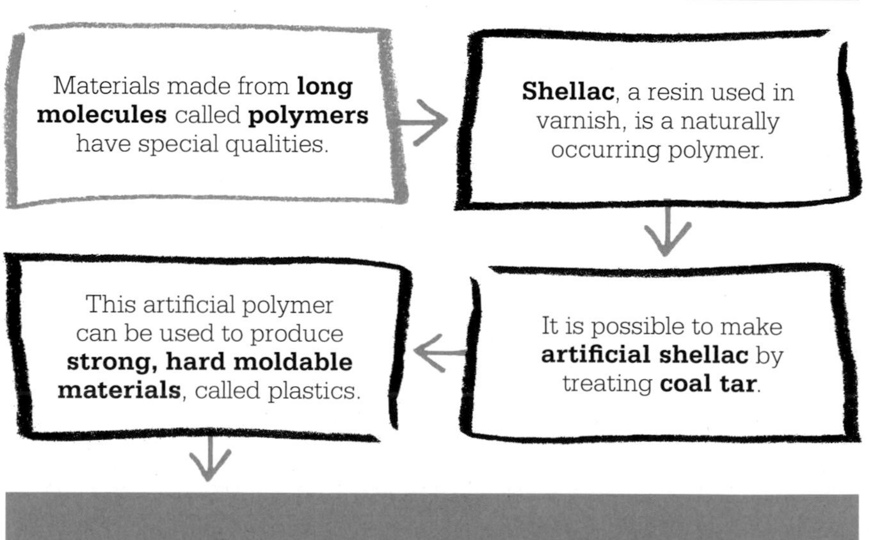

See also: Friedrich Wöhler 124–25 ▪ August Kekulé 160–65 ▪ Linus Pauling 254–59 ▪ Harry Kroto 320–21

I was trying to make something really hard, but then I thought I should make something really soft instead, that could be moulded into different shapes. That was how I came up with the first plastic.
Leo Baekeland

Although the molecules of natural polymers were far too complex to figure out in the 1800s, some scientists began to explore ways of making them synthetically from chemical reactions. In 1862, British chemist Alexander Parkes created a synthetic form of cellulose, which he called Parkesine. A few years later, American John Hyatt developed another, which became known as celluloid.

Imitating nature

After developing the world's first photographic paper in the 1890s, Baekeland sold the idea to Kodak and used the money to buy a house equipped with its own laboratory. Here, he experimented with ways of creating synthetic shellac. Shellac is a resin secreted by the female lac beetle. It is a natural polymer that was used to give furniture and other objects a tough, shiny coat. Baekeland found that by treating phenol resin made from coal tar with formaldehyde, he could make a kind of shellac. In 1907, he added various kinds of powder to this resin and found that he could create a remarkable hard, moldable plastic.

Chemically this plastic is known as polyoxybenzylmethylenglycolanhydride, but Baekeland called it simply Bakelite. Bakelite was a "thermoset" plastic—plastic that holds its shape after being heated. Due to its properties of electrical insulation and heat resistance, Bakelite was soon being used to make radios, telephones, and electrical insulators. Many more uses were quickly found for it.

Today, there are thousands of synthetic plastics, including Plexiglass, polythene, low-density polyethylene, and cellophane, each with its own properties and uses. The majority are based on hydrocarbons (chemicals made from hydrogen and carbon) derived from oil or natural gas. However, in recent decades, carbon fibers, nanotubes and other materials have been added to create superlight, superstrong plastic materials such as Kevlar. ■

Heat-resistant and nonconductive of electricity, Bakelite was an ideal material to use for the casings of electrical goods such as telephones and radios.

Leo Baekeland

Leo Baekeland was born in Ghent in Belgium and studied at the university there. In 1889, he became associate professor of chemistry and married Celine Swarts. While the young couple were on honeymoon in New York, Baekeland met Richard Anthony, head of a well-known photographic company. Anthony was so impressed by Baekeland's work with photographic processes that he hired him as a consulting chemist. Baekeland moved to the US and was soon in business for himself.

Baekeland invented the first photographic papers, known as Velox, before developing Bakelite, which made him rich. He is credited with many inventions besides plastic, registering more than 50 patents in total. In later life, he became an eccentric recluse, eating food only from tin cans. He died in 1944 and is buried in Sleepy Hollow Cemetery, New York.

Key work

1909 *Paper on Bakelite read to the American Chemical Society*

I HAVE CALLED THIS PRINCIPLE NATURAL SELECTION

CHARLES DARWIN (1809–1882)

IN CONTEXT

BRANCH
Biology

BEFORE
1794 Erasmus Darwin (Charles's grandfather) recounts his vision of evolution in *Zoonomia*.

1809 Jean-Baptiste Lamarck proposes a form of evolution through the inheritance of acquired characteristics.

AFTER
1937 Theodosius Dobzhansky publishes his experimental evidence for the genetic basis of evolution.

1942 Ernst Mayr defines the concept of species through populations that reproduce only with one another.

1972 Niles Eldredge and Stephen Jay Gould propose that evolution occurs mainly in short bursts interspersed with periods of relative stability.

Most organisms produce **more offspring than can survive** due to constraints such as lack of food and living space.

Offspring vary from each other in many ways.

Variation means **some offspring are better suited** or adapted to the struggle for survival.

If these individuals **pass on the advantageous traits** to their offspring, these also survive.

I have called this principle "natural selection."

The British naturalist Charles Darwin was by no means the first scientist to suggest that plants, animals, and other organisms are not fixed and unchanging—or, to use the popular word of the time, "immutable." Like others before him, Darwin proposed that species of organisms change, or evolve, through time. His great contribution was to show how evolution took place by a process he termed natural selection. He laid out his central idea in his book *On the Origin of Species by Means of Natural Selection, or the Preservation of Favoured Races in the Struggle for Life*, published in London in 1859. Darwin described the book as "one long argument."

"Confessing a murder"

On the Origin of Species met with academic and popular opposition. It made no mention of religious doctrine, which insisted that species were indeed fixed and immutable and designed by God. But gradually the ideas in the book changed the scientific perspective on the natural world. Its core notion forms the basis for all modern biology, providing a simple, but immensely powerful, explanation of life forms both past and present.

Darwin was acutely aware of the potential blasphemy in his work during the decades he spent writing it. Fifteen years before publication, he explained to his confidant, the botanist Joseph Hooker, that his theory required no God or unchanging species: "At last gleams of light have come, & I am almost convinced (quite contrary to opinion I started with) that species are not (it is like confessing a murder) immutable."

See also: James Hutton 96–101 ▪ Jean-Baptiste Lamarck 118 ▪ Gregor Mendel 166–71 ▪ Thomas Henry Huxley 172–73 ▪ Thomas Hunt Morgan 224–25 ▪ Barbara McClintock 271 ▪ James Watson and Francis Crick 276–83 ▪ Michael Syvanen 318–19

Creation is not an event that happened in 4004 BCE; it is a process that began some 10 billion years ago and is still under way.
Theodosius Dobzhansky

Darwin's approach to evolution, like the rest of his wide-ranging work in natural history, was cautious, careful, and deliberate. He proceeded step by step, amassing great quantities of evidence along the way. Over almost 30 years, he integrated his extensive knowledge of fossils, geology, plants, animals, and selective breeding, with concepts from demography, economics, and many other fields. The resulting theory of evolution by natural selection is regarded as one of the greatest scientific advances ever.

The role of God

In the early 19th century, fossils were widely discussed in Victorian society. Some regarded them as naturally formed rock shapes, and nothing to do with living organisms. Others saw them as the handiwork of the Creator, put on Earth to test believers. Or they thought that they were the remains of organisms still alive somewhere in the world, since God had created living things in perfection.

In 1796, the French naturalist Georges Cuvier recognized that certain fossils, such as those of mammoths or giant sloths, were the remains of animals that had become extinct. He reconciled this with his religious belief by invoking catastrophes such as the Flood depicted in the Bible. Each disaster swept away a whole assortment of living things; God then replenished Earth with new species. Between each disaster, each species remained fixed and immutable. This theory was known as "catastrophism" and it became widely known following the publication of Cuvier's *Preliminary Discourse* in 1813.

However, at the time Cuvier was writing, various ideas based on evolution were already in circulation. Erasmus Darwin, the free-thinking grandfather of Charles, proposed an early, idiosyncratic theory. More influential were the ideas of Jean-Baptiste Lamarck, professor of zoology at France's National Museum of Natural History. His *Philosophie Zoologique* of 1809 articulated what was perhaps the first reasoned theory of evolution. He theorized that living beings evolved from simple beginnings through increasingly sophisticated stages, due to a "complexifying force." They faced environmental challenges on their body physiques, and from this came the idea of use and disuse in an individual: "More frequent and continuous use of any organ gradually strengthens, develops and enlarges that organ… while the permanent disuse of any organ imperceptibly weakens and deteriorates it…until it finally disappears." The organ's greater power was then passed to offspring, a phenomenon that became known as inheritance of acquired characteristics.

Although his theory came to be largely discounted, Lamarck was later praised by Darwin for having opened up the possibility that change did not occur as a result of what Darwin disparagingly termed "miraculous interposition."

Adventures of the Beagle

Darwin had plenty of time to muse on the immutability of species during an around-the-world voyage aboard the survey ship HMS *Beagle*, in 1831–36, under captain Robert FitzRoy. As expedition scientist, Darwin was charged with collecting all types of fossil, plant, and animal specimens, and sending them back to Britain from each port of call. »

By studying the fossil record, Georges Cuvier established that species had become extinct. But he believed that the evidence pointed to a series of catastrophes, not gradual change.

This epic voyage opened the eyes of the young Darwin, still only in his twenties, to the incredible variety of life. Wherever the *Beagle* docked, Darwin keenly observed all aspects of nature. In 1835, he described and collected a group of small, insignificant birds on the Galápagos Islands, a Pacific Ocean archipelago 560 miles (900 km) west of Ecuador. He thought there were nine species, six being finches.

After his return to England, Darwin organized his mass of data and oversaw a multivolume, multiauthor report, *The Zoology of the Voyage of HMS* Beagle. In the volume on birds, the renowned ornithologist John Gould declared that there were in fact 13 species in Darwin's specimens, all of them finches. Within the group, however, were birds with differently shaped beaks, adapted to different diets.

In his own, bestselling account of his adventure, *The Voyage of the* Beagle, Darwin wrote, "Seeing this gradation and diversity of structure in one small, intimately related group of birds, one might really fancy that from an original paucity of birds in this archipelago, one species had been taken and modified for different ends." This was one of the first clear, public formulations of where his thoughts on evolution were heading.

Comparing species

Darwin's finches, as the Galápagos specimens became known, were not the only trigger for his work on evolution. In fact, his thoughts had been mounting throughout the *Beagle*'s voyage, and especially during his visit to the Galápagos. He was fascinated by the giant tortoises he saw, and by the way the shapes of their shells differed subtly from island to island. He was also impressed by the species of mockingbirds. They, too, varied between the islands, yet they also had similarities not only among themselves, but with species that lived on the South American mainland.

Darwin suggested that the various mockingbirds might have evolved from a common ancestor that had somehow crossed the Pacific from the mainland; then each group of birds evolved by adapting to the particular environment on each island and its available food. Observing giant tortoises, Falkland Island foxes, and other species supported these early conclusions. But Darwin was sensitive about where such blasphemous ideas would lead: "Such facts would undermine the stability of species."

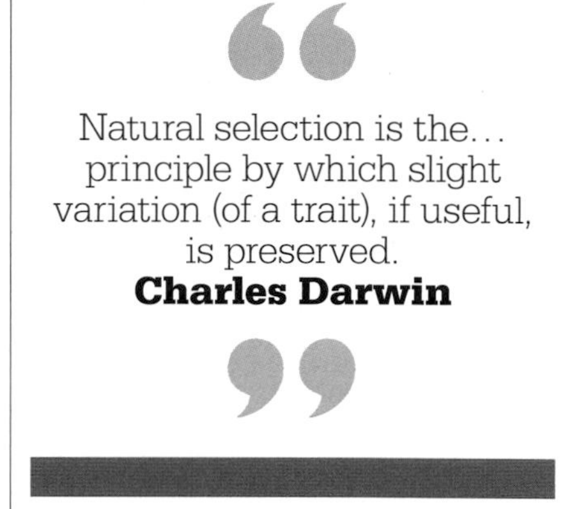

Natural selection is the… principle by which slight variation (of a trait), if useful, is preserved.
Charles Darwin

Other parts of the jigsaw

On his way to South America in 1831, Darwin had read the first volume of Charles Lyell's *Principles of Geology*. Lyell argued against Cuvier's catastrophism history and his theory of fossil formation. Instead, he adapted the ideas of geological renewal put forward by James Hutton into a theory known as "uniformitarianism." Earth was continually being formed, altered, and reformed over immense time periods by processes such as wave erosion and volcanic upheaval that were the same as those happening today. There was no need to invoke disastrous interventions by God.

Lyell's ideas transformed the way Darwin interpreted the landscape formations, rocks, and

This giant tortoise is only found on the Galápagos Islands, where unique subspecies have developed on each island. Darwin gathered evidence here for his theory of evolution.

The finches of the Galápagos have evolved differently shaped beaks adapted to specific diets.

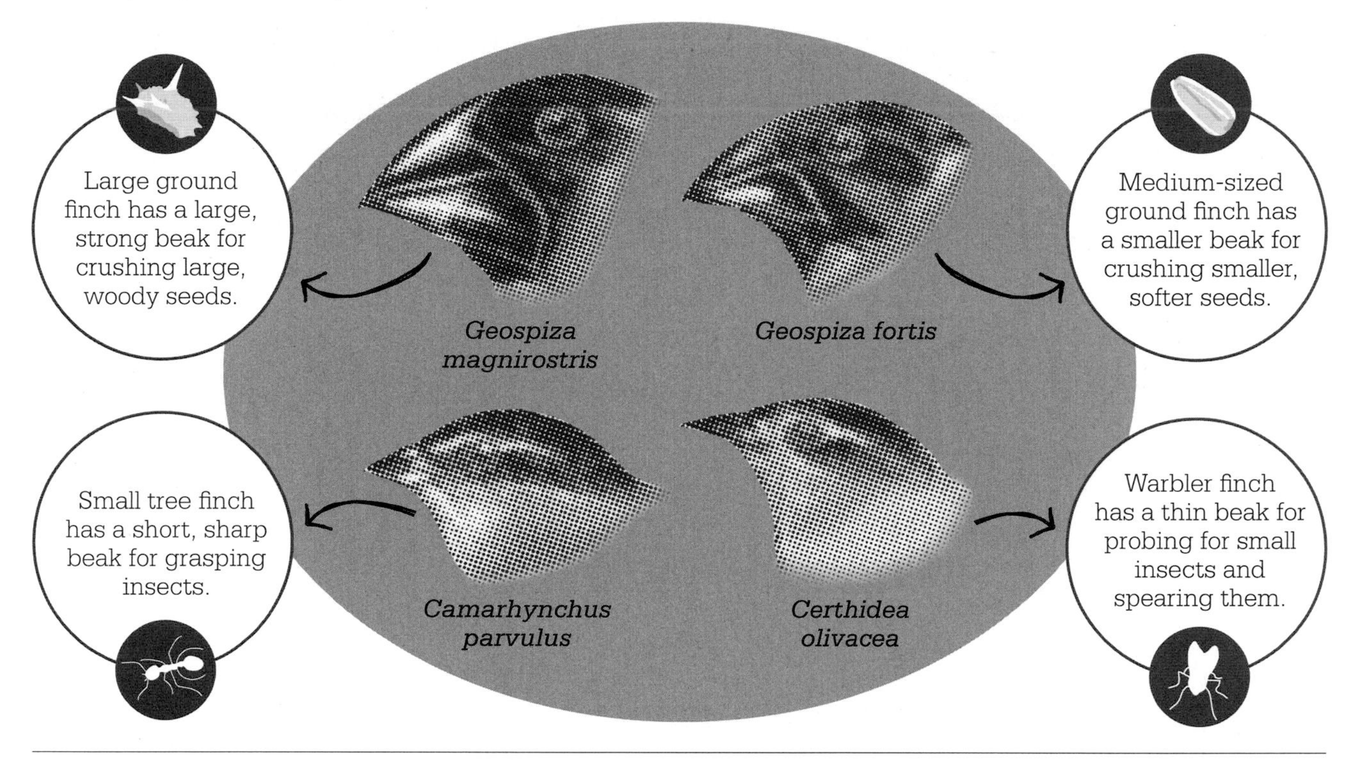

fossils he found on his explorations, which he now saw "through Lyell's eyes." However, while he was in South America, volume two of *Principles of Geology* arrived. In it, Lyell rejected ideas of gradual evolution of plants and animals, including Lamarck's theories. Instead, he invoked the concept of "centres of Creation" to explain species' diversity and distribution. Although Darwin admired Lyell as a geologist, he had to discount this latest concept as the evidence for evolution mounted.

Another piece of the jigsaw was revealed in 1838 when Darwin read *An Essay on the Principle of Population* by the English demographer Thomas Malthus, which had been published 40 years earlier. Malthus described how human populations can increase in an exponential way, with the potential to double after one generation of 25 years, then double again in the next generation, and so on. However, food supplies could not expand in the same way, and the result was a struggle for existence. Malthus's ideas were one of the main inspirations for Darwin's theory of evolution.

The quiet years

Even before the *Beagle* had returned to England, the interest generated by the specimens Darwin had sent back had made him a celebrity. After his return, his scientific and popular accounts of the voyage increased his fame. However, his health deteriorated and gradually he withdrew from the public eye.

In 1842, Darwin moved to the peace and quiet of Down House in Kent, where he continued to amass evidence to support his theory of evolution. Scientists around the world sent him specimens and data. He studied the domestication of animals and plants, and the role of selective breeding, or artificial selection, especially in pigeons. In 1855, he started breeding varieties of *Columbia livia*, or rock doves, and they would feature prominently in the first two chapters in *On the Origin of Species*.

Through his work on pigeons, Darwin began to understand the extent and relevance of variation among individuals. He rejected the accepted wisdom that environmental factors were responsible for such differences, insisting that reproduction was the cause, with variation somehow inherited from parents. He added this to the ideas of Malthus and applied them to the natural world. »

Much later, in his autobiography, Darwin recalled his reaction when he first read Malthus back in 1838. "Being well prepared to appreciate the struggle for existence…it at once struck me that under these circumstances favourable variations would tend to be preserved, and unfavourable ones to be destroyed. The result of this would be the formation of new species…I had at last got a theory by which to work."

Knowing more about the role of variation, by 1856 Darwin the pigeon breeder could imagine not humans but nature doing the choosing. From the term "artificial selection" he derived "natural selection."

Jolt into action

On June 18, 1858, Darwin received a short essay by a young British naturalist named Alfred Russel Wallace. Wallace described a flash of insight in which he had suddenly understood how evolution occurred, and asked Darwin for his opinion. Darwin was startled to read that Wallace's insight replicated almost exactly the same ideas he himself had been working on for more than 20 years.

Alfred Russel Wallace, like Darwin, developed his theory of evolution in the light of extensive field work, conducted first in the Amazon River Basin and later in the Malay Archipelago.

Worried about precedence, Darwin consulted Charles Lyell. They agreed to a joint presentation of Darwin's and Wallace's papers at the Linnaean Society in London on July 1, 1858. Neither author would attend in person. The audience's response was polite, with no outcry about blasphemy. Encouraged, Darwin now finished his book. Published on November 24, 1859, *On the Origin of Species* sold out on its first day.

Darwin's theory

Darwin states that species are not immutable. They change, or evolve, and the main mechanism for this change is natural selection. The process relies on two factors. First, more offspring are born than can survive when faced with the challenges of climate, food supply, competition, predators, and diseases; this leads to a struggle for existence. Second, there is variation, sometimes tiny but nonetheless present, among the offspring within a species. For evolution, these variations must fulfill two criteria. One: they should have some effect on the struggle to survive and breed, that is, they should help to confer reproductive success. Two: they should be inherited, or passed to offspring, where they would confer the same evolutionary advantage.

Darwin describes evolution as a slow and gradual process. As a population of organisms adapts to a new environment, it becomes a new species, different from

Charles Darwin

Born in Shrewsbury, England, in 1809, Darwin was originally destined to follow his father into medicine, but his childhood was filled with pursuits such as beetle collecting, and with little inclination to become a physician, he trained for the clergy. A chance appointment in 1831 placed him as expedition scientist on HMS *Beagle*'s around-the-world trip.

Following the voyage, Darwin was under the scientific spotlight, gaining fame as a perceptive observer, reliable experimenter, and talented writer. He wrote on the formation of coral reefs and on marine invertebrates, especially barnacles, which he studied for almost 10 years. He also wrote works on fertilization, of orchids, insect-eating plants, movement in plants, and variation among domesticated animals and plants. Later in life, he tackled the origin of humans.

Key works

1839 *The Voyage of the* Beagle
1859 *On the Origin of Species by Means of Natural Selection*
1871 *The Descent of Man, and Selection in Relation to Sex*

I think I have found out (here's presumption!) the simple way by which species become exquisitely adapted to various ends.
Charles Darwin

its ancestors. Meanwhile, those ancestors may remain the same, or they may evolve in response to their own changing environment, or they may lose the struggle for survival and become extinct.

Aftermath

Faced with such a thorough, reasoned, evidence-based exposition of evolution by natural selection, most scientists soon accepted Darwin's concept of "survival of the fittest." Darwin's book was careful to avoid any mention of humans in connection with evolution, other than the single sentence, "Light will be shed on the origin of man, and his history." However, there were protests from the Church, and the clear implication that humans had evolved from other animals was ridiculed in many quarters.

Darwin, as ever avoiding the limelight, remained engrossed in his studies at Down House. As controversy mounted, numerous scientists sprang to his defense. The biologist Thomas Henry Huxley was vociferous in supporting the theory—and arguing the case for human descent from apes—and dubbed himself "Darwin's bulldog."

However, the mechanism by which inheritance occurred—how and why some traits are passed on, others not—remained a mystery. Coincidentally, at the same time that Darwin published his book, a monk named Gregor Mendel was experimenting with pea plants in Brno (in the present-day Czech Republic). His work on inherited characteristics, reported in 1865, formed the basis of genetics, but was overlooked by mainstream science until the 20th century, when new discoveries in genetics were integrated into evolutionary theory, providing a mechanism for heredity. Darwin's principle of natural selection remains key to understanding the process. ■

This cartoon ridiculing Darwin appeared in 1871, the year in which he applied his theory of evolution to humans—something he had been careful to avoid in earlier works.

FORECASTING THE WEATHER

ROBERT FITZROY (1805–1865)

IN CONTEXT

BRANCH
Meteorology

BEFORE
1643 Evangelista Torricelli invents the barometer, which measures air pressure.

1805 Francis Beaufort develops the Beaufort scale of wind force.

1847 Joseph Henry proposes a telegraph link to warn the eastern United States of storms coming from the west.

AFTER
1870 The US Army Signal Corps begins creating weather maps for the whole US.

1917 The Bergen School of Meteorology in Norway develops the notion of weather fronts.

2001 Systems of Unified Surface Analysis use powerful computers to give highly detailed local weather.

A century and a half ago, notions of weather prediction were deemed little more than folklore. The man who changed that and gave us modern weather forecasting was British naval officer and scientist Captain Robert FitzRoy.

FitzRoy is better known today as the captain of the *Beagle*, the ship that carried Charles Darwin on the voyage that led to his theory of evolution by natural selection. Yet FitzRoy was a remarkable scientist in his own right.

FitzRoy was just 26 when he sailed from England with Darwin in 1831. Yet he had already served more than a decade at sea, and had studied at the Royal Naval College at Greenwich, where he was the first candidate to pass the lieutenant's exam with perfect marks. He had even commanded the *Beagle* on an earlier survey trip around South America, where the importance of studying the weather was impressed upon him. His ship almost met with disaster in a violent wind off the coast of Patagonia after he had ignored the warning signs of falling pressure on the ship's barometer.

With a barometer, two or three thermometers, some brief instructions, and an attentive observation, not of instruments only, but the sky and atmosphere, one may utilise Meteorology.
Robert FitzRoy

Naval weather pioneers

It was no coincidence that many of the first breakthroughs in weather forecasting came from naval officers. Knowing what weather lay ahead was crucial in the days of sailing ships. Missing a good wind could have huge financial consequences—and being caught at sea in a storm could be disastrous.

Two naval officers in particular had already made significant contributions. One was Irish

Robert FitzRoy

Born in 1805 in Suffolk, England, to an aristocratic family, Robert FitzRoy joined the Navy at just 12 years old. He went on to serve many years at sea as an outstanding sea captain. He captained the *Beagle* on two major survey voyages to South America, including the around-the-world voyage with Charles Darwin. FitzRoy was, however, a devout Christian who opposed Darwin's theory of evolution. After leaving active service in the Navy, FitzRoy became governor of New Zealand, where his even-handed treatment of the Maori earned him the resentment of the settlers. He returned to England in 1848 to command the Navy's first steamship, and was appointed head of the British Meteorological Office when it was established in 1854. There he developed the methods that became the foundation of scientific weather forecasting.

Key works

1839 *Narrative of the Voyages of the* Beagle
1860 *The Barometer Manual*
1863 *The Weather Book*

See also: Robert Boyle 46–49 ▪ George Hadley 80 ▪ Gaspard-Gustave de Coriolis 126 ▪ Charles Darwin 142–49

mariner Francis Beaufort, who created a standard scale showing the wind speed or "force" linked to particular conditions at sea, and later on land. This allowed the severity of storms to be recorded and compared methodically for the first time. The scale ranged from 1, indicating "light air" to 12, "hurricane." The first time the Beaufort scale was used was by FitzRoy on the *Beagle* voyage. Thereafter it became standard in all naval ships' logs.

Another naval weather pioneer was American Matthew Maury. He created wind and current charts for the North Atlantic, which resulted in dramatic improvements for sailing times and certainty. He also advocated the creation of an international sea and land weather service, and led a conference in Brussels in 1853 that began to coordinate observations on conditions at sea from all around the world.

Before FitzRoy began his weather reporting systems, mariners had already observed that winds form cyclonic patterns in hurricanes, and that wind direction could be used to predict the storm's path.

The Meteorological Office

In 1854, FitzRoy, encouraged by Beaufort, was given the task of setting up the British contribution at the Meteorological Office. But with characteristic zeal and insight, FitzRoy went much further than his brief. He began to see that a system of simultaneous weather observations from around the world could not only reveal hitherto undiscovered patterns, but actually be used to make weather predictions.

Observers already knew that in tropical hurricanes, for example, the winds blow in a circular or "cyclonic" pattern around a central area of low air pressure or "depression." It was soon realized »

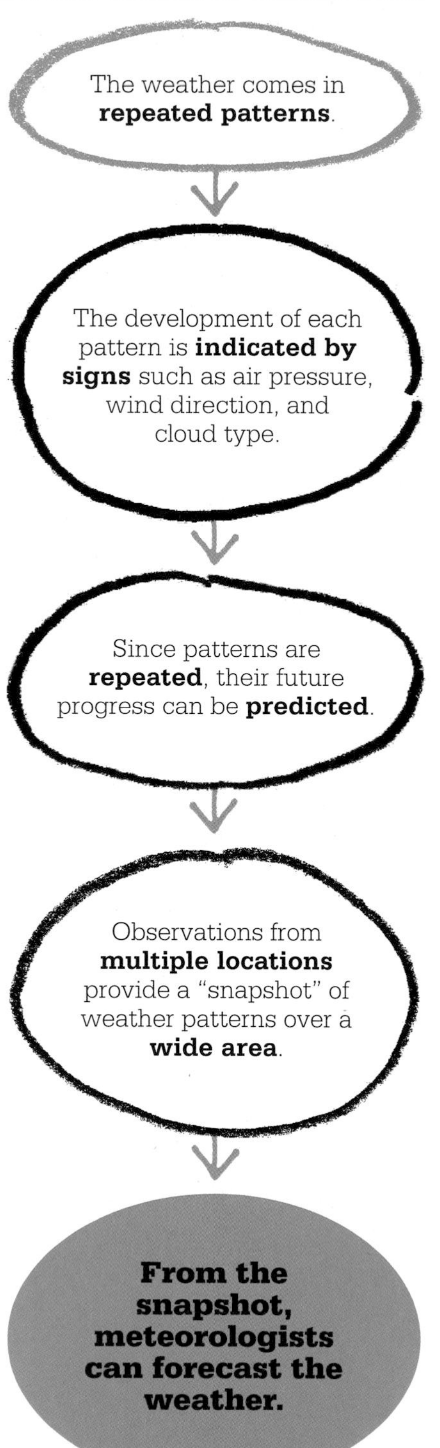

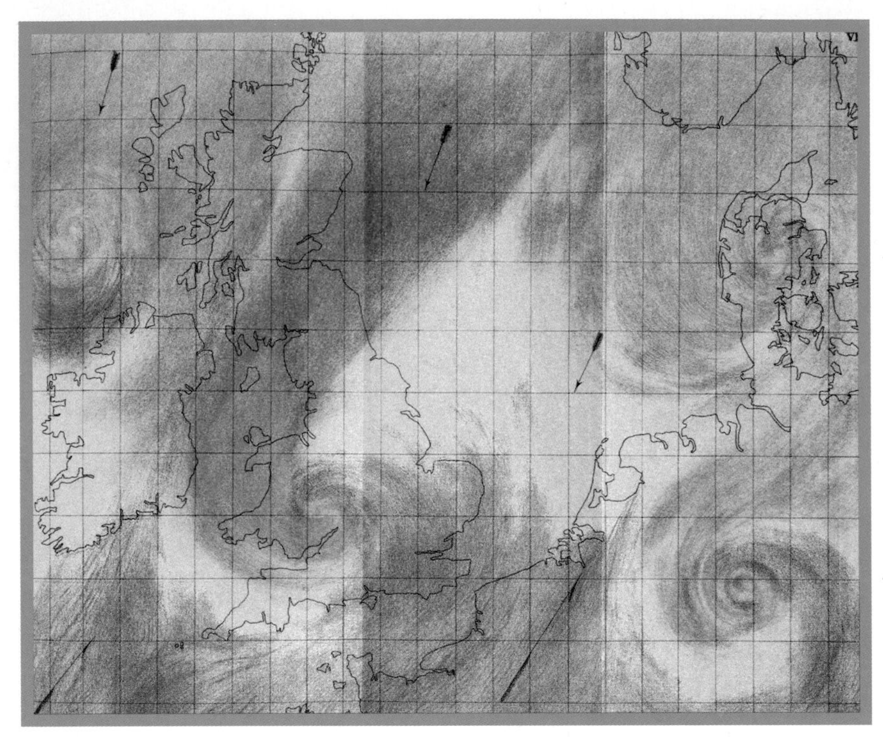

FitzRoy colored his daily "synoptic" charts in crayon. This one, made in 1863, shows a low-pressure front bringing storms toward northern Europe from the west. The lower right of the chart reveals a cyclone forming.

that most of the large storms that blow in the mid-latitudes show this cyclonic depression shape. So the direction of the wind gives a clue as to whether the storm is approaching or receding.

In the 1850s, better records of weather events, and the use of the new electric telegraph to communicate over long distances, almost instantly revealed that cyclonic storms, which form over land, move eastward. In contrast, hurricanes (tropical North Atlantic storms) form over water and migrate westward. So in North America when a storm hit one place far inland, a telegraph could be sent to warn places farther east that a storm was on its way. Observers already knew that a drop in air pressure on the barometer gave warning of a storm to come. The telegraph allowed such readings to be relayed rapidly over great distances and therefore gave warnings much further in advance.

Synoptic weather

FitzRoy understood that the keys to weather prediction were systematic observations of air pressure, temperature, and wind speed and direction taken at set times from widely spread locations. When these observations were sent instantly by telegraph to his coordinating office in London, he could build up a picture or "synopsis" of weather conditions over a vast area.

This synopsis gave such a complete picture of the weather conditions that it not only revealed current weather patterns on a wide scale, it also enabled weather patterns to be tracked. FitzRoy realized that weather patterns were repeated. From this, it was clear to him that he could figure out how weather patterns may develop over a short time in the future, from how they have developed in the past. This provides the basis for a detailed forecast of the weather at any and every point within the region covered. This was a remarkable insight that formed the basis of modern forecasting.

The observation figures alone were enough, but FitzRoy also used them to create the first modern meteorological chart, the "synoptic" chart that revealed the swirling shapes of cyclonic storms as clearly as satellite pictures do today. FitzRoy's ideas were summed up in his book, titled simply *The Weather Book* (1863), which introduced the word "forecast" and laid out the principles of modern forecasting for the first time.

A crucial step was to divide the British Isles into weather areas, collate current weather conditions, and use past weather data from each area to help make forecasts. FitzRoy recruited a network of observers, particularly at sea and in ports in the British Isles. He also obtained data from France and Spain, where the idea of constant weather observation was catching on. Within a few years, his network

> I try, by my warnings of probable bad weather, to avoid the need for a life-boat.
> **Robert FitzRoy**

was operating so effectively that he could get a daily snapshot of weather patterns right across Western Europe. Patterns in the weather were revealed so clearly that he could forecast how it was likely to change over the next day at least—and so produce the first national forecasts.

Daily weather forecasts

Every morning, weather reports would come to FitzRoy's office from scores of weather stations across Western Europe, and within an hour, the synoptic picture was figured out. Instantly, forecasts were despatched to *The Times* newspaper to be published for all to read. The first weather forecast was published by the newspaper on August 1, 1861.

FitzRoy set up a system of signaling cones in highly visible places at ports to warn if a storm was on the way and from which direction. This system worked so well that it saved countless lives.

Some shipowners, however, resented the system when their captains began to delay setting sail if warned of a storm. There were also problems disseminating the forecasts in time. It took 24 hours to distribute the newspaper, so FitzRoy had to make forecasts for not just one day ahead but two—otherwise the weather would have happened by the time people read his forecasts. He was aware that longer-range forecasts were far more unreliable, and was frequently exposed to ridicule, particularly when *The Times* disassociated itself from mistakes.

FitzRoy's legacy

Faced with a barrage of ridicule and criticism from vested interests, the forecasts were suspended and FitzRoy committed suicide in 1865. When it was discovered that he had spent his fortune on his research at the Meteorological Office, the government compensated his family. But within a few years, pressure from mariners ensured that his storm warning system was again in widespread use. Picking up the detailed forecasts and storm warnings for particular shipping areas is now an essential part of every mariner's day.

As communications technology improved and added ever more detail to the observational data, the value of FitzRoy's system came into its own in the 20th century.

Having collated and duly considered the Irish telegrams [or from any other weather area], the first forecast for that district is drawn…and forthwith sent out for immediate publication.
Robert FitzRoy

Modern forecasting

Today, the world is dotted with a network of more than 11,000 weather stations, in addition to the numerous satellites, aircraft, and ships—all continuously feeding information into a global meteorological data bank. Powerful number-crunching supercomputers churn out weather forecasts that are, in the short-term at least, highly accurate, and a huge range of activities, from air travel to sports events, rely on them. ■

This weather station, located in the remote mountains of Ukraine, sends data on temperature, humidity, and wind speed via satellite to weather supercomputers.

OMNE VIVUM EX VIVO— ALL LIFE FROM LIFE

LOUIS PASTEUR (1822–1895)

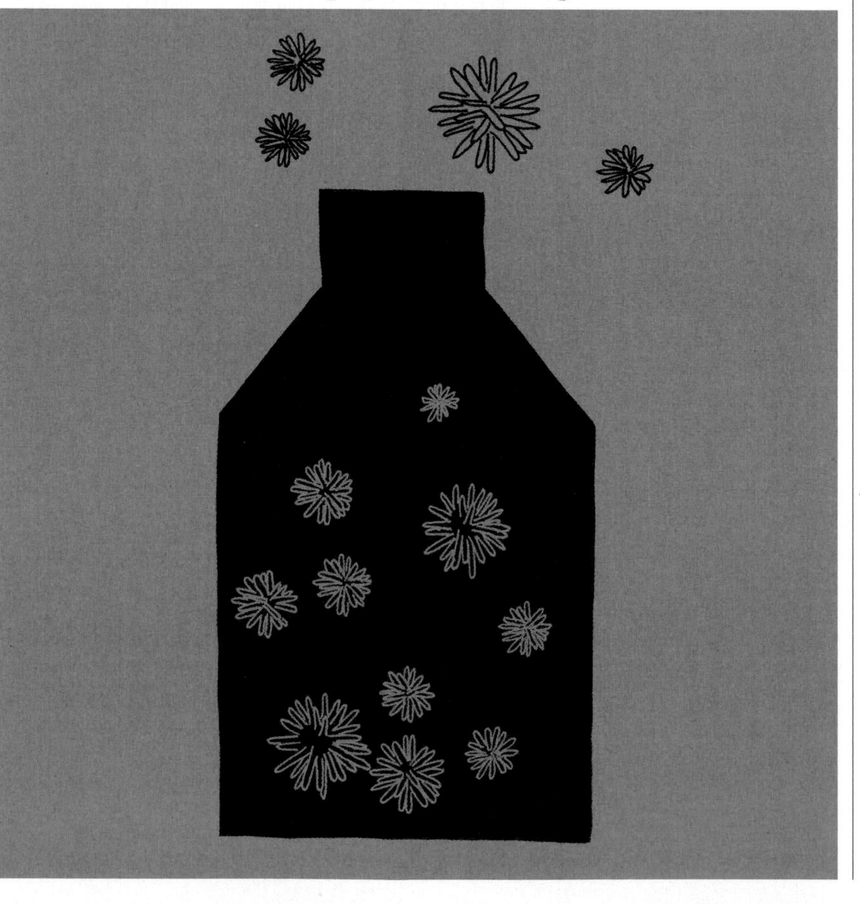

IN CONTEXT

BRANCH
Biology

BEFORE
1668 Francesco Redi demonstrates that maggots arise from flies—and not spontaneously.

1745 John Needham boils broth to kill microbes, and believes that spontaneous generation has occurred when they grow back.

1768 Lazzaro Spallanzani shows that microbes do not grow in boiled broth when air is excluded.

AFTER
1881 Robert Koch isolates microbes that cause disease.

1953 Stanley Miller and Harold Urey create amino acids—essential to life—in an experiment that simulates origin-of-life conditions.

Modern biology teaches that living things can only arise from other living things by a process of reproduction. This may seem self-evident today, but when the basic principles of biology were in their infancy, many scientists adhered to a notion called "abiogenesis"—the idea that life could spontaneously generate itself. Long after Aristotle claimed that living organisms could emerge from decaying matter, some even believed in methods that purported to make creatures from inanimate objects. In the 17th century, for example, Flemish physician Jan Baptista von Helmont wrote that sweaty

See also: Robert Hooke 54 ▪ Antonie van Leeuwenhoek 56–57 ▪ Thomas Henry Huxley 172–73 ▪ Harold Urey and Stanley Miller 274–75

Many **living organisms are microscopic**, and are suspended in the air around us.

→ **Some** of these microbes cause **spoilage of food** or **infectious disease**.

→ Spoilage or infection do not occur if **microbes** are **prevented from contaminating and reproducing**.

→ **Microbes cannot arise by spontaneous generation. All life comes from life.**

underwear and some wheat grain left in a jar in the open would spawn adult mice. Spontaneous generation had its advocates until well into the 19th century. In 1859, however, a French microbiologist named Louis Pasteur devised a clever experiment that disproved it. In the course of his studies, he also proved that infections were caused by living microbes—germs.

Before Pasteur, the link between disease or decay and organisms had been suspected but never substantiated. Until microscopes could prove otherwise, the notion that there were such things as tiny living entities that were invisible to the naked eye seemed fanciful. In 1546, Italian physician Girolamo Fracastoro described "seeds of contagion," and came close to the truth of the matter. But he fell short of explicitly stating that they were living, reproducible things, and his theory made little impact. Instead, people believed that infectious disease was caused by "miasma"—or noxious air—that came from rotting matter. Without a clear idea of the nature of germs as microbes, no one could properly appreciate that the transmission of infection and the propagation of life were in effect two sides of the same coin.

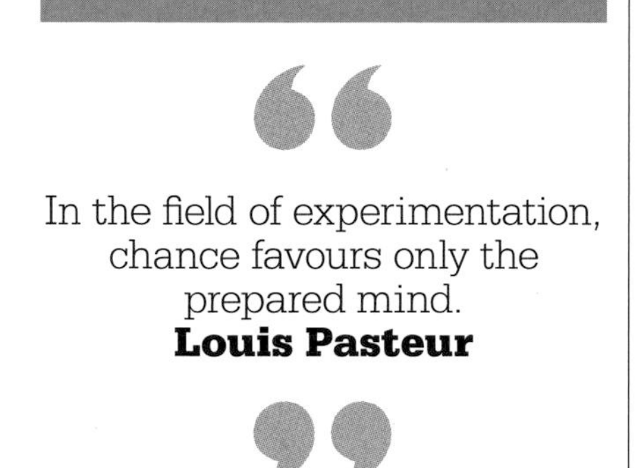

In the field of experimentation, chance favours only the prepared mind.
Louis Pasteur

First scientific observations

In the 17th century, scientists attempted to trace the origins of larger creatures by studying reproduction. In 1661, English physician William Harvey (known for his discovery of the circulation of blood) dissected a pregnant deer in an effort to discover the origin of a fetus, and proclaimed *"Omne vivum ex ovo"*—all life from eggs. He failed to find the deer's egg in question, but it was at least a hint of things to come.

Italian physician Francesco Redi was the first to offer experimental evidence for the impossibility of spontaneous generation—at least in so far as creatures visible to the human eye were concerned. In 1668, he studied the process by which meat becomes riddled with maggots. He covered one piece of meat with parchment and left another exposed. Only the exposed meat became infected with maggots, because it attracted flies, which deposited their eggs on it. Redi repeated the experiment with cheesecloth—which absorbed the meat's odor and attracted flies—and showed that flies' eggs taken from the cheesecloth could then be used to "seed" uninfected meat with maggots. Redi argued that maggots could only arise from »

This drawing by Francesco Redi shows maggots turning into flies. His work showed not only that flies come from maggots, but also that maggots come from flies.

flies, rather than spontaneously. However, the significance of Redi's experiment was not appreciated, and even Redi himself did not fully reject abiogenesis, believing that it did occur in certain circumstances.

Among the first makers and users of the microscope for detailed scientific study, Dutch scientist Antonie van Leeuwenhoek showed that some living things were so small that they could not be seen with the naked eye—and also that the reproduction of larger creatures depended upon similar microscopic living entities, such as sperm.

Yet the idea of abiogenesis was so deeply entrenched in the minds of scientists that many still thought that these microscopic organisms were too small to contain reproductive organs and must therefore arise spontaneously. In 1745, English naturalist John Needham set out to prove it. He knew that heat could kill microbes, so he boiled some mutton gravy in a flask—thereby killing its microbes—and then allowed it to cool. After observing the broth for a time, he saw that the microbes had come back. He concluded that they had arisen spontaneously from the sterilized broth. Two decades later, Italian physiologist Lazzaro Spallanzani repeated Needham's experiment, but showed that the microbes did not grow back if he removed air from the flask. Spallanzani thought that the air had "seeded" the broth, but his critics proposed instead that air was actually a "vital force" for the new generation of microbes.

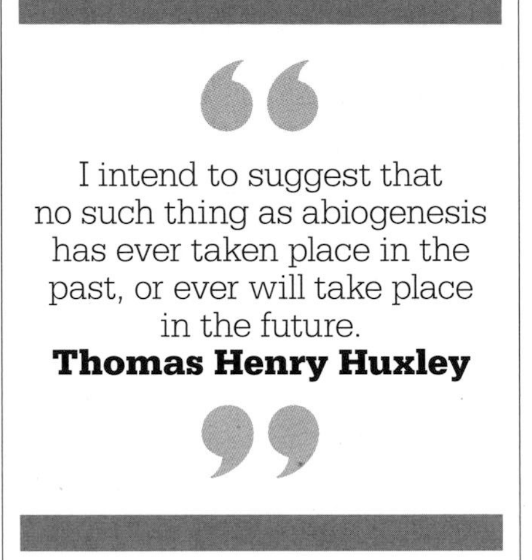

I intend to suggest that no such thing as abiogenesis has ever taken place in the past, or ever will take place in the future.
Thomas Henry Huxley

Viewed in the context of modern biology, the results of Needham's and Spallanzani's experiments can be easily explained. Although heat does indeed kill most microbes, some bacteria, for example, can survive by turning into dormant, heat-resistant spores. And most microbes, as with most life, need oxygen from the air in order to derive energy from their nutrition. Most importantly, however, these sorts of experiments were always vulnerable to contamination—microscopic airborne microbes can easily colonize a growth medium, even after a brief exposure to the atmosphere. So in fact, neither of these experiments had addressed conclusively the question of abiogenisis, one way or another.

Conclusive proof

A century later, microscopes and microbiology had advanced far enough for it to became possible to settle the matter. Louis Pasteur's experiment demonstrated that there were microbes suspended in air, ready to infect any exposed surface. First, he filtered air through cotton. Then he analyzed the contaminated cotton filters and examined the trapped dust

Pasteur's swan-neck experiment proved that a sterilized broth will remain free of microorganisms as long as they are prevented from falling into it from the air.

Air can get in through tube

Microorganisms get trapped in the curve

The broth is boiled to kill any microoganisms in it.

When the broth cools it remains free of microorganisms.

Tilting the tube allows microorganisms back into the broth.

The microorganisms quickly multiply again.

with a microscope. He found it to be teeming with the type of microbes that had been linked with the decay and spoilage of food. It looked as though infection was caused when microbes literally fell out of the air. This was the critical information Pasteur needed to succeed in the next step, when he took up a challenge laid down by the French Academy of Sciences—to disprove the idea of spontaneous generation once and for all.

For his experiment, Pasteur boiled nutrient-rich broth—just as Needham and Spallanzani had done a century before—but this time made a critical modification to the flask. He heated the flask's neck to soften the glass, then drew the glass outward and downward to form a tube in the shape of a swan's neck. When the setup had cooled, the tube was part-way directed downward so that microbes could not fall onto the broth, even though the temperature was now suitable for their growth and there was plentiful oxygen since the tube communicated with the outside air. The only way microbes could grow back in the flask was spontaneously—and this did not happen.

As final proof that microbes needed to contaminate the broth from the air, Pasteur repeated the experiment, but snapped off the swan-necked tube. The broth became infected: he had finally disproved spontaneous generation, and had shown that all life came from life. It was clear that microbes could no more spontaneously appear in a flask of broth than mice could appear in a dirty jar.

Abiogenesis returns

In 1870, English biologist Thomas Henry Huxley championed Pasteur's work in a lecture entitled "biogenesis and abiogenesis." It was a crushing blow to the last devotees of spontaneous generation, and marked the birth of a new biology solidly founded on the disciplines of cell theory, biochemistry, and genetics. By the 1880s, German physician Robert Koch had shown that the disease anthrax was transmitted by infectious bacteria.

Nevertheless, nearly a century after Huxley's address, abiogenesis would once again focus the minds of a new generation of scientists as they began to ask questions about the origin of the very first life on Earth. In 1953, American chemists Stanley Miller and Harold Urey sent electrical sparks through a mix of water, ammonia, methane, and hydrogen to simulate the atmospheric conditions at the dawn of life on Earth. Within weeks, they had created amino acids—the building blocks of proteins and key chemical constituents of living cells. Miller and Urey's experiment triggered a resurgence of work directed at showing that living organisms can emerge from nonliving matter, but this time scientists were equipped with the tools of biochemistry and an understanding of processes that took place billions of years ago. ■

> I observe facts alone; I seek but the scientific conditions under which life manifests itself.
> **Louis Pasteur**

Louis Pasteur

Born to a poor French family in 1822, Louis Pasteur became such a towering figure in the world of science that, upon his death, he was given a full state funeral. After training in chemistry and medicine, his professional career included academic positions at the French universities of Strasbourg and Lille.

His first research was on chemical crystals, but he is better known in the field of microbiology. Pasteur showed that microbes turned wine into vinegar and soured milk, and developed a heat-treating process that killed them—known as pasteurization. His work on microbes helped to develop modern germ theory: the idea that some microbes caused infectious disease. Later in his career, he developed several vaccines, and established the Institut Pasteur devoted to the study of microbiology, which thrives to this day.

Key works

1866 *Studies on Wine*
1868 *Studies on Vinegar*
1878 *Microbes: Their Roles in Fermentation, Putrefaction, and Contagion*

ONE OF THE SNAKES GRABBED ITS OWN TAIL

AUGUST KEKULÉ (1829–1896)

IN CONTEXT

BRANCH
Chemistry

BEFORE
1852 Edward Frankland introduces the idea of valency—the number of bonds an atom can form with other atoms.

1858 Archibald Couper suggests that carbon atoms can link directly to one another, forming chains.

AFTER
1858 Italian chemist Stanislao Cannizzaro explains the difference between atoms and molecules, and publishes atomic and molecular weights.

1869 Dmitri Mendeleev lays out the periodic table.

1931 Linus Pauling elucidates the structure of the chemical bond in general, and that of the benzene molecule in particular, using the ideas of quantum mechanics.

The early years of the 19th century saw huge developments in chemistry that fundamentally changed the scientific view of matter. In 1803, John Dalton suggested that each element was made of atoms that are unique to that element, and used the concept of atomic weight to explain how elements always combine with each other in whole-number proportions. Jöns Jakob Berzelius studied 2,000 compounds to investigate these proportions. He invented the naming system we use today—H for hydrogen, C for carbon, and so on—and compiled a list of atomic weights for all 40 elements that were then known. He also coined the term "organic chemistry" for the chemistry of living organisms—the term later came to mean most chemistry involving carbon. In 1809, French chemist Joseph Louis Gay-Lussac explained how gases combine in simple proportions by volume, and two years later the Italian Amedeo Avogadro suggested that equal volumes of gas contain equal numbers of molecules. It was clear that there were strict rules governing the combination of the

> I spent a part of the night putting at least sketches of those musings down on paper. This is how the structural theory came into being.
> **Friedrich August Kekulé**

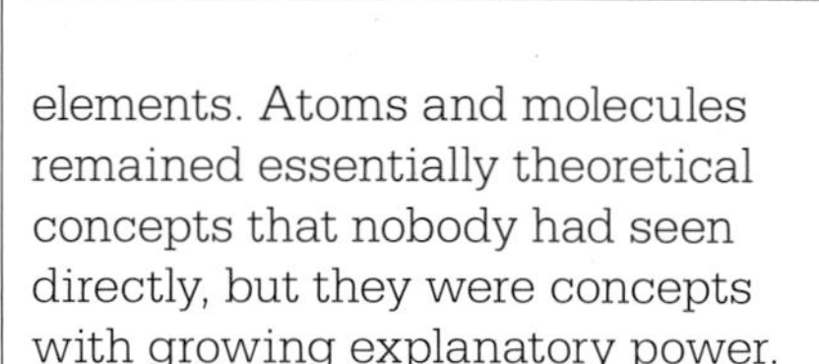

elements. Atoms and molecules remained essentially theoretical concepts that nobody had seen directly, but they were concepts with growing explanatory power.

Valency

In 1852, the first step toward an understanding of how atoms combine with each another was taken by English chemist Edward Frankland, who introduced the idea of valency—which is the number of atoms each atom of an element can combine with. Hydrogen has a valency of one; oxygen has

The **atoms** of each element can **combine with other atoms** in a set number of ways. This is called **valency**.

Carbon atoms have a valency of **four**.

In the molecules of benzene, **carbon atoms** bond with each other to **form rings**, onto which hydrogen atoms bond.

This structure came to Kekulé in a vision of a snake grabbing its own tail.

See also: Robert Boyle 46–49 ▪ Joseph Black 76–77 ▪ Henry Cavendish 78–79 ▪ Joseph Priestley 82–83 ▪ Antoine Lavoisier 84 ▪ John Dalton 112–13 ▪ Humphry Davy 114 ▪ Linus Pauling 254–59 ▪ Harry Kroto 320–21

a valency of two. Then, in 1858, British chemist Archibald Couper suggested that bonds were formed between self-linking carbon atoms, and that molecules were chains of atoms bonded together. So water, which was known to consist of two parts of hydrogen to one of oxygen, could be represented as H_2O, or H–O–H, where "–" signifies a bond. Carbon has a valency of four, making it tetravalent, so a carbon atom can form four bonds, as in methane (CH_4), where the hydrogen atoms are arranged in a tetrahedron around the carbon. (Today, chemists think of a bond as representing a pair of electrons shared between the two atoms, and the symbols H, O, and C as representing the central part of the appropriate atom.)

Couper was working at the time at a laboratory in Paris. Meanwhile, in Heidelberg, Germany, August Kekulé had come up with the same idea, announcing in 1857 that carbon has a valency of four, and early in 1858 that carbon atoms can bond to one another. Publication of Couper's paper had been delayed, allowing Kekulé to publish a month before him and claim priority for the idea of self-bonding carbon atoms. Kekulé called the bonds between atoms "affinities," and explained his ideas in greater detail in his popular *Textbook of Organic Chemistry*, which first appeared in 1859.

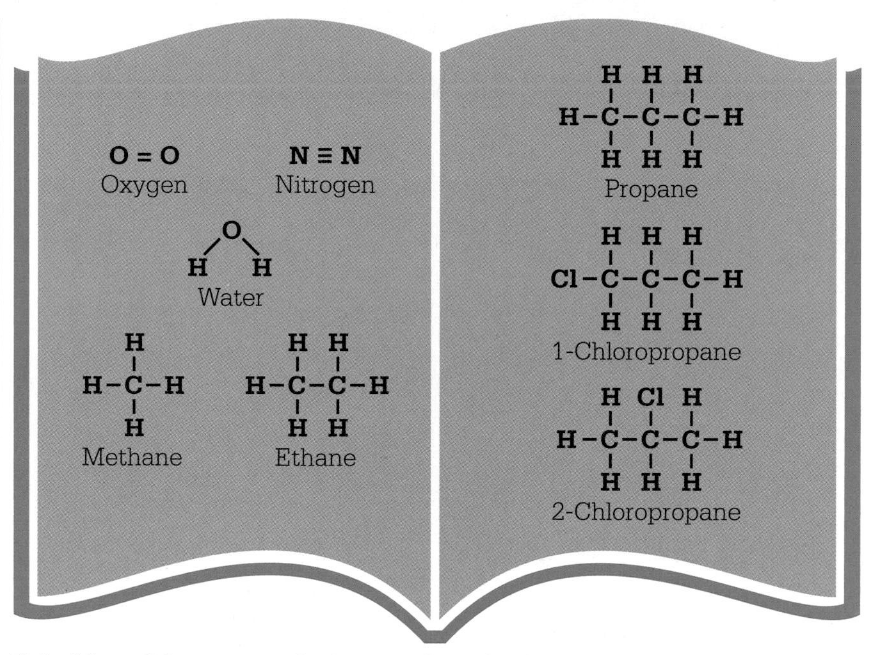

Kekulé used the concept of valency to describe the bonds that are formed between atoms to make various molecules. Here, each bond is represented by a line.

Carbon compounds

Figuring out theoretical models based on evidence from chemical reactions, Kekulé declared that tetravalent carbon atoms could link together to form what he called a "carbon skeleton," to which other atoms with other valencies (such as hydrogen, oxygen, and chlorine) could bond. Suddenly, organic chemistry began to make sense, and chemists assigned structural formulae to all kinds of molecules.

Simple hydrocarbons such as methane (CH_4), ethane (C_2H_6), and propane (C_3H_8) were now seen to be chains of carbon atoms where the spare valencies were occupied by hydrogen atoms. Reacting such a compound with, say, chlorine (Cl_2) produced compounds in which one or more of the hydrogen atoms were replaced by chlorine atoms, making compounds such as chloromethane or chloroethane. One feature of this substitution was that chloropropane came in two distinct forms, either 1-chloropropane or 2-chloropropane, depending on whether the chlorine was attached to the middle carbon atom or one of the end carbon atoms (see the diagram above). Some compounds need double bonds to satisfy the valencies of the atoms: the oxygen molecule (O_2), for example, and the molecule of ethylene (C_2H_4). Ethylene reacts with chlorine, and the result is not substitution but addition. The chlorine adds across the double bond, to make 1,2 dichloroethane ($C_2H_4Cl_2$). Some compounds even have triple bonds, including the nitrogen molecule (N_2) and acetylene (C_2H_2), which is highly reactive, and used in oxyacetylene welding torches.

Benzene, however, remained a puzzle. It turned out to have the formula C_6H_6, but is much less reactive than acetylene, even though both compounds have equal numbers of carbon and hydrogen atoms. Devising a » linear structure that was not highly reactive was a real conundrum. There clearly had to be double bonds, but how they were arranged was a mystery.

Furthermore, benzene reacts with chlorine not by addition (like ethylene) but by substitution: a chlorine atom replaces a hydrogen atom. When one of benzene's hydrogen atoms is substituted by a chlorine atom, the result is only a single compound C_6H_5Cl, chlorobenzene. This seemed to show that all the carbon atoms were equivalent, since the chlorine atom might be attached to any one of them.

Benzene rings

The solution to the puzzle of benzene's structure came to Kekulé in 1865 in a dream. The answer was a ring of carbon atoms, a ring in which all six atoms were equal, with a hydrogen atom bonded to each one. This meant that the chlorine in chlorobenzene could be attached anywhere around the ring.

Further support for this theory came from substituting hydrogen twice, to make dichlorobenzene ($C_6H_4Cl_2$). If benzene is a six-membered ring with all the carbon atoms equal, there should be three distinct forms, or "isomers," of this compound—the two chlorine atoms could be on adjacent carbon atoms, on carbon atoms separated by one other carbon, or at opposite ends of the ring. This turned out to be the case, and the three isomers were named ortho-, meta-, and para-dichlorobenzene respectively.

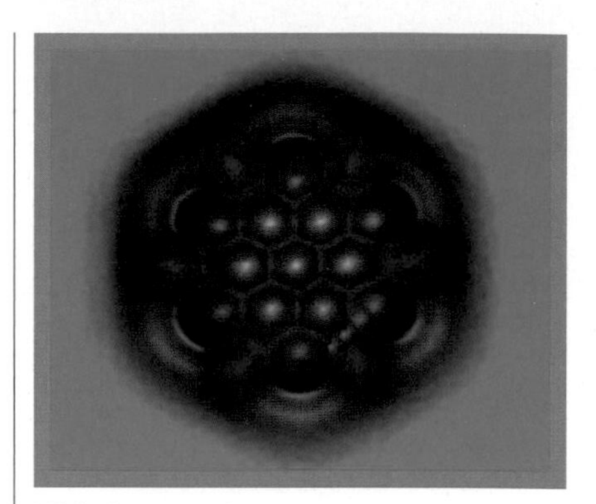

This image of a hexabenzocoronene molecule was captured using an atomic force microscope. It is 1.4 nanometers in diameter and shows carbon–carbon bonds of different lengths.

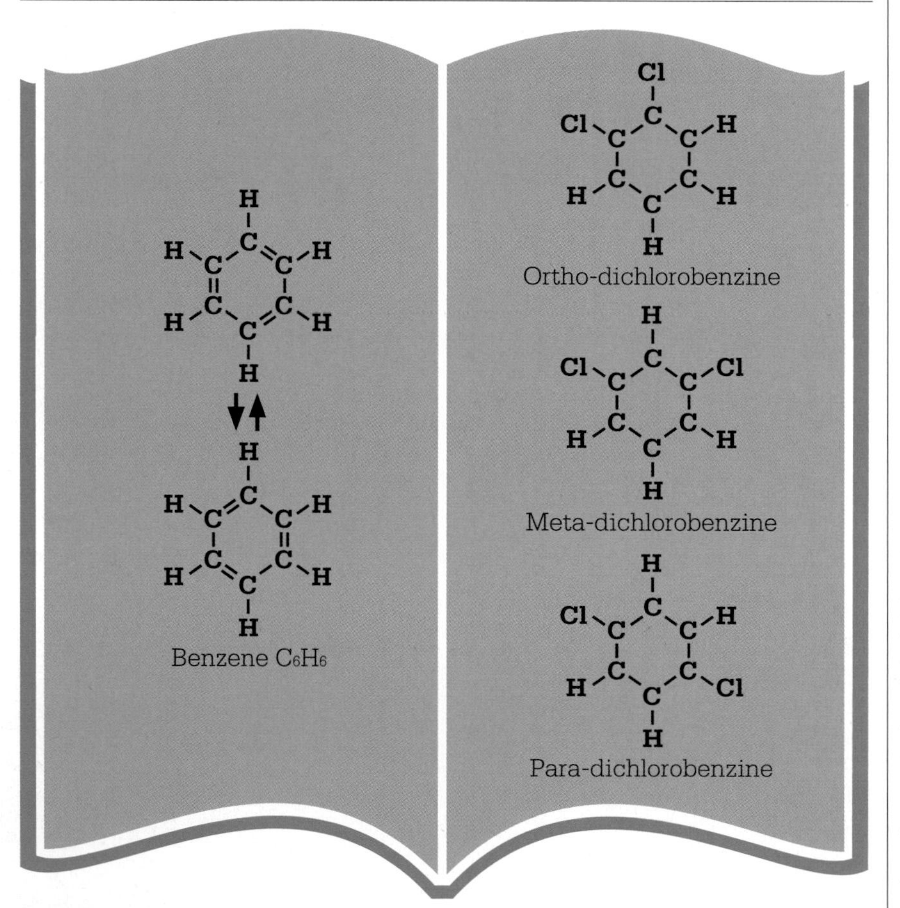

Kekulé suggested that double and single bonds between carbon atoms in a benzene ring alternated (left). Two chlorine atoms can substitute for two of the hydrogen atoms in three different ways (right).

Establishing symmetry

An unsolved mystery still remained over the observed symmetry of the benzene ring. To satisfy its tetravalency, each carbon atom should have four bonds to other atoms. This meant that they all had a "spare" bond. At first, Kekulé drew alternating single and double bonds around the ring, but when it became apparent that the ring had to be symmetrical, he suggested that the molecule oscillated between the two structures.

The electron was not discovered until 1896. The idea that bonds form through the sharing of electrons was first proposed by American chemist G. N. Wilson in 1916. In the 1930s, Linus Pauling then used quantum mechanics to explain that the six spare electrons in the benzene ring are not localized in double bonds, but

Kekulé described the moment that he formulated his theory of benzene rings as a dreamlike vision, in which he saw a snake biting its own tail as in the ancient symbol of the ouroboros, which is depicted here as a dragon.

are delocalized around the ring, and shared equally between the carbon atoms, so that the carbon-carbon bonds are neither single nor double, but 1.5 (see pp.254–59). It would take these new ideas from physics to finally solve the puzzle of the structure of the benzene molecule.

Dream of inspiration

Kekulé's report of his dream is the most cited personal account of a flash of inspiration in all of science. It seems that he was in a hypnagogic state—on the edge of going to sleep: that state where realities and imagination slide into each another. He described it as *Halbschlaf,* or half-sleep. In fact he describes two such reveries: the first, probably in 1855, on top of a bus in south London, heading for Clapham Road. "Atoms fluttered before my eyes. I had always seen these tiny particles in motion, but I had never succeeded in fathoming the manner of their motion. Today I saw how frequently two smaller ones merged into a pair; how larger ones engulfed two smaller ones, still larger ones bonded three and even four of the small ones."

The second occasion was in his study in Ghent in Belgium, possibly inspired by the ancient ouroboros symbol of a snake biting its own tail: "The same thing happened with the benzene ring theory… I turned the chair to face the fireplace and slipped into a languorous state…atoms fluttered before my eyes….Long rows, frequently linked more densely; everything in motion, winding and turning like snakes. And lo, what was that? One of the snakes grabbed its own tail and the image whirled mockingly before my eyes." ■

August Kekulé

Friedrich August Kekulé, who called himself August, was born on September 7, 1829 in Darmstadt, now in the German state of Hesse. While at the University of Giessen, he abandoned the study of architecture and switched to chemistry after hearing the lectures of Justus von Liebig. He eventually became professor of chemistry at Bonn University.

In 1857 and the following years, Kekulé published a series of papers on the tetravalence of carbon, the bonding in simple organic molecules, and the structure of benzene, which made him the principal architect of the theory of molecular structure. In 1895, he was ennobled by Kaiser Wilhelm II, and became August Kekulé von Stradonitz. Three of the first five Nobel prizes in chemistry were won by his students.

Key works

1859 *Textbook of Organic Chemistry*
1887 *The Chemistry of Benzene Derivatives or Aromatic Substances*

THE DEFINITELY EXPRESSED AVERAGE PROPORTION OF THREE TO ONE

GREGOR MENDEL (1822–1884)

IN CONTEXT

BRANCH
Biology

BEFORE
1760 German botanist Josef Kölreuter describes experiments in breeding tobacco plants, but fails to explain his results correctly.

1842 Swiss botanist Carl von Nägeli studies cell division and describes threadlike bodies that are later identified as chromosomes.

1859 Charles Darwin publishes his theory of evolution by natural selection.

AFTER
1900 Botanists Hugo de Vries, Carl Correns, and William Bateson concurrently "rediscover" Mendel's laws.

1910 Thomas Hunt Morgan corroborates Mendel's laws and confirms the chromosomal basis for heredity.

In the history of scientific understanding, one of the greatest of all the natural mysteries was the mechanism of inheritance. The fact of heredity had been known ever since people noticed that family members were recognizably similar. Practical implications were everywhere—from the breeding of crops and livestock in agriculture, to the knowledge that some diseases, such as hemophilia, could be passed on to children. But no one knew how it happened.

Greek philosophers thought that there was some sort of essence or material "principle" that was passed from parents to offspring. Parents conveyed the principle to the next generation during sexual intercourse; it was supposed to have originated in the blood, and paternal and maternal principles blended to make a new person. This idea persisted for centuries—mainly because no one came up with anything better—but when it reached Charles Darwin, its fundamental weakness became all too clear. Darwin's theory of evolution by natural selection proposed that species changed over many generations—and in doing so gave rise to biological diversity. But if inheritance relied on the blending of chemical principles, surely the biological diversity would be diluted out of existence? It would be like mixing paints of different colors, and ending up with gray. The adaptations and novelties upon which Darwin's theory rested would not persist.

Inherited characteristics had been observed for millennia before Mendel, but the biological mechanism that produced phenomena such as identical twins was unknown.

Gregor Mendel

Born Johann Mendel in 1822 in Silesia in the Austrian Empire, Mendel initially trained in mathematics and philosophy before entering the priesthood as a way of furthering his education—changing his name to Gregor and becoming an Augustinian monk. He completed his studies at the University of Vienna and returned to teach at the abbey in Brno (now in Czech Republic). Here, Mendel developed his interest in inheritance—and at various times studied mice, bees, and peas. Under pressure from the bishop, he abandoned work on animals and concentrated on breeding peas. It was this work that led him to devise his laws of heredity and develop the critical idea that inherited characteristics are controlled by discrete particles, later called genes. He became abbot of the monastery in 1868 and stopped his scientific work. On his death, his scientific papers were burned by his successor.

Key work

1866 *Experiments in Plant Hybrizidation*

See also: Jean-Baptiste Lamarck 118 ▪ Charles Darwin 142–49 ▪ Thomas Hunt Morgan 224–25 ▪ James Watson and Francis Crick 276–83 ▪ Michael Syvanen 318–19 ▪ William French Anderson 322–23

Mendel's discovery

The breakthrough in understanding inheritance came nearly a century before the chemical structure of DNA was established—and less than a decade after Darwin published *On the Origin of Species*. Gregor Mendel, an Augustinian monk in Brno, was a teacher, scientist, and mathematician who succeeded where many better-known naturalists had failed. It was, perhaps, Mendel's skills in mathematics and probability theory that proved the difference.

Mendel conducted experiments with the common pea, *Pisum sativum*. This plant varies in several identifiable ways, such as height, flower color, seed color, and seed shape. Mendel started looking at the inheritance of one characteristic at a time and applied his mathematical mind to the results. By breeding pea plants, which were easily cultivated in the monastery grounds, he could conduct a series of experiments to obtain meaningful data.

Mendel took critical precautions in his work. Recognizing that characteristics can skip and hide through generations, he was careful to start with pea plants of "pure" stock—such as white-flowered plants that only produced white-flowered offspring. He crossed pure white-flowered plants with pure purple-flowered ones, pure tall with pure short, and so on. In each case, he also precisely controlled the fertilization: using tweezers, he transferred pollen from unopened flower buds to stop them from scattering indiscriminately. He performed these breeding experiments many times and documented the numbers and characteristics of plants in the next generation, and the generation after that. He found that alternate varieties (such as purple flower and white flower) were inherited in fixed proportions. In the first generation, only one variety, such as purple flower, came through; in the second generation, this variety accounted for three-quarters of the offspring. Mendel called this the dominant variety. He called the other variety the recessive variety. In this case, white flower was recessive, and made up a quarter of the second generation plants. For each characteristic—tall/short; seed color; flower color; and seed shape—it was possible to identify dominant and recessive varieties according to these proportions.

The key conclusion

Mendel went further and tested the inheritance of two characteristics simultaneously—such as flower color and seed color. He found that offspring ended up with different combinations of traits and—once again—these combinations occurred in fixed proportions. In the first generation, all plants had both dominant traits (purple flower, yellow seed), but in the second generation there was a mixture of combinations. »

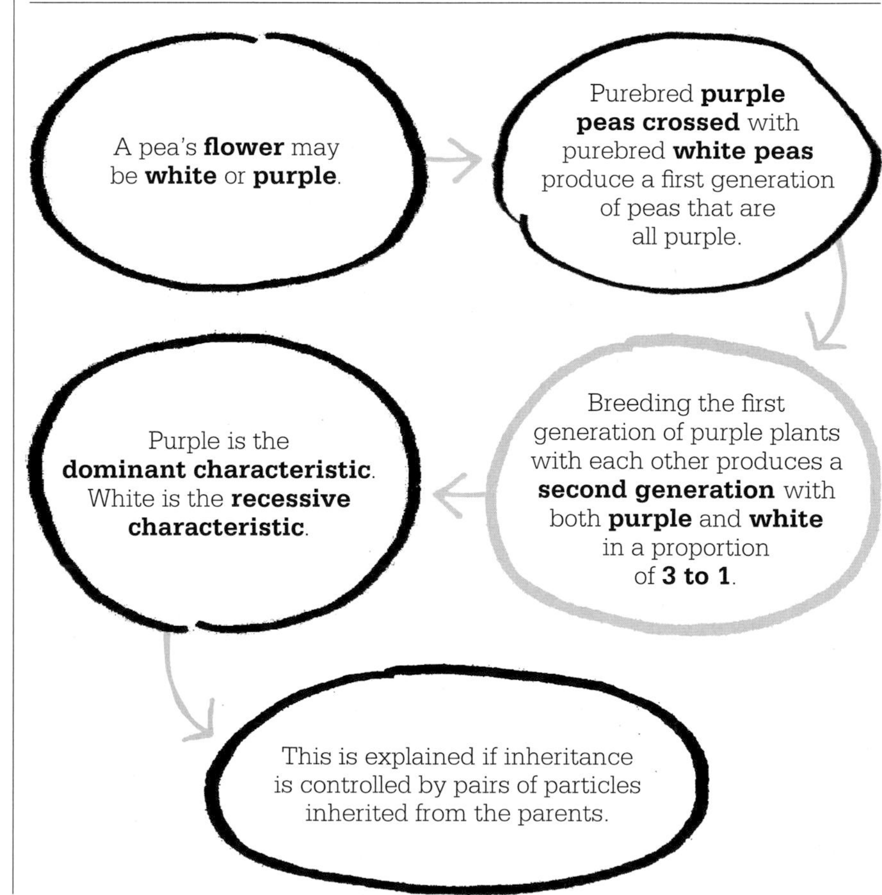

For example, one-sixteenth of the plants had the combination with both recessive traits (white flower, green seed). Mendel concluded that the two characteristics were inherited independently of one another. In other words, inheritance of flower color had no effect on inheritance of seed color and vice versa. The fact that heredity was precisely proportional in this way led Mendel to conclude that it was not due to the blending of vague chemical principles after all, but happened because of discrete "particles." There were particles controlling flower color, particles for seed color, and so on. These particles were transferred from parents to offspring intact. This explained why recessive traits could hide their effects and skip a generation: a recessive trait would only show through if a plant inherited two identical doses of the particle concerned. Today we recognize these particles as genes.

Genius recognized

Mendel published the results of his findings in a journal of natural history in 1866, but his work failed to make an impact in the wider scientific world. The esoteric nature of his title—*Experiments in Plant Hybridization*—might have restricted the readership but, in any case, it took more than 30 years for Mendel to be properly appreciated for what he had done. In 1900, Dutch botanist Hugo de Vries published the results of plant breeding experiments similar to those of Mendel—including a corroboration of the three-to-one ratio. De Vries followed up with an acknowledgment that Mendel had got there first.

Traits disappear entirely in the hybrids, but reappear unchanged in their progeny.
Gregor Mendel

A few months later, German botanist Carl Correns explicitly described Mendel's mechanism for inheritance. Meanwhile, in England—spurred on after reading the papers of de Vries and Correns—Cambridge biologist William Bateson read Mendel's original paper for the first time and immediately recognized its significance. Bateson would become a champion of Mendelian ideas, and he ended up coining the term "genetics" for this new field of biology. Posthumously, the Augustinian monk had at last been appreciated.

By then, work of a different kind—in the fields of cell biology and biochemistry—was guiding biologists down new avenues of research. Microscopes were replacing plant breeding experiments as scientists searched for clues by looking right inside cells. Nineteenth-century biologists had a hunch that the key to heredity lay in the cell's nucleus. Unaware of Mendel's work, in 1878, German Walther Flemming identified the threadlike structures inside cell nuclei that moved around during cell division. He named them chromosomes, meaning "colored body." Within a few years of the rediscovery

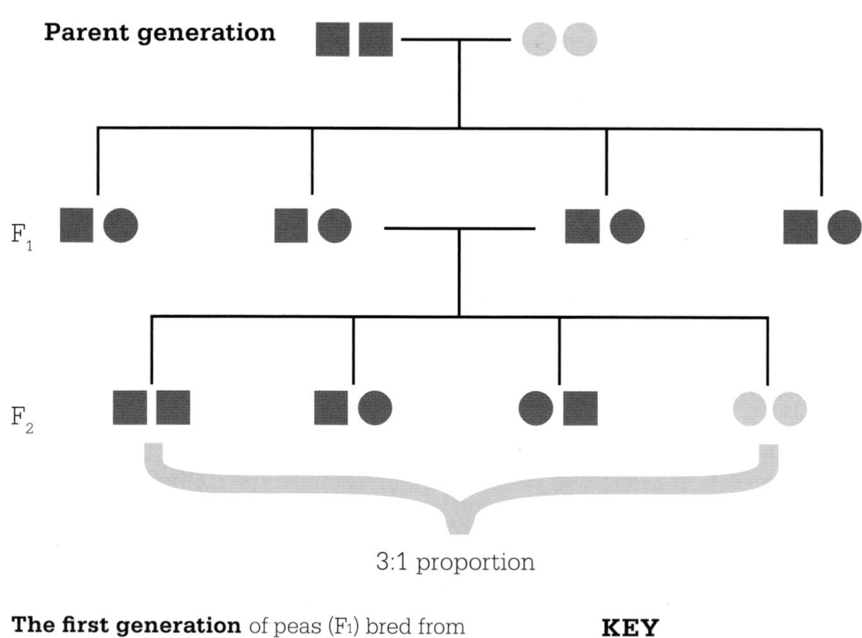

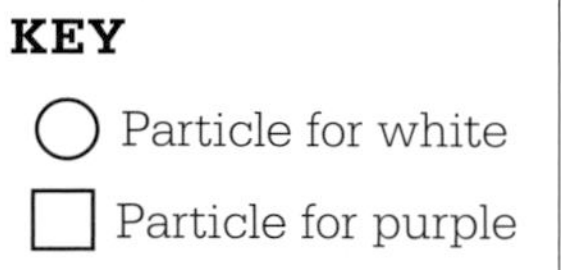

The first generation of peas (F_1) bred from "pure" white- and purple-flowered plants all have one particle from each parent. Purple is dominant, so all the F_1 flowers are purple. In the second generation (F_2), one plant in four will inherit two "white" particles and produce white flowers.

Hugo de Vries discovered the 3:1 ratio of characteristics in experiments with a variety of plants in the 1890s. He would later concede that Mendel had a claim to priority in the discovery.

of Mendel's work, biologists had demonstrated that Mendel's "particles of inheritance" were real and that they were carried on chromosomes.

Laws of inheritance refined

Mendel had established two laws of inheritance. First, the fixed proportions of characteristics in offspring led him to conclude that the particles of inheritance came in pairs. There was a particle pair for flower color, a pair for seed color, and so on. Pairs were formed at fertilization because one particle came from each parent—and separated again when the new generation reproduced to form its own sex cells. If the particles coming together were different varieties (such as those for purple and white flower), only the dominant particle would be expressed.

In modern terms, the different varieties of genes are called alleles. Mendel's first law became known as the Law of Segregation because the alleles segregated to form sex cells. Mendel's second law arose when he considered two characteristics. The Law of Independent Assortment suggests that the relevant genes for each trait are inherited independently.

Mendel's choice of plant species was, it turns out, fortuitous. We now know that the characteristics of *Pisum sativum* follow the simplest pattern of inheritance. Each characteristic—such as flower color—is under the control of a single type of gene that comes in different varieties (alleles). However, many biological characteristics—such as human height—are the outcome of the interactions of many different genes.

Furthermore, the genes Mendel studied were inherited independently. Later work would show that genes can sit side-by-side on the same chromosome. Each chromosome carries hundreds or thousands of genes on a string of DNA. Chromosome pairs separate to create sex cells, and the chromosome is then passed on whole. This means that the inheritance of traits controlled by different genes on the same chromosome is not independent. Each pea characteristic studied by Mendel is due to a gene on a separate chromosome. If they had been on the same chromosome, his results would have been more complex and harder to interpret.

In the 20th century, research would reveal the exceptions to Mendel's laws. As scientists probed more deeply into the behavior of genes and chromosomes, they confirmed that inheritance can happen in more complicated ways than Mendel had found. However, these discoveries build on, rather than contradict, Mendel's findings, which laid the foundation for modern genetics. ■

> I suggest…the term Genetics, which sufficiently indicates that our labours are devoted to the elucidation of the phenomena of heredity and variation.
>
> **William Bateson**

AN EVOLUTIONARY LINK BETWEEN BIRDS AND DINOSAURS

THOMAS HENRY HUXLEY (1825–1895)

IN CONTEXT

BRANCH
Biology

BEFORE
1859 Charles Darwin publishes *On the Origin of Species*, describing his theory of evolution.

1860 The first *Archaeopteryx* fossil, discovered in Germany, is sold to London's Natural History Museum.

AFTER
1875 The "Berlin specimen" of *Archaeopteryx*, with teeth, is found.

1969 US paleontologist John Ostrom's study of microraptor dinosaurs highlights new similarities with birds.

1996 *Sinosauropteryx*, the first known feathered dinosaur, is discovered in China.

2005 US biologist Chris Organ shows the similarity between the DNA of birds and that of *Tyrannosaurus rex*.

In 1859, Charles Darwin described his theory of evolution by natural selection. In the heated debates that followed, Thomas Henry Huxley was the most formidable champion of Darwin's ideas, earning himself the nickname "Darwin's bulldog." More significantly, the British biologist did pioneering work on a key tenet in the evidence for Darwin's theories—the idea that birds and dinosaurs are closely related.

If Darwin's theory that species gradually changed into others was true, then the fossil record should show how species that were very different had diverged from ancestors that were very similar. In 1860, a remarkable fossil was found in limestone in a German quarry. It dated from the Jurassic period, and was named *Archaeopteryx lithographica*. With wings and feathers like a bird's, yet from the time of the dinosaurs, it seemed to be an example of the kind of missing link between species that Darwin's theory predicted.

One sample, however, was not nearly enough to prove the connection between birds and dinosaurs, and *Archaeopteryx*

Eleven fossils of *Archaeopteryx* have been discovered. This birdlike dinosaur lived in the Late Jurassic period, about 150 million years ago, in what is now southern Germany.

could simply have been one of the earliest birds, rather than a feathered dinosaur. But Huxley began to study closely the anatomy of both birds and dinosaurs, and for him, the evidence was compelling.

A transitional fossil

Huxley made detailed comparisons between *Archaeopteryx* and various other dinosaurs, and found that it was very similar to the small dinosaurs *Hypsilophodon* and *Compsognathus*. The discovery, in 1875, of a more complete *Archaeopteryx* fossil, this time with dinosaur-like teeth, seemed to confirm the connection.

See also: Mary Anning 116–17 ▪ Charles Darwin 142–49

Detailed studies of **fossils of small dinosaurs** show many features in common with **birds**.

Birdlike *Archaeopteryx* fossils have **teeth**, like dinosaurs.

The **similarities between** the anatomy of **birds and dinosaurs** are too great to be a coincidence.

There is an evolutionary link between birds and dinosaurs.

Huxley came to believe that there was an evolutionary link between birds and dinosaurs, but he did not imagine a common ancestor would ever be found. What mattered to him were the very clear similarities. Like reptiles, birds have scales—feathers are simply developments of scales—and they lay eggs. They also have a host of similarities in bone structure.

Nevertheless, the link between dinosaurs and birds remained disputed for another century. Then, in the 1960s, studies of the sleek, agile raptor *Deinonychus* (a relative of *Velociraptor*) began finally to convince many paleontologists of the link between birds and these microraptors (small predatory dinosaurs). In recent years, a host of finds of fossils of ancient birds and birdlike dinosaurs in China has strengthened the link—including the discovery in 2005 of a small dinosaur with feathered legs, *Pedopenna*. Also that year, a groundbreaking study of DNA extracted from the fossilized soft tissue of a *Tyrannosaurus rex* showed that dinosaurs are genetically more similar to birds than to other reptiles. ■

Birds are essentially similar to Reptiles…these animals may be said to be merely an extremely modified and aberrant Reptilian type.
Thomas Henry Huxley

Thomas Henry Huxley

Born in London, Huxley became an apprentice doctor at 13 years old. At 21, he was a surgeon aboard a Royal Navy ship assigned to chart the seas around Australia and New Guinea. During the voyage, he wrote papers on the marine invertebrates he collected, and these so impressed the Royal Society that he was elected a fellow in 1851. On his return in 1854, Huxley became a lecturer in natural history at the Royal School of Mines.

After meeting Charles Darwin in 1856, Huxley became a strong advocate of Darwin's theories. In a debate on evolution held in 1860, Huxley won the day against Samuel Wilberforce, Bishop of Oxford, who argued for God's creation. Along with his work showing similarities between birds and dinosaurs, he gathered evidence on the subject of human origins.

Key works

1858 *The Theory of the Vertebrate Skull*
1863 *Evidence as to Man's Place in Nature*
1880 *The Coming of Age of the Origin of Species*

AN APPARENT PERIODICITY OF PROPERTIES

DMITRI MENDELEEV (1834–1907)

IN CONTEXT

BRANCH
Chemistry

BEFORE
1803 John Dalton introduces the idea of atomic weights.

1828 Johann Döbereiner attempts first classification.

1860 Stanislao Cannizzaro publishes an extensive table of atomic and molecular weights.

AFTER
1913 Lothar Meyer shows the periodic relationship between elements by plotting atomic weight against volume.

1913 Henry Moseley redefines the periodic table using atomic numbers—the number of protons in an atom's nucleus.

1913 Niels Bohr suggests a model for the structure of the atom. It includes shells of electrons that explain the relative reactivity of the different groups of elements.

In 1661, Anglo-Irish physicist Robert Boyle defined elements as "certain primitive and simple, or perfectly unmingled bodies; which not being made of any other bodies, or of one another, are the ingredients of which all those called perfectly mixt bodies are immediately compounded, and into which they are ultimately resolved." In other words, an element cannot be broken down by chemical means into simpler substances. In 1803, British chemist John Dalton introduced the idea of atomic weights (now called relative atomic masses) for these elements. Hydrogen is the lightest element, and he gave it the value 1, which we still use today.

Law of eight

In the first half of the 19th century, chemists gradually isolated more elements, and it became clear that certain groups of elements had similar properties. For example, sodium and potassium are silvery solids (alkali metals) that react violently with water, liberating hydrogen gas. In fact, they are so similar that British chemist Humphry Davy did not distinguish between them when he first discovered them. Similarly, the halogen elements chlorine and bromine are both pungent, poisonous oxidizing agents, even though chlorine is a gas and bromine a liquid. British chemist John Newlands noticed that when the known elements were listed in order of increasing

The first to attempt a classification of the elements was German chemist Johann Döbereiner. By 1828, he had found that some elements formed groups of three with related properties.

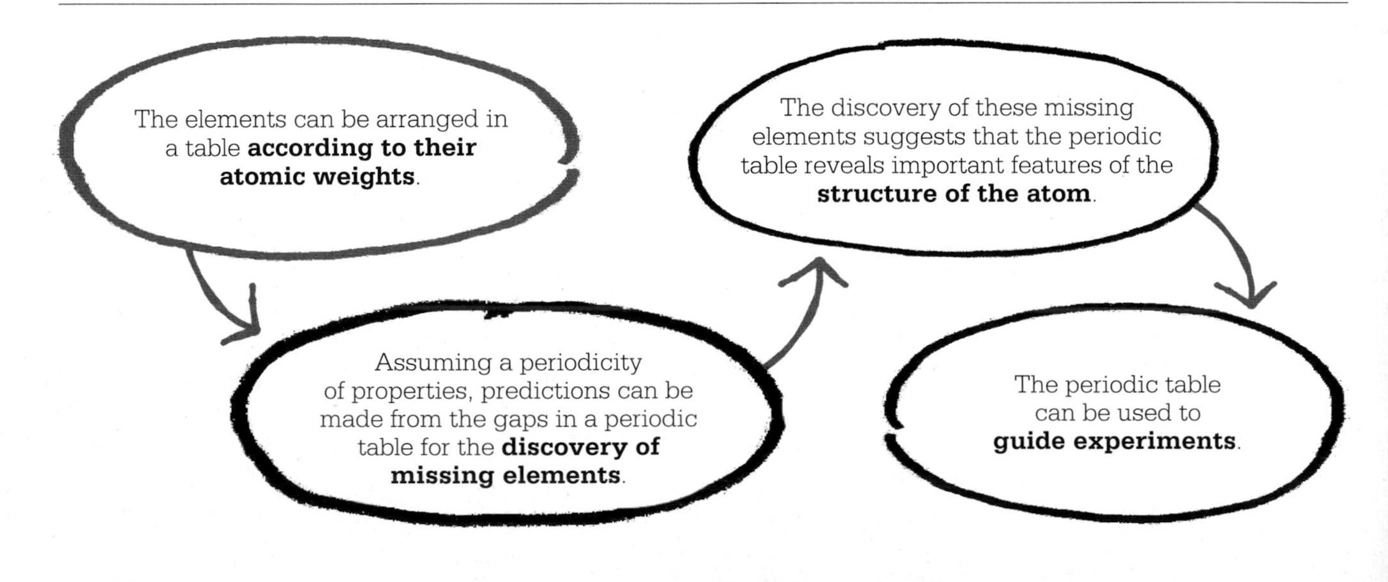

See also: Robert Boyle 46–49 ▪ John Dalton 112–13 ▪ Humphry Davy 114 ▪ Marie Curie 190–95 ▪ Ernest Rutherford 206–13 ▪ Linus Pauling 254–59

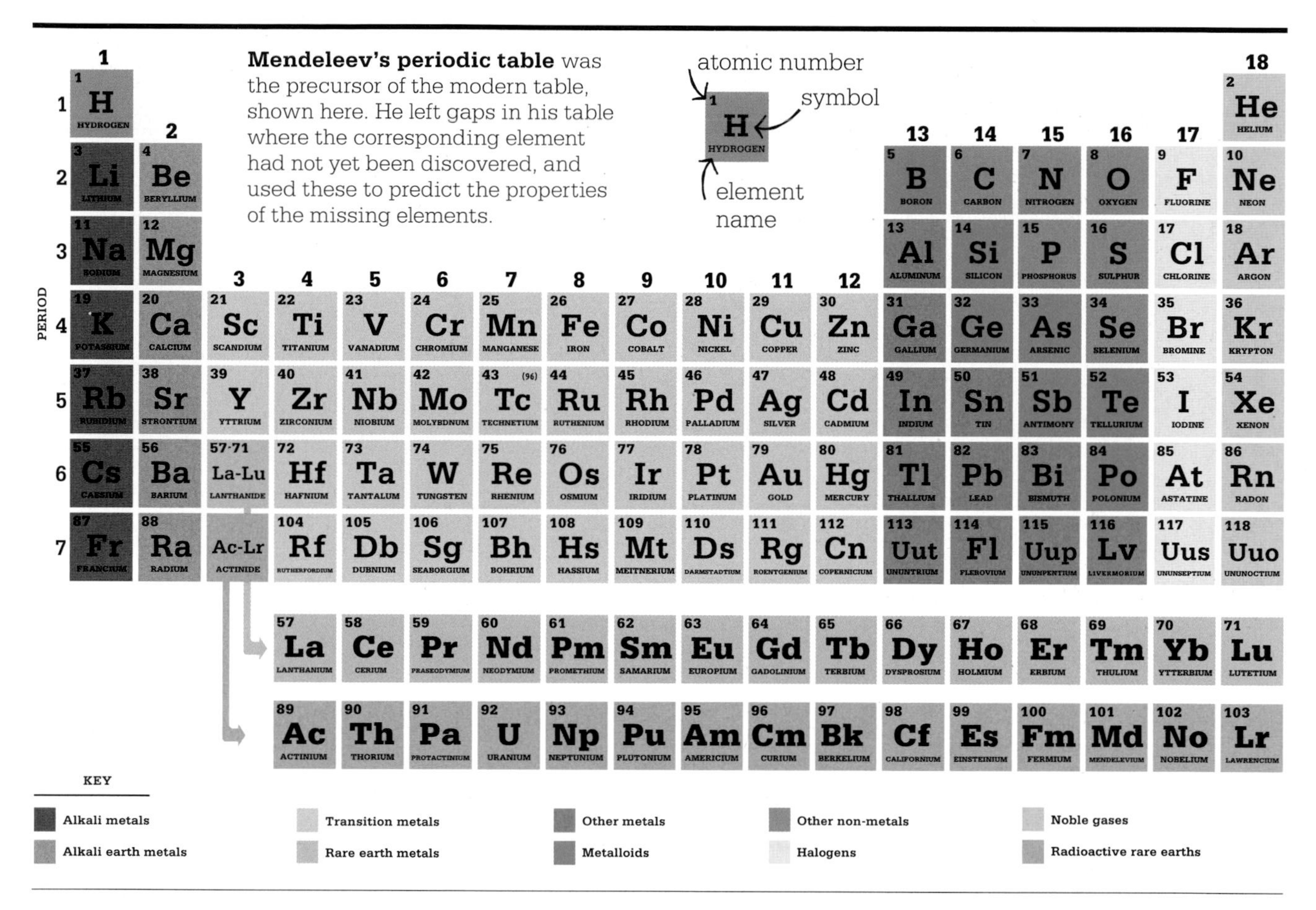

Mendeleev's periodic table was the precursor of the modern table, shown here. He left gaps in his table where the corresponding element had not yet been discovered, and used these to predict the properties of the missing elements.

atomic weight, similar elements occurred every eighth place. He published his findings in 1864.

In the journal *Chemical News* Newlands wrote: "Elements belonging to the same group appear in the same horizontal line. Also the numbers of similar elements differ by seven or multiples of seven…This peculiar relationship I propose to call The Law of Octaves." The patterns in his table make sense as far as calcium, but then go haywire. On March 1, 1865, Newlands was ridiculed by the Chemical Society, who said that he might as well list the elements in alphabetical order, and refused to publish his paper. The significance of Newlands' achievement would not be recognized for more than 20 years. Meanwhile, French mineralogist Alexandre-Émile Béguyer de Chancourtois had also noticed the patterns, publishing his ideas in 1862, but few people noticed.

Card puzzle

Around the same time, Dmitri Mendeleev was struggling with the same problem as he wrote his book *Principles of Chemistry* in St. Petersburg, Russia. In 1863, there were 56 known elements, and new ones were being discovered at a rate of about one a year. Mendeleev was convinced that there must be a pattern to them. In an effort to solve the puzzle, he made a set of 56 playing cards, each labeled with the name and major properties of one element.

Mendeleev is said to have made his breakthrough as he was about to embark on a winter journey in 1868. Before setting out, he laid out his cards on the table and began to ponder the puzzle, as though playing a game of solitare. When his coachman came to the door for the luggage, Mendeleev waved him away, saying he was busy. He moved the cards back and forth until finally he managed to arrange all 56 elements to his satisfaction, with the similar groups running » vertically. The following year, Mendeleev read a paper at the Russian Chemical Society stating that: "The elements, if arranged according to their atomic weight, exhibit an apparent periodicity of properties." He explained that elements with similar chemical properties have atomic weights that are either of nearly the same value (such as potassium, iridium, and osmium) or that increase regularly (such as potassium, rubidium, and cesium). He further explained that the arrangement of the elements into groups in the order of their atomic weights corresponds to their valency, which is the number of bonds the atoms can form with other atoms.

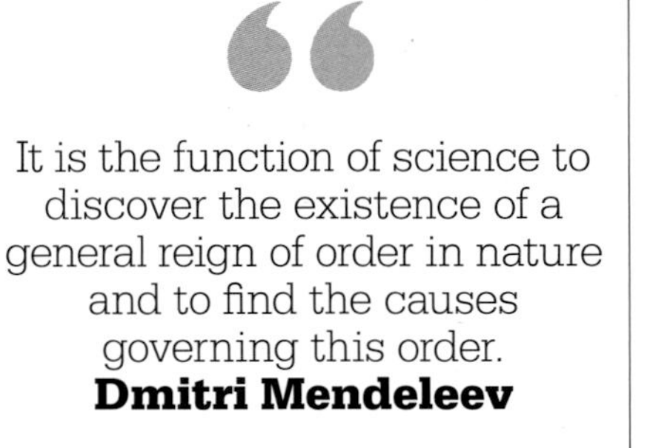

It is the function of science to discover the existence of a general reign of order in nature and to find the causes governing this order.
Dmitri Mendeleev

Predicting new elements

In his paper, Mendeleev made a bold prediction: "We must expect the discovery of many yet unknown elements—for example, two elements, analogous to aluminum and silicon, whose atomic weights would be between 65 and 75."

Mendeleev's arrangement included crucial improvements over Newlands' Octaves. Below boron and aluminum, Newlands had placed chromium, which made little sense. Mendeleev reasoned that there must exist an as-yet undiscovered element, and predicted that one would be found with an atomic weight of about 68. It would form an oxide (a compound formed by an element with oxygen) with a chemical formula of M_2O_3, where "M" is the symbol for the new element. This formula meant that two atoms of the new element would combine with three oxygen atoms to make the oxide. He predicted two more elements to fill other spaces: one with an atomic weight of about 45, forming the oxide M_2O_3, and the other with an atomic weight of 72, forming the oxide MO_2.

Critics were sceptical, but Mendeleev had made very specific claims, and one of the most powerful ways to support a scientific theory is to make predictions that are proved true. In this case, the element gallium (atomic weight 70, forming the oxide Ga_2O_3) was discovered in 1875; scandium (weight 45, Sc_2O_3) in 1879; and germanium (weight 73, GeO_2) in 1886. These discoveries made Mendeleev's reputation.

The six alkali metals are all soft, highly reactive metals. The outer layer of this lump of pure sodium has reacted with the oxygen in the air to give it a coating of sodium oxide.

Mistakes in the table

Mendeleev did make some mistakes. In his 1869 paper, he asserted that the atomic weight of tellurium must be incorrect: it should lie between 123 and 126, because the atomic weight of iodine is 127, and iodine should clearly follow tellurium in the table,

The six noble gases that occur naturally (listed in group 18 of the table) are helium, neon, argon, krypton, xenon, and radon. They have very low chemical reactivity because they each have a full valence shell—a shell of electrons surrounding the atom's nucleus. Helium has just one shell containing two electrons, while the other elements have outer shells of eight electrons. Radioactive radon is unstable.

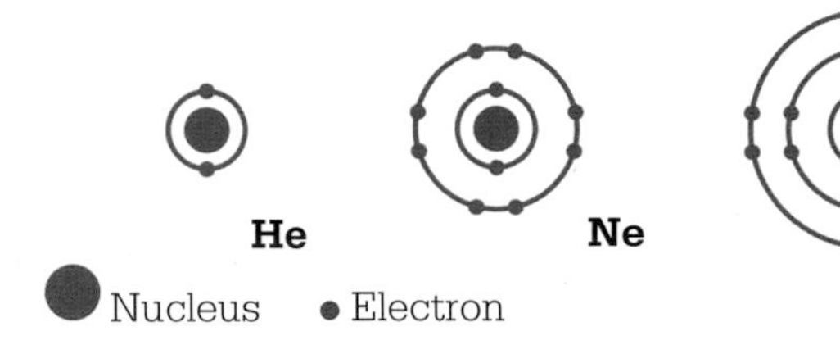

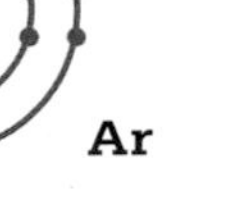

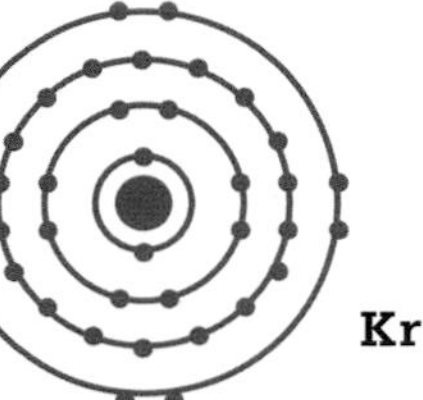

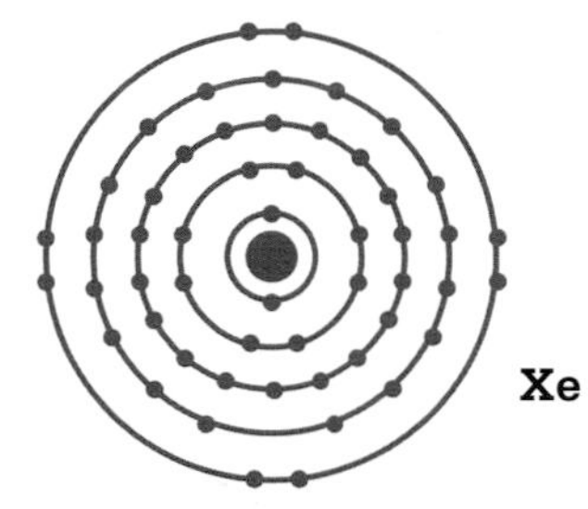

according to its properties. He was wrong—the relative atomic weight of tellurium is in fact 127.6; it is greater than that of iodine. A similar anomaly occurs between potassium (weight 39) and argon (weight 40), where argon clearly precedes potassium in the table—but Mendeleev was not aware of these problems in 1869, because argon was not discovered until 1894. Argon is one of the noble gases, which are colorless, odorless, and hardly react with other elements. Difficult to detect, none of the noble gases were known at that time, so there were no spaces for them in Mendeleev's table. Once argon had turned up, however, there were several more holes to fill, and by 1898, Scottish chemist William Ramsay had isolated helium, neon, krypton, and xenon. In 1902, Mendeleev incorporated the noble gases into his table as Group 18, and this version of the table forms the basis of the periodic table we use today.

The anomaly of the "wrong" atomic weights was solved in 1913 by British physicist Henry Moseley, who used X-rays to determine the number of protons in the nucleus of each atom of a particular element.

> We must expect the discovery of elements analogous to aluminum and silicon—whose atomic weight would be between 65 and 75.
> **Dmitri Mendeleev**

This came to be called the atomic number of the element, and it is this number that determines the element's position on the periodic table. The fact that atomic weights had given a close approximation followed from the fact that for the lighter elements, the atomic weight is roughly (though not exactly) twice the atomic number.

Using the table

The periodic table of the elements may look like just a cataloguing system—a neat way of ordering the elements—but it has far greater importance in both chemistry and physics. It allows chemists to predict the properties of an element, and to try variations in processes; for example, if a particular reaction does not work with chromium, perhaps it works with molybdenum, the element below chromium in the table.

The table was also crucial in the search for the structure of the atom. Why did the properties of elements repeat in these patterns? Why were the Group 18 elements so unreactive, while the elements in the groups on either side were the most reactive of all? Such questions led directly to the picture of the structure of the atom that has been accepted ever since.

Mendeleev was to some extent lucky to have been credited for his table. Not only did he publish his ideas after Béguyer and Newlands, but also German chemist Lothar Meyer, who plotted atomic weight against atomic volume to show the periodic relationship between elements, was ahead of him, too, publishing in 1870. As so often in science, the time had been ripe for a particular discovery, and several people had reached the same conclusion independently, without knowing about each other's work. ■

Dmitri Mendeleev

The youngest of at least 12 children, Dmitri Mendeleev was born in 1834 in a village in Siberia. When his father went blind and lost his teaching post, Mendeleev's mother supported the family with a glass factory business. When that burned down, she took her 15-year-old son across Russia to St. Petersburg to receive a higher education.

In 1862, Mendeleev married Feozva Nikitichna Leshcheva, but in 1876 he became obsessed with Anna Ivanova Popova, and married her before his divorce from his first wife was final.

In the 1890s, Mendeleev organized new standards for producing vodka. He investigated the chemistry of oil, and helped to set up Russia's first oil refinery. In 1905, he was elected a member of the Royal Swedish Academy of Science, who recommended him for a Nobel Prize, but his candidacy was blocked, possibly due to his bigamy. The radioactive element 101 mendelevium is named in his honor.

Key work

1870 *Principles of Chemistry*

LIGHT AND MAGNETISM ARE AFFECTATIONS OF THE SAME SUBSTANCE

JAMES CLERK MAXWELL (1831–1879)

IN CONTEXT

BRANCH
Physics

BEFORE
1803 Thomas Young's double-slit experiments appear to show that light is a wave.

1820 Hans Christian Ørsted demonstrates a link between electricity and magnetism.

1831 Michael Faraday shows that a changing magnetic field produces an electric field.

AFTER
1900 Max Planck suggests that in some circumstances, light can be treated as if it were composed of tiny "wave packets," or quanta.

1905 Albert Einstein shows that light quanta, today known as photons, are real.

1940s Richard Feynman and others develop quantum electrodynamics (QED) to explain the behavior of light.

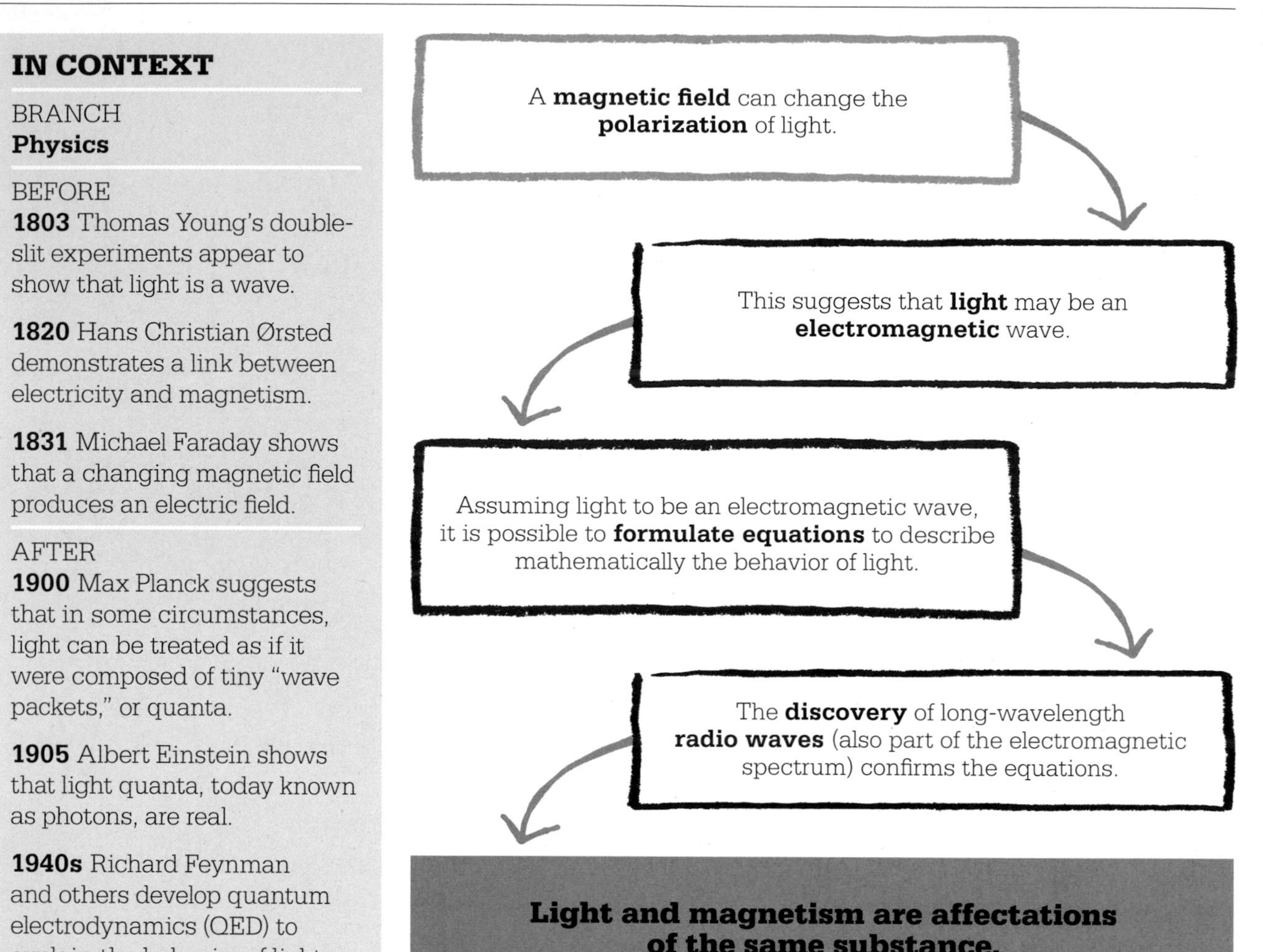

The series of differential equations describing the behavior of electromagnetic fields developed by Scottish physicist James Clerk Maxwell through the 1860s and 1870s are rightly considered one of the towering achievements in the history of physics. A truly transformative discovery, they not only revolutionized the way that scientists viewed electricity, magnetism, and light, but also laid the ground rules for an entirely new style of mathematical physics. This would have far-reaching consequences in the 20th century, and today offers hope for unifying our understanding of the universe into a comprehensive "Theory of Everything."

The Faraday effect

Danish physicist Hans Christian Ørsted's discovery, in 1820, of a link between electricity and magnetism set the stage for a century of attempts to discover the links and interconnections between seemingly unconnected phenomena. It also inspired a significant breakthrough by Michael Faraday. Today, Faraday is perhaps best known for his invention of the electric motor and the discovery of electromagnetic induction, but it was a less celebrated discovery that provided Maxwell's departure point.

For two decades, Faraday had been attempting, on and off, to find a link between light and electromagnetism. Then, in 1845, he devised an ingenious experiment that answered the question once and for all. It involved passing a beam of polarized light (one in which the waves oscillate

See also: Alessandro Volta 90–95 ▪ Hans Christian Ørsted 120 ▪ Michael Faraday 121 ▪ Max Planck 202–05 ▪ Albert Einstein 214–21 ▪ Richard Feynman 272–73 ▪ Sheldon Glashow 292–93

The special theory of relativity owes its origins to Maxwell's equations of the electromagnetic field.
Albert Einstein

in a single direction, easily created by bouncing a beam of light off a smooth reflecting surface) through a strong magnetic field, and testing the angle of polarization on the other side using a special eyepiece. He found that by rotating the orientation of the magnetic field, he was able to affect the angle of polarization of the light. Based on this discovery, Faraday argued for the first time that light waves were some kind of undulation in the lines of force by which he interpreted electromagnetic phenomena.

Theories of electromagnetism

However, while Faraday was a brilliant experimentalist, it took the genius of Maxwell to put this intuitive idea onto sound theoretical footing. Maxwell came to the problem from the opposite direction, discovering the link between electricity, magnetism, and light almost by accident.

Maxwell's main concern was to explain just how the electromagnetic forces involved in phenomena such as Faraday's induction—where a moving magnet induces an electric current—were operating. Faraday had invented the ingenious idea of "lines of force," spreading in concentric rings around moving electric currents, or emerging and reentering the poles of magnets. When electrical conductors moved in relation to these lines, currents flowed within them. The density of the lines of force and the speed of relative motion both influenced the strength of the current.

But while lines of force were a useful aid to understanding the phenomenon, they did not have a physical existence—electrical and magnetic fields make their presence felt at every point in space that lies within their range of influence, not just when certain lines are cut. Scientists who attempted to describe the physics of electromagnetism tended to fall into one of two schools: those who saw electromagnetism as some form of "action at a distance" similar to Newton's model of gravity, and those who believed that electromagnetism was propagated through space by waves. In general, the supporters of "action at a distance" hailed from continental Europe and followed the theories of electrical pioneer André-Marie Ampère (p.120), while the believers in waves tended to be British. One clear way of distinguishing between the two basic theories was that action at a distance would take place instantaneously, while waves would inevitably take some time to propagate through space. »

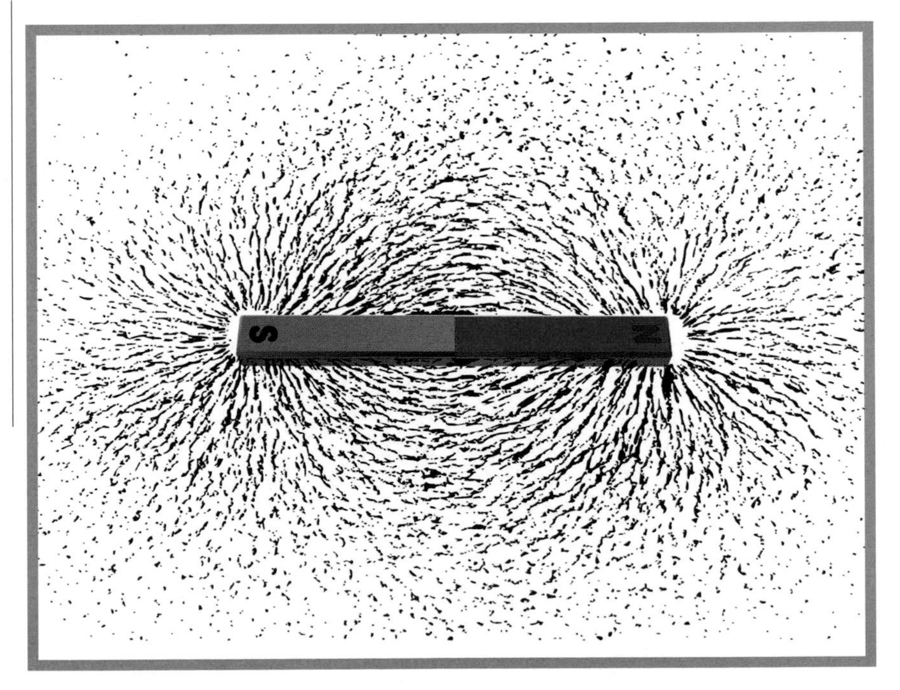

The pattern of iron filings around a magnet would seem to suggest the lines of force described by Faraday. In fact they show the direction of the force experienced by a charge at a given point in an electromagentic field, as represented in Maxwell's equations.

Maxwell's models

Maxwell began to develop his theory of electromagnetism in a pair of papers published in 1855 and 1856. These were an attempt to model Faraday's lines of force geometrically in terms of the flow in a (hypothetical) incompressible fluid. He had limited success and in subsequent papers tried an alternative approach, modeling the field as a series of particles and rotating vortices. By analogy, Maxwell was able to demonstrate Ampère's circuital law, which relates the electric current passing through a conducting loop to the magnetic field around it. Maxwell also showed that in this model, changes in the electromagnetic field would propagate at a finite (if high) speed.

Maxwell derived an approximate value for the speed of propagation, at about 193,060 miles/s (310,700 km/s). This value was so suspiciously close to the speed of light as measured in numerous experiments that he immediately realized that Faraday's intuition about the nature of light must be correct. In the final paper of the series, Maxwell described how magnetism could affect the orientation of an electromagnetic wave as seen in the Faraday effect.

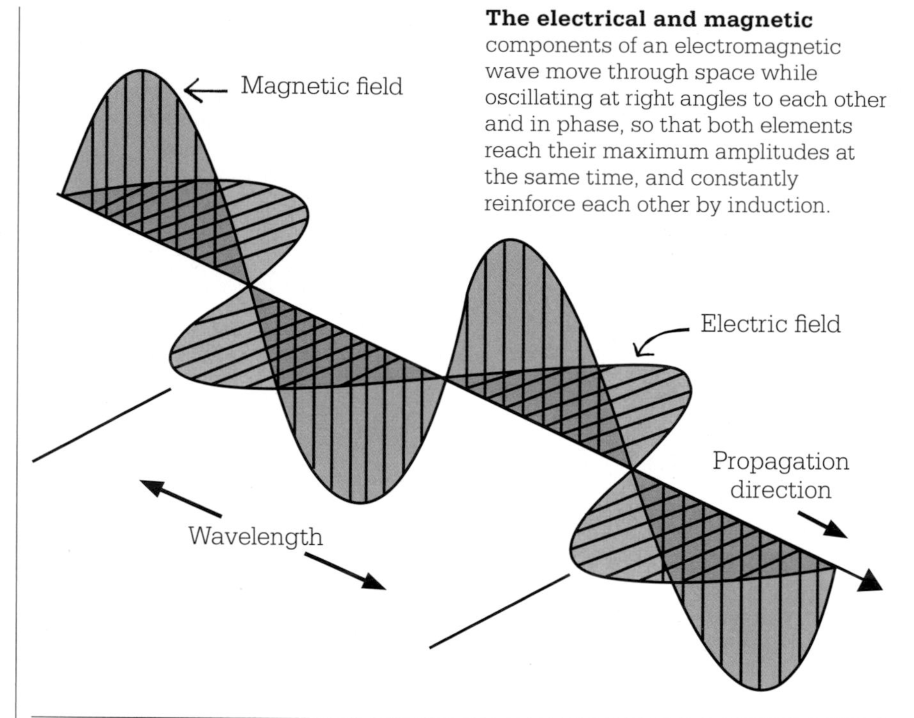

The electrical and magnetic components of an electromagnetic wave move through space while oscillating at right angles to each other and in phase, so that both elements reach their maximum amplitudes at the same time, and constantly reinforce each other by induction.

From a long view of the history of mankind…there can be little doubt that the most significant event of the 19th century will be judged as Maxwell's discovery of the laws of electrodynamics.

Richard Feynman

Developing the equations

Satisfied that the essentials of his theory were correct, Maxwell set out in 1864 to put it on a sound mathematical footing. In *A Dynamical Theory of the Electromagnetic Field*, he described light as a pair of electrical and magnetic transverse waves, oriented perpendicular to each other and locked in phase in such a way that changes to the electric field reinforce the magnetic field, and vice versa (the orientation of the electrical wave is the one that normally determines the wave's overall polarization). In the last part of his paper, he laid out a series of 20 equations that offered a complete mathematical description of electromagnetic phenomena in terms of electrical and magnetic potentials—in other words, the amount of electrical or magnetic potential energy a point charge would experience at a specific point in the electromagnetic field.

Maxwell went on to show how electromagnetic waves moving at the speed of light arose naturally from the equations, apparently settling the debate about the nature of electromagnetism once and for all.

He summed up his work on the subject in the 1873 *Treatise on Electricity and Magnetism*, but, convincing as the theory was, it remained unproven at the time of Maxwell's death, since the short wavelength and high frequency of light waves made their properties impossible to measure. However, eight years later, in 1887, German physicist Heinrich Hertz provided the final piece of the puzzle (and made an enormous technological breakthrough) when he succeeded in producing a very different form of electromagnetic wave with low

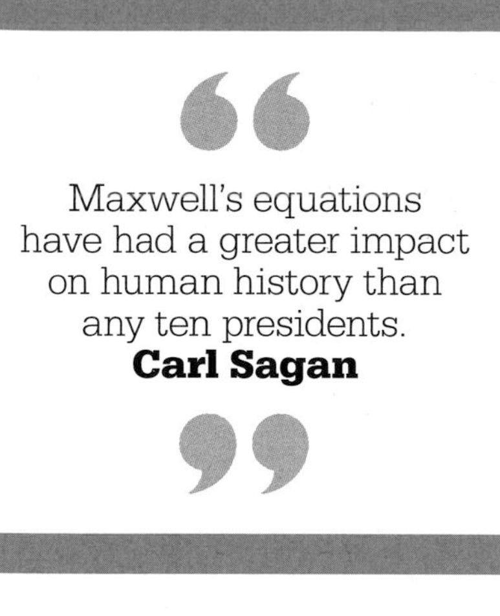

Maxwell's equations have had a greater impact on human history than any ten presidents.
Carl Sagan

frequencies and long wavelengths, but with the same overall speed of propagation—the form of electromagnetism known today as radio waves.

Heaviside weighs in

By the time of Hertz's discovery, there had been one other important development that finally produced Maxwell's equations in the form we know today.

In 1884, a British electrical engineer, mathematician, and physicist named Oliver Heaviside—a self-trained genius who had already patented the coaxial cable for the efficient transmission of electrical signals—devised a way of transforming the potentials of Maxwell's equations into vectors. These were values that described both the value and the direction of the force that was experienced by a charge at a given point in an electromagnetic field. By describing the direction of charges across the field rather than simply its strength at individual points, Heaviside reduced a dozen of the original equations to a mere four, and in doing so made them much more useful for practical applications. Heaviside's contribution is largely forgotten today, but it is his set of four elegant equations that now bear Maxwell's name.

While Maxwell's work settled many questions about the nature of electricity, magnetism, and light, it also served to highlight outstanding mysteries. Perhaps the most significant of these was the nature of the medium through which electromagnetic waves moved—for surely light waves, like all others, required such a medium? The quest to measure this so-called luminiferous ether was to dominate physics in the late 19th century, leading to the development of some ingenious experiments. The continued failure to detect it created a crisis in physics that would pave the way for the twin 20th-century revolutions of quantum theory and relativity. ■

The Maxwell-Heaviside equations, although couched in the abstruse mathematical grammar of differential equations, actually provide a concise description of the structure and effect of electrical and magnetic fields.

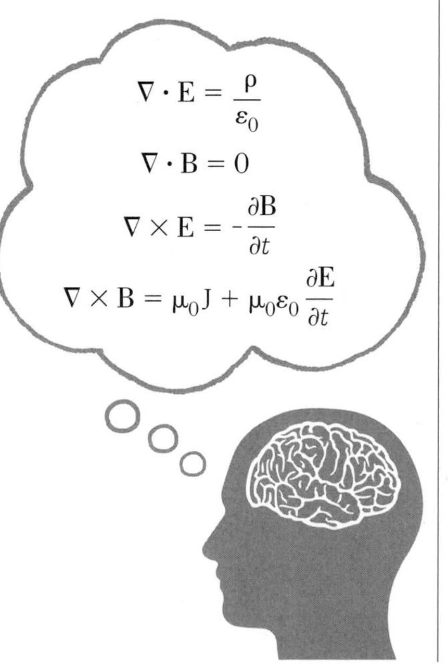

James Clerk Maxwell

Born in Edinburgh, Scotland, in 1831, James Clerk Maxwell showed genius from an early age, publishing a scientific paper on geometry at 14 years old. Educated at the universities of Edinburgh and Cambridge, he became a professor at Marischal College in Aberdeen, Scotland, at 25 years old. It was there that he began his work on electromagnetism.

Maxwell was interested in many other scientific problems of the age: in 1859, he was the first to explain the structure of Saturn's rings; between 1855 and 1872, he did important work on the theory of color vision, and from 1859 to 1866 he developed a mathematical model for the distribution of particle velocities in a gas.

A shy man, Maxwell was also fond of writing poetry and remained devoutly religious all his life. He died of cancer at 48.

Key works

1861 *On Physical Lines of Force*
1864 *A Dynamical Theory of the Electromagnetic Field*
1872 *Theory of Heat*
1873 *Treatise on Electricity and Magnetism*

RAYS WERE COMING FROM THE TUBE

WILHELM RÖNTGEN (1845–1923)

IN CONTEXT

BRANCH
Physics

BEFORE
1838 Michael Faraday passes an electrical current through a partially evacuated glass tube, producing a glowing electric arc.

1869 Cathode rays are observed by Johann Hittorf.

AFTER
1896 First clinical use of X-rays in diagnosis, producing an image of a bone fracture.

1896 First clinical use of X-rays in cancer treatment.

1897 J. J. Thomson discovers that cathode rays are in fact streams of electrons. X-rays are produced when a stream of electrons hits a metal target.

1953 Rosalind Franklin uses X-rays to help her to determine the structure of DNA.

When an electric current is passed through a sealed glass tube, **cathode rays** cause part of the tube to glow.

Fluorescent screens near the tube also **glow**, even when it is covered in black cardboard.

Some **unknown type of ray** must have **passed through the cardboard** to make the screen glow.

Invisible rays are coming from the tube.

Like many scientific discoveries, X-rays were first observed by scientists studying something else—in this case, electricity. An artificially produced electric arc (a glowing discharge jumping between two electrodes) was first observed in 1838 by Michael Faraday. He passed an electrical current through a glass tube that had been partially evacuated of air. The arc stretched from the negative electrode (the cathode) to the positive electrode (the anode).

Cathode rays

This arrangement of electrodes inside a sealed container is called a discharge tube. By the 1860s, British physicist William Crookes had developed discharge tubes with hardly any air in them. German physicist Johann Hittorf used these tubes to measure the electricity carrying capacity of charged atoms and molecules. There was no glowing arc between the electrodes in Hittorf's tubes, but the glass tubes themselves glowed. Hittorf concluded that the

See also: Michael Faraday 121 ▪ Ernest Rutherford 206–13 ▪ James Watson and Francis Crick 276–83

"rays" must have come from the cathode, or negative electrode. They were named cathode rays by Hittorf's colleague Eugen Goldstein, but in 1897, British physicist J. J. Thomson showed that they are streams of electrons.

Discovering X-rays

During his experiments, Hittorf noticed that photographic plates in the same room were becoming fogged, but he did not investigate this effect any further. Others observed similar effects, but Wilhelm Röntgen was the first to investigate their cause—finding that it was a ray that could pass right through many opaque substances. At his request, his laboratory notes were burned after his death, so we cannot be sure exactly how he discovered these "X-rays," but he may have first observed them when he noticed that a screen near his discharge tube was glowing even though the tube was covered in black cardboard. Röntgen abandoned his original experiment and spent the next two months investigating the properties of these invisible rays, which are still called Röntgen rays in many countries. We now know that X-rays are a form of short-wavelength electromagnetic radiation. They have a wavelength ranging from 0.01–10 nanometers (billionths of a meter). In contrast, visible light falls between the range of 400–700 nanometers.

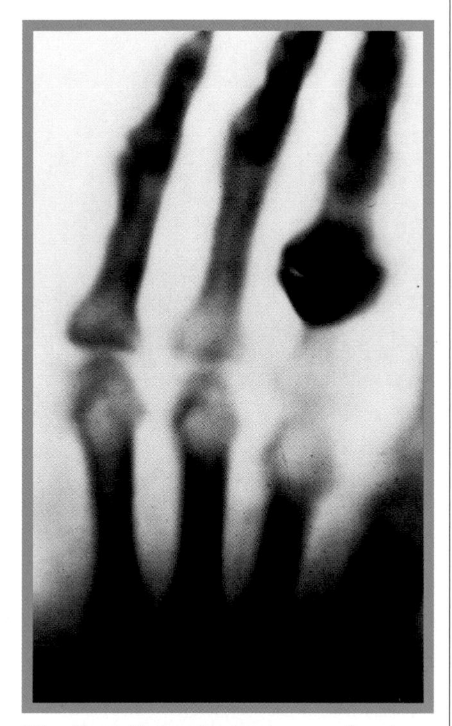

The first X-ray image was taken by Röntgen of his wife Anna's hand. The dark circle is her wedding ring. On seeing the image, Anna is said to have exclaimed: "I have seen my own death."

Using X-rays today

Today, X-rays are produced by firing a stream of electrons at a metal target. They pass through some materials better than others, and can be used to form images of the insides of the body or to detect metals in closed containers. In CT (computed tomography) scans, a computer combines a series of X-ray images to form a 3D image of the inside of the body.

X-rays can also be used to form images of very small objects, and X-ray microscopes were developed in the 1940s. The image resolution that is possible when using light microscopes is limited by the wavelengths of visible light. With their much shorter wavelengths, X-rays can be used to form images of much smaller objects. Diffraction of X-rays can be used to figure out how atoms in crystals are arranged—a technique that proved crucial in elucidating the structure of DNA. ■

Wilhelm Röntgen

Wilhelm Röntgen was born in Germany, but lived in the Netherlands for part of his childhood. He studied mechanical engineering in Zurich before becoming a lecturer in physics at Strasbourg University in 1874, and a professor two years later. He took senior positions at several universities during his career.

Röntgen studied many different areas of physics, including gases, heat transfer, and light. However, he is best known for his research into X-rays, and in 1901 he was awarded the first Nobel Prize in Physics for this work. He refused to limit the potential uses of X-rays by taking out patents, saying that his discoveries belonged to humanity, and gave away his Nobel Prize money. Unlike many of his contemporaries, Röntgen used lead protective shields in his work with radiation. He died from an unrelated cancer at 77 years old.

Key works

1895 *On a New Kind of Rays*
1897 *Additional Observations on the Properties of X-rays*

SEEING INTO THE EARTH

RICHARD DIXON OLDHAM (1858–1936)

IN CONTEXT

BRANCH
Geology

BEFORE
1798 Henry Cavendish publishes his calculations of the density of Earth. The value is greater than the density of the surface rocks, showing that Earth must contain denser materials.

1880 British geologist John Milne invents the modern seismograph.

1887 Britain's Royal Society funds 20 earthquake observatories worldwide.

AFTER
1909 Croatian seismologist Andrija Mohorovicic identifies the seismic boundary between Earth's crust and the mantle.

1926 Harold Jeffreys claims that the core of Earth is liquid.

1936 Inge Lehmann argues that Earth has a solid inner core and a molten outer core.

There are different types of **seismic wave**.

P waves are not detected at certain distances from an earthquake…

…therefore **rocks** inside Earth **must be deflecting** the paths of the waves.

Earth's core has **properties** that are **different** from those in Earth's upper layers.

The shaking caused by earthquakes spreads out in the form of seismic waves, which we can detect using seismographs. While working for the Geological Survey of India between 1879 and 1903, Richard Dixon Oldham wrote a survey of an earthquake that struck Assam in 1897. In it he made his greatest contribution to plate tectonic theory. Oldham noted that the quake had three phases of motion, which he took to represent three different types of wave. Two of these were "body" waves, which traveled through Earth. The third type was a wave that traveled around the surface of Earth.

Wave effects

The body waves Oldham identified are today known as P waves and S waves (primary and secondary—the order in which they arrive at a seismograph). P waves are longitudinal waves; as the wave passes, rocks are moved backward and forward in the same direction as the waves are traveling. S waves are transverse waves (like the waves on the surface of water); the rocks are moved sideways to the direction of the wave. P waves

See also: James Hutton 96–101 ▪ Nevil Maskelyne 102–03 ▪ Alfred Wegener 222–23

travel faster than S waves, and can travel through solids, liquids, or gases. S waves can travel only through solid materials.

Shadow zones

Later, Oldham studied seismograph records for many earthquakes around the world, and noticed that there was a P-wave "shadow zone" extending partway around Earth from the earthquake location. Hardly any P waves from an earthquake were detected in this zone. Oldham knew that the speed at which seismic waves travel inside Earth depends on the density of the rocks. He concluded that properties of the rocks change with depth, and the resulting changes in speed cause refraction (the waves followed curved paths). The shadow zone is therefore caused by a sudden change in the properties of rocks deep within Earth.

Today, we know that there is a much larger shadow zone for S waves, which extends across most of the hemisphere opposite the focus of the earthquake. This indicates an Earth interior that has very different properties than those of the mantle. In 1926, American geophysicist Harold Jeffreys used this evidence from S waves to suggest that Earth's core is liquid, since S waves cannot pass through liquids. The P-wave shadow zone is not completely "shadowed," since some P waves are detected there. In 1936, Danish seismologist Inge Lehmann interpreted these P waves as reflections from an inner, solid core. This is the model of Earth we use today: a solid inner core surrounded by liquid, then the mantle with crustal rocks on top. ■

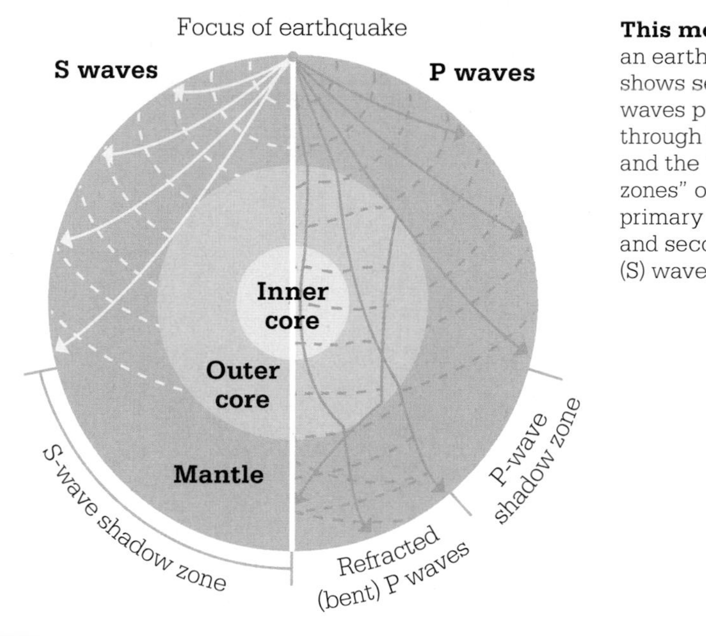

This model of an earthquake shows seismic waves passing through Earth and the "shadow zones" of the primary (P) waves and secondary (S) waves.

Richard Dixon Oldham

Born in Dublin in 1858, the son of the superintendent of the Geological Survey of India (GSI), Richard Dixon Oldham studied at the Royal School of Mines, before joining the GSI himself and became superintendent as well.

The GSI's main work involved mapping the rock strata, but it also compiled detailed reports on earthquakes in India, and it is for this aspect of his work that Oldham is best known. He retired on health grounds in 1903 and returned to the United Kingdom, publishing his ideas about Earth's core in 1906. He was awarded the Lyell Medal by the Geological Society of London, and was made a Fellow of the Royal Society.

Key works

1899 *Report of the Great Earthquake of 12th June 1897*
1900 *On the Propagation of Earthquake Motion to Great Distances*
1906 *The Constitution of the Interior of the Earth*

> The seismograph, recording the unfelt motion of distant earthquakes, enables us to see into the earth and determine its nature.
> **Richard Dixon Oldham**

RADIATION IS AN ATOMIC PROPERTY OF THE ELEMENTS

MARIE CURIE (1867–1934)

IN CONTEXT

BRANCH
Physics

BEFORE
1895 Wilhelm Röntgen investigates the properties of X-rays.

1896 Henri Becquerel discovers that uranium salts emit penetrating radiation.

1897 J. J. Thomson discovers the electron while exploring the properties of cathode rays.

AFTER
1904 Thomson proposes the "plum pudding" model of the atom.

1911 Ernest Rutherford and Ernest Marsden propose the "nuclear model" of the atom.

1932 British physicist James Chadwick discovers the neutron.

Like many major scientific discoveries, radiation was found by accident. In 1896, French physicist Henri Becquerel was investigating phosphorescence, which occurs when light falls on a substance that then emits light of a different color. Becquerel wanted to know whether phosphorescent minerals also emitted X-rays, which had been discovered by Wilhelm Röntgen a year earlier. To find out, he placed one of these minerals on top of a photographic plate that was wrapped in thick black paper and exposed both to the Sun. The experiment worked—the plate darkened; the mineral appeared to have emitted X-rays. Becquerel also showed that metals would block the "rays" that caused the plate to darken. The next day was cloudy so he could not repeat the experiment. He left the mineral on a photographic plate in a drawer, but the plate still darkened, even without the sunshine. He realized that the mineral must have an internal source of energy, which turned out to be the result of the breakdown of atoms of uranium in the mineral he was using. He had detected radioactivity.

It was necessary at this point to find a new term to define this new property of matter manifested by the elements of uranium and thorium. I proposed the word radioactivity.
Marie Curie

Rays produced by atoms

Following Becquerel's discovery, his Polish doctoral student, Marie Curie, decided to investigate these new "rays." Using an electrometer—a device for measuring electrical currents—she found that air around a sample of a uranium-containing mineral was conducting electricity. The level of electrical activity depended only on the amount of uranium present, not on the total mass of the mineral (which included elements

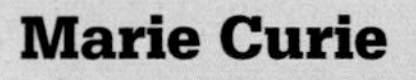

Marie Curie

Maria Salomea Skłodowska was born in Warsaw in 1867. At that time Poland was under Russian rule and women were not allowed into higher education. She worked to help finance her sister's medical studies in Paris, France, and in 1891 moved there herself to study mathematics, physics, and chemistry. There, she married her colleague, Pierre Curie, in 1895. When her daughter was born in 1897, she began teaching to help support the family, but continued to research with Pierre in a converted shed. After Pierre's death, she accepted his chair at the University of Paris, the first woman to hold this position. She was also the first woman to be awarded a Nobel Prize, and the first to be awarded a second Nobel. During World War I, she helped set up radiology centers. She died in 1934 of anemia, probably caused by her long exposure to radiation.

Key works

1898 *Emissions of Rays by Uranium and Thorium Compounds*
1935 *Radioactivity*

See also: Wilhelm Röntgen 186–87 ▪ Ernest Rutherford 206–13 ▪ J. Robert Oppenheimer 260–65

other than uranium). This led her to the belief that the radioactivity came from the uranium atoms themselves, and not from any reactions between uranium and other elements.

Curie soon found that some minerals that contained uranium were more radioactive than uranium itself, and wondered whether these minerals contained another substance—one that was more active than uranium. By 1898, she had identified thorium as another radioactive element. She rushed to present her findings in a paper to the Académie des Sciences, but the discovery of thorium's radioactive properties had already been published.

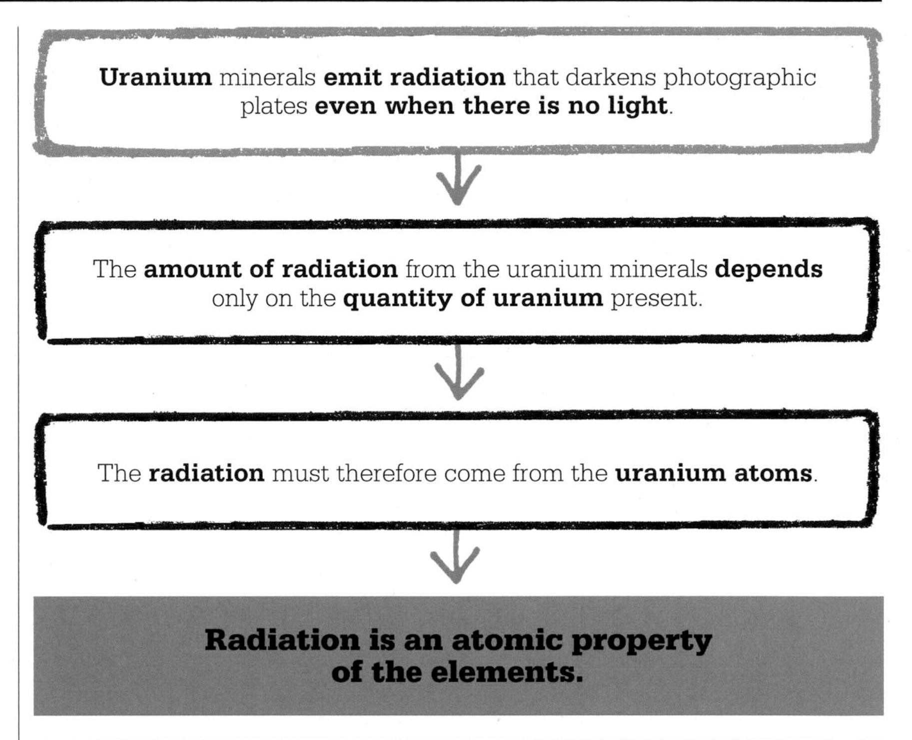

Science double

Curie and her husband Pierre worked together to discover the additional radioactive elements responsible for the high activity of the uranium-rich minerals pitchblende and chalcolite. By the end of 1898 they had announced the discovery of two new elements, which they called polonium (after her native country, Poland) and radium. They attempted to prove their discoveries by obtaining pure samples of the two elements, but it was not until 1902 that they obtained 0.003 oz (0.1 g) of radium chloride from a metric ton of pitchblende.

During this time, the Curies published dozens of scientific papers, including one outlining their discovery that radium could help to destroy tumors. They did not patent these discoveries, but in 1903, they were jointly awarded the Nobel Prize in Physics, along with Becquerel. Marie continued her scientific work after her husband's death in 1906, and succeeded in isolating a pure sample of radium in 1910. In 1911, she was awarded the Nobel Prize in Chemistry, becoming the first person to win or share in two prizes.

New model of the atom

The Curies' discovery of radiation paved the way for the two New Zealand-born physicists Ernest Rutherford and Ernest Marsden to formulate their new model of the atom in 1911, but it was not until 1932 that English physicist James Chadwick discovered neutrons and the process of radiation could be fully explained. Neutrons and positively charged protons are subatomic particles that make up the nucleus of an atom, which also has negatively charged electrons buzzing around it. The protons and neutrons contribute almost all the mass of the atom. Atoms of a particular element always have the same number of protons but may have different numbers of neutrons. Atoms with different numbers of neutrons are called isotopes of the element. For example, an atom of uranium always has 92 protons in its nucleus, but may have between 140 and 146 neutrons. These »

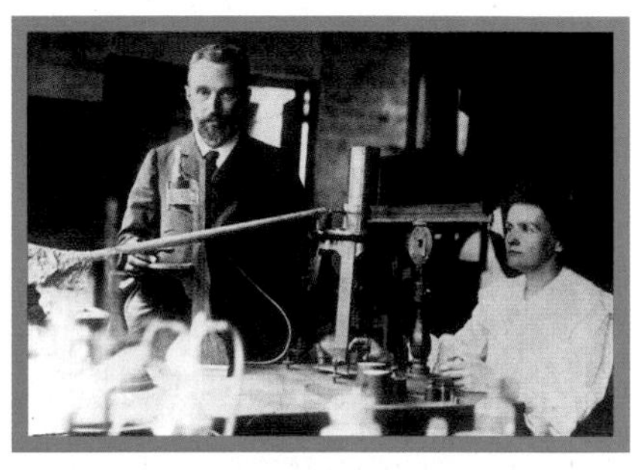

Marie and Pierre Curie did not have a dedicated laboratory. Most of their work was done in a leaking shed next to the University of Paris's School of Physics and Chemistry.

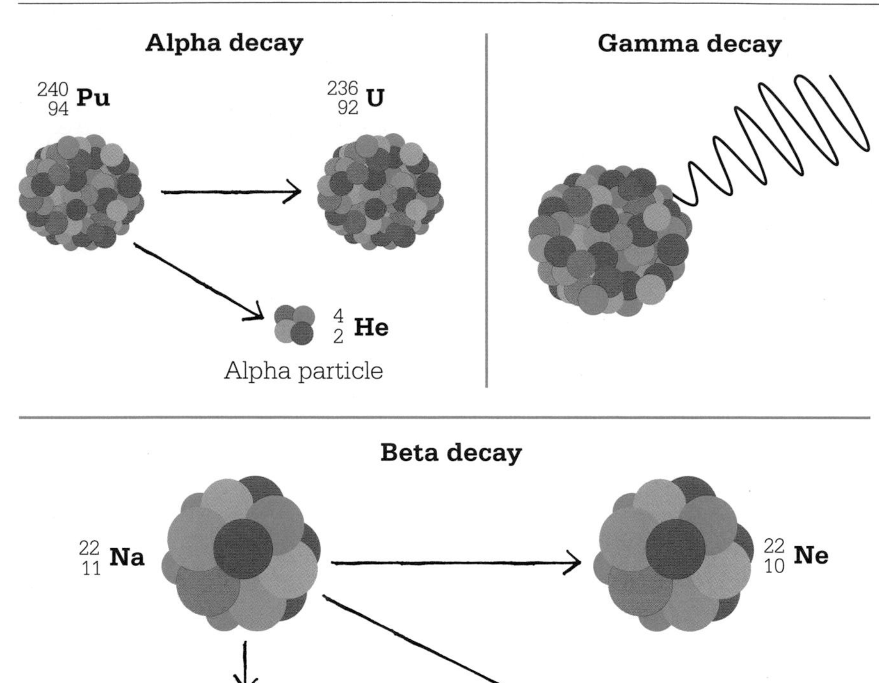

Radioactive decay can happen in three ways. Plutonium-240 (top left) decays to make uranium and an alpha particle. This is an example of alpha decay. During beta decay, sodium-22 decays to make neon, a beta particle (in this case a positron), and a neutrino. With gamma decay, a high-energy nucleus gives off gamma radiation but no particles.

isotopes are named after the total number of protons and neutrons, so the most common isotope of uranium, with 146 neutrons, is written as uranium-238 (i.e. 92 + 146).

Many heavy elements, such as uranium, have nuclei that are unstable, and this leads to spontaneous radioactive decay. Rutherford named the emissions from radioactive elements alpha, beta, and gamma rays. The nucleus becomes more stable by emitting an alpha particle, a beta particle, or gamma radiation. An alpha particle consists of two protons and two neutrons. Beta particles can be electrons or their opposites, positrons, emitted from the nucleus when a proton turns into a neutron or vice versa. Alpha and beta decay both change the number of protons in the nucleus of the decaying atom so that it becomes an atom of a different element. Gamma rays are a form of high-energy short-wave electromagnetic radiation and do not change the nature of the element.

Radioactive decay is different from the fission process that takes place inside nuclear reactors, and the fusion process that powers the Sun. In fission, unstable nuclei such as uranium-235 are bombarded with neutrons and break up to form much smaller atoms, releasing energy in the process. In fusion, two small nuclei combine to form a larger one. Fusion also releases energy, but the great temperatures and pressures required to start the process explain why scientists have only achieved fusion in the form of nuclear weapons. So far, attempts to use nuclear fusion to generate electricity consume more energy than is released.

Half-life

As a radioactive material decays, the atoms of the radioactive element change to other elements, and so the number of unstable atoms reduces with time. The fewer unstable atoms there are, the less radioactivity will be produced. The reduction in activity of a radioactive isotope is measured by its half-life. This is the time it takes for the activity to halve, which is the same as saying the time for the number of unstable atoms in a sample to halve. For example, the isotope technetium-99m is widely used in medicine, and has a half-life of 6 hours. This means that 6 hours after a dose is injected into a patient, the activity will be half of its original level; 12 hours after injection, the activity will be one quarter of the original level, and so on. By contrast, uranium-235 has a half-life of over 700 million years.

Radioactive dating

This idea of half-life can be used to date minerals or other materials. Many different radioactive elements with known half-lives can be used to do this, but one of the best known is carbon. The most common isotope of carbon is carbon-12, with 6 protons and 6 neutrons in each atom. Carbon-12 makes up 99 percent of the carbon found on Earth, and has a stable nucleus. A tiny proportion of the carbon is carbon-14, which has two extra neutrons. This unstable isotope has a half-life of 5,730 years. Carbon-14 is constantly being produced in the upper atmosphere as nitrogen atoms are bombarded with cosmic rays. This means there is a relatively constant ratio of carbon-12 to carbon-14 in the atmosphere. Since photosynthesizing plants take in carbon dioxide from the atmosphere, and our food consists of plants (or animals that have eaten plants), there is also a relatively constant proportion in plants and animals while they are alive, even though the carbon-14 is constantly decaying. When an organism dies, no more carbon-14 is taken into its body, while the carbon-14 already there continues to decay. By measuring the ratio of carbon-12 to carbon-14 in the body, scientists can figure out how long ago the organism died.

This radiometric method is used to date wood, charcoal, bone, and shells. There are natural variations in the ratios of the carbon isotopes, but dates can be cross-checked with other dating methods such as tree rings, and the corrections applied to objects of similar age.

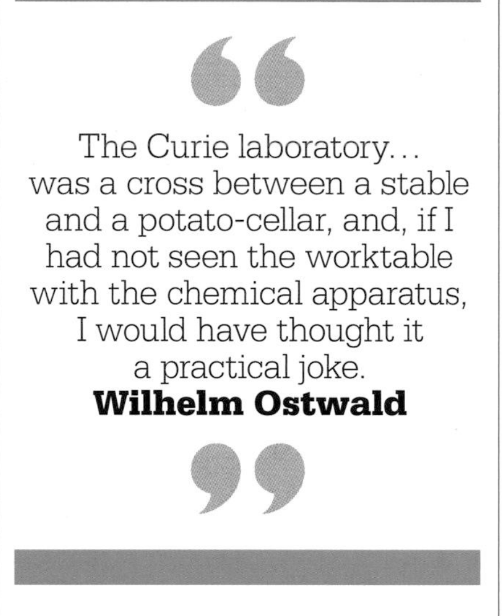

> The Curie laboratory... was a cross between a stable and a potato-cellar, and, if I had not seen the worktable with the chemical apparatus, I would have thought it a practical joke.
> **Wilhelm Ostwald**

A wonder treatment

Curie realized that radioactivity had medicinal uses. During World War I, she used the small amount of radium she had extracted to produce radon gas (a radioactive gas produced when radium decays). This was sealed into glass tubes and inserted into patients' bodies to kill diseased tissue. It was seen as a wonder cure, and even marketed in beauty treatments to help firm up aging skin. It was only later that the importance of using materials with a short half-life was recognized.

Radioactive isotopes are also widely used in medical imaging to diagnose disease, and in treatment of cancer. Gamma rays are used to sterilize surgical instruments, and even food, to increase its shelf life. Gamma ray emitters can be used for the internal inspection of metal objects, to detect cracks, or to inspect the contents of cargo containers to identify contraband. ■

The erection of Ale's stones in Sweden was dated to 600 CE by the radiometric dating of wooden tools found at the site. The actual stones are hundreds of millions of years older.

A CONTAGIOUS LIVING FLUID

MARTINUS BEIJERINCK (1851–1931)

IN CONTEXT

BRANCH
Biology

BEFORE
1870s and 80s Robert Koch and others identify bacteria as the cause of diseases such as tuberculosis and cholera.

1886 German plant biologist Adolf Mayer shows tobacco mosaic disease can be transferred between plants.

1892 Dmitri Ivanovsky demonstrates that tobacco plant sap passing through the finest unglazed porcelain filters still carries infection.

AFTER
1903 Ivanovsky reports light-microscope "crystal inclusions" in infected host cells, but suspects they are very small bacteria.

1935 US biochemist Wendell Stanley studies the structure of the tobacco mosaic virus, and realizes that viruses are large chemical molecules.

Tobacco mosaic disease shows **features of an infection**, but...

...**filters** that catch bacteria **do not catch** and remove **the contagion**, so it cannot be bacteria.

Also, unlike bacteria, **the infectious agent grows only in a living host**, not in laboratory gels or broths.

So the causative agent must be different and even smaller, deserving a new name—**virus**.

These days, the word "virus" is all too familiar as a medical term, and many people understand the idea that viruses are just about the smallest of the harmful agents, or germs, that cause infections in humans, other animals, plants, and fungi.

Yet at the end of the 19th century, the term virus was only just making its way into science and medicine. It was suggested in 1898 by Dutch microbiologist Martinus Beijerinck for a new category of contagious disease-causing agents. Beijerinck had a special interest in plants and a skilled talent for microscopy. He experimented with tobacco plants that were suffering from mosaic disease, a discoloring mottled effect on the leaves that was costly

See also: Friedrich Wöhler 124–25 ▪ Louis Pasteur 156–59 ▪ Lynn Margulis 300–01 ▪ Craig Venter 324–25

for the tobacco industry. His results led him to apply the term virus—already in occasional use for substances that were toxic or poisonous—to the contagious agents that caused the disease.

At the time, most of Beijerinck's contemporaries in science and medicine were still grappling with understanding bacteria. Louis Pasteur and German physician Robert Koch had first isolated and identified them as disease-causing in the 1870s, and more were being discovered constantly.

A common method of testing for bacteria at the time was to pass liquid containing the suspected contagions through various sets of filters. One of the best known was the Chamberland filter, invented in 1884 by Pasteur's colleague Charles Chamberland. It used minute pores in unglazed porcelain to capture particles as small as bacteria.

Too small to filter

Several researchers had suspected that there was a class of infectious agents even tinier than bacteria that could pass on disease. In 1892, Russian botanist Dmitri Ivanovsky performed tests on tobacco mosaic disease, showing that its infection agent passed through the filters. He established that the agent in this case could not be bacteria, but did not investigate further to discover what the agent might be.

Beijerinck repeated Ivanovsky's experiment. He, too, established that even after juice pressed from the leaves was filtered, tobacco mosaic disease was still present. Indeed, at first he thought that the cause was the fluid itself, which he called *contagium vivum fluidium* (contagious living fluid). He further demonstrated that the contagion carried in the fluid could not be grown in laboratory nutrient gels or broths, nor in any host organism. It had to infect its own specific living host in order to multiply and spread the disease.

Even though viruses could not be seen by light microscopes of the time, grown with the usual laboratory culture methods, or detected by any of the standard microbiological techniques, Beijerinck figured out that they really did exist. He insisted that they caused disease, propelling microbiology and medical science into a new era. It would not be until 1939, with the aid of electron microscopes, that tobacco mosaic virus became the first virus to have its photograph taken. ■

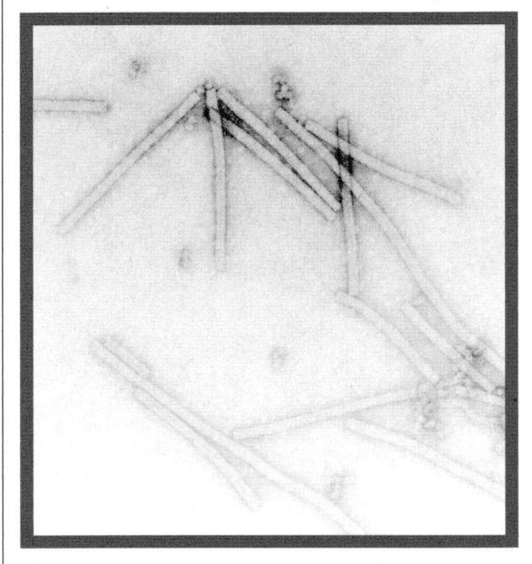

This electron micrograph image shows particles of the tobacco mosaic virus at 160,000x magnification. The particles have been stained to enhance their visibility.

Martinus Beijerinck

Something of a recluse, Martinus Beijerinck spent many solitary hours experimenting in the laboratory. He was born in Amsterdam in 1851, and studied chemistry and biology in Delft, graduating in 1872 from Leiden University. Focusing on soil and plant microbiology at Delft, he performed his famous filtering experiments on the tobacco mosaic virus in the 1890s. He also studied how plants capture nitrogen from the air and incorporate it into their tissues—a kind of natural fertilizer system that enriches the soil—as well as working on plant galls, fermentation by yeasts and other microbes, the nutrition of microbes, and sulfur bacteria. By the end of his life, he was internationally recognized. The Beijerinck Virology Prizes, set up in 1965, are awarded every two years in the field of virology.

Key works

1895 *On Sulphate Reduction by* Spirillum desulfuricans
1898 *Concerning a* contagium vivum fluidium *as a Cause of the Spot-disease of Tobacco Leaves*

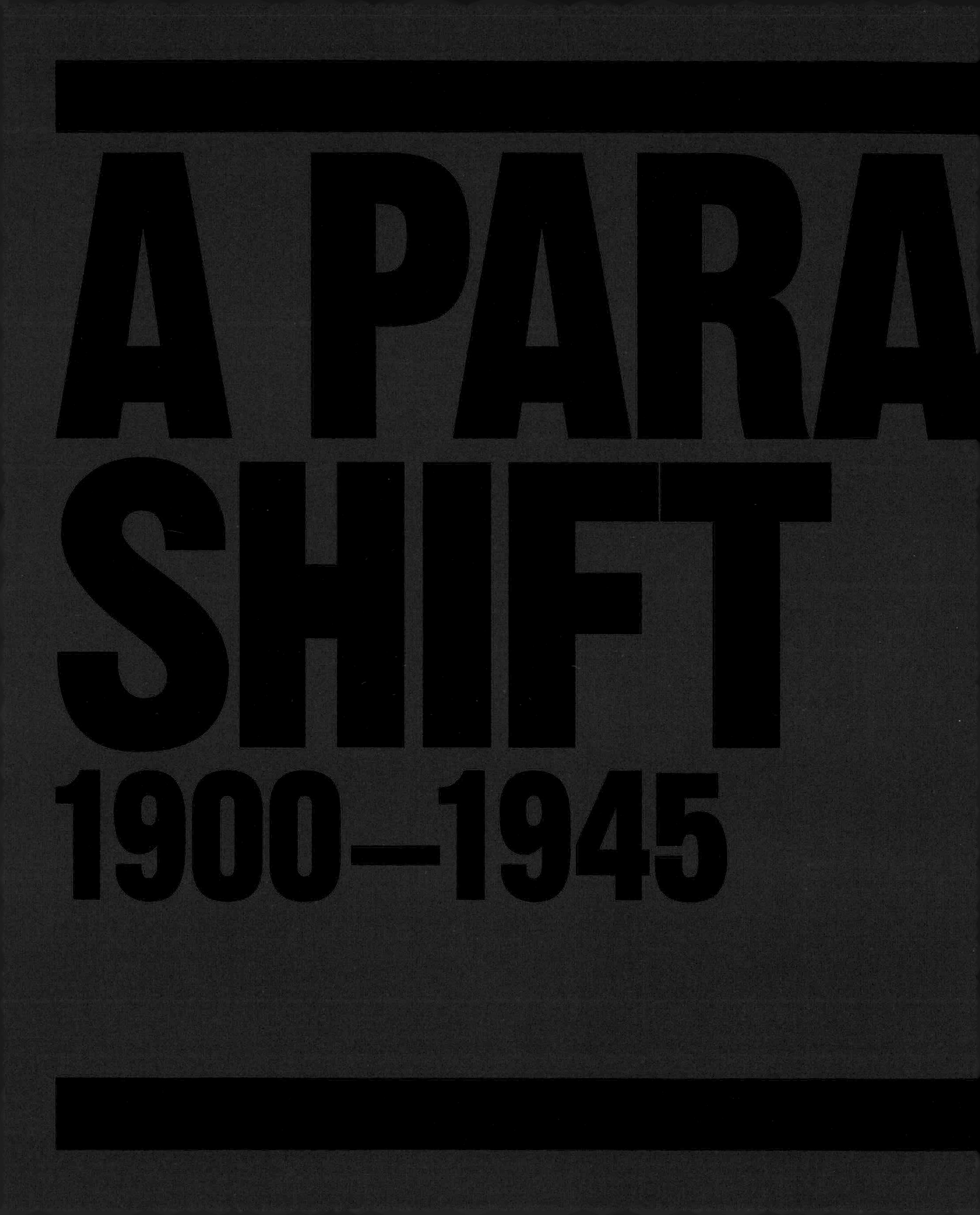
A PARA
SHIFT
1900–1945

DIGM

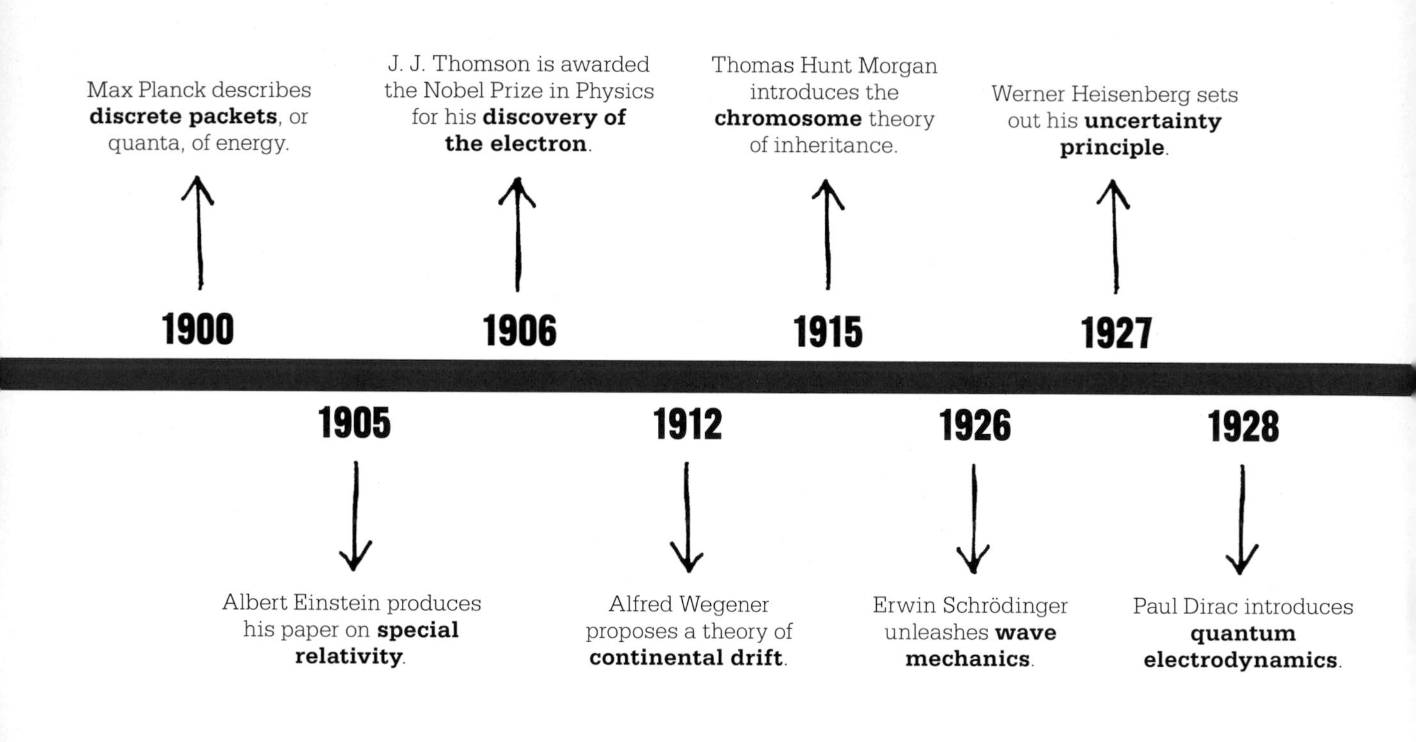

While the 19th century had seen a fundamental change in the way scientists view life processes, the first half of the 20th would prove even more of a shock. The old certainties of classical physics, largely unchanged since Isaac Newton, were about to be thrown away, and nothing short of a new way to view space, time, and matter was to replace it. By 1930, the old idea of a predictable universe had been shattered.

A new physics

Physicists were finding that the equations of classical mechanics were producing some nonsensical results. It was clear that something was fundamentally wrong. In 1900, Max Planck solved the puzzle of the spectrum of radiation emitted by a "black box," which had stubbornly resisted classical equations, by imagining that electromagnetism traveled not in continuous waves, but in discrete packets, or "quanta." Five years later, Albert Einstein, a clerk working at the Swiss Patent Office, produced his paper on special relativity, asserting that the speed of light is constant and independent of the movement of source or observer. After working through the implications of general relativity, Einstein had found by 1916 that notions of an absolute time and space independent of the observer had gone, to be replaced by a single space-time, which was warped by the presence of mass to produce gravity. Einstein had further demonstrated that matter and energy should be considered aspects of the same phenomenon, capable of being converted from one to the other, and his equation describing their relation—$E = mc^2$—hinted at an enormous potential energy locked inside atoms.

Wave–particle duality

Worse was to follow for the old picture of the universe. At Cambridge, English physicist J. J. Thomson discovered the electron, showing that it has a negative charge and is at least a thousand times smaller and lighter than any atom. Studying the properties of the electron was to produce new puzzles. Not only did light have particle-like properties, but particles had wavelike properties, too. Austrian Erwin Schrödinger drew up a series of equations that described

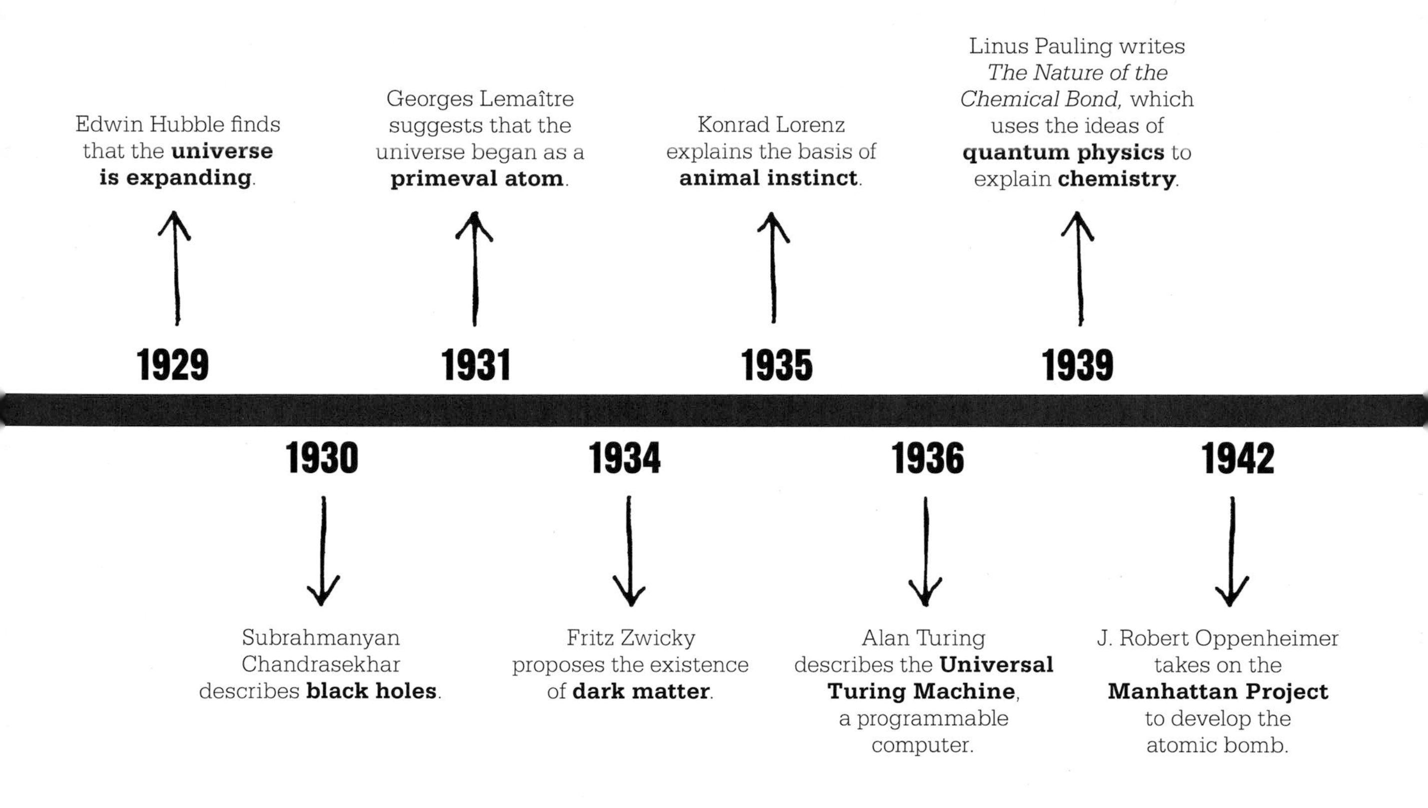

the probability of finding a particle in a particular place and state. His German colleague Werner Heisenberg showed that there was an inherent uncertainty to the values of place and momentum, which was initially thought to be a problem of measurement, but later found to be fundamental to the structure of the universe. A strange picture was emerging of a warped, relative space-time with particles of matter smeared across it in the form of probability waves.

Splitting the atom

New Zealander Ernest Rutherford first showed that an atom is made mostly of space, with a small, dense nucleus and electrons in orbit around it. He explained certain forms of radioactivity as the splitting of this nucleus. Chemist Linus Pauling took this new picture of an atom and used the ideas of quantum physics to explain how atoms bonded to each another. In the process, he showed how the discipline of chemistry was, in reality, a subsection of physics. By the 1930s, physicists were working on ways to unlock the energy in the atom, and in the US, J. Robert Oppenheimer led the Manhattan Project, which was to produce the first nuclear weapons.

The universe expands

Up to the 1920s, nebulae were thought to be clouds of gas or dust within our own galaxy, the Milky Way, which comprised the entire known universe. Then American astronomer Edwin Hubble discovered that these nebulae were in fact distant galaxies. The universe was suddenly enormously bigger than anyone had thought. Hubble further found that the universe was expanding in all directions. Belgian priest and physicist Georges Lemaître proposed that the universe had expanded from a "primeval atom." This was to become the Big Bang theory. A further puzzle was uncovered when astronomer Fritz Zwicky coined the term "dark matter" to explain why the Coma galaxy cluster appeared to contain 400 times as much mass (as seen from its gravity) as he could explain from the observable stars. Not only was matter not quite what it had been thought to be, but much of it was not even directly detectable. It was clear that there were still major holes in scientific understanding. ■

QUANTA ARE DISCRETE PACKETS OF ENERGY

MAX PLANCK (1858–1947)

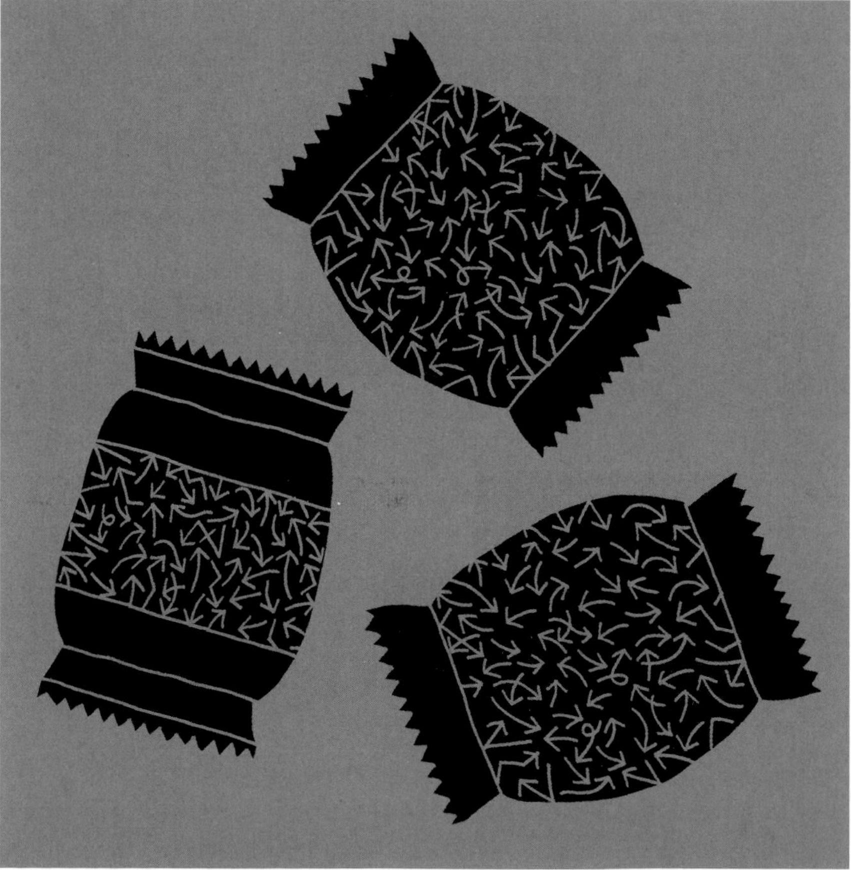

IN CONTEXT

BRANCH
Physics

BEFORE
1860 The distribution of so-called black-body radiation fails to match predictions made by theoretical models.

1870s Austrian physicist Ludwig Boltzmann's analysis of entropy (disorder) introduces a probabilistic interpretation of quantum mechanics.

AFTER
1905 Albert Einstein proposes that the quantum is a real entity, using Planck's concept of quantized light to introduce the idea of the photon.

1924 Louis de Broglie proves that matter behaves both as a particle and as a wave.

1926 Erwin Schrödinger formulates an equation for the wave behavior of particles.

In December 1900, the German theoretical physicist Max Planck presented a paper setting out his method for resolving a long-standing theoretical conflict. In doing so, he made one of the most important conceptual leaps in the history of physics. Planck's paper marked the turning point between the classical mechanics of Newton and quantum mechanics. The certainty and precision of Newtonian mechanics was to give way to an uncertain, probabilistic description of the universe.

Quantum theory has its roots in the study of thermal radiation, the phenomenon that explains why we feel heat from a fire, even when the

See also: Ludwig Boltzmann 139 ▪ Albert Einstein 214–21 ▪ Erwin Schrödinger 226–33

Classical mechanics treats radiation as if it were emitted across a **continuous range**.

But **nonsense results** are reached for the **distribution of black-body radiation**, assuming a continuous range.

The **problem is solved** by treating radiation as if it were produced in **discrete "quanta."**

Radiation is not continuous, but is emitted in discrete quanta of energy.

air in between it and us is cold. Every object absorbs and emits electromagnetic radiation. If its temperature rises, the wavelength of the radiation it emits decreases while its frequency increases. For example, a lump of coal at room temperature emits energy below the frequency of visible light, in the infrared spectrum. We cannot see the emission, so the coal appears black. Once we set the coal alight, however, it emits higher-frequency radiation, glowing a dull red as the emissions break into the visible spectrum, then white-hot and finally a brilliant blue. Extremely hot objects, such as stars, radiate even shorter-wavelength ultraviolet light and X-rays, which again we cannot see. Meanwhile, in addition to producing radiation, a body also reflects radiation, and it is this reflected light that gives objects color even when they do not glow.

In 1860, German physicist Gustav Kirchhoff thought of an idealized concept he called a "perfect black body." This is a theoretical surface that, when at thermal equilibrium (not heating up or cooling down), absorbs every frequency of electromagnetic radiation that falls on it, and does not itself reflect any radiation. The spectrum of thermal radiation coming off this body is "pure," since it is not mixed with any reflections—it will only be the result of the body's own temperature. Kirchhoff believed that such "black-body radiation" is fundamental in nature—the Sun, for example, comes close to being a black-body object whose emitted spectrum is almost entirely a result of its own temperature. Studying the distribution of a black body's light would show that emission of radiation depended only on a body's temperature, and not its physical shape or chemical composition. Kirchhoff's hypothesis kick-started a new experimental program designed to find a theoretical framework that would describe black-body radiation.

Entropy and black bodies

Planck arrived at his new quantum theory through the failure of classical physics to explain the experimental results of black-body radiation distribution. Much of Planck's work focussed on the second law of thermodynamics, which he had identified as an "absolute." This law states that isolated systems move over time toward a state of thermodynamic equilibrium (meaning that all parts of the system are at the same temperature). Planck attempted to »

> A new scientific truth does not triumph by convincing its opponents and making them see the light, but rather because…a new generation grows up that is familiar with it.
>
> **Max Planck**

explain the thermal radiation pattern of a black body by figuring out the entropy of the system. Entropy is a measure of disorder, though more strictly it is defined as a count of the number of ways a system can organize itself. The higher the entropy of a system, the more ways the system has of organizing and producing the same overall pattern. For instance, imagine a room where all the molecules of air start off bunched up in the top corner. There are far more ways for the molecules to organize themselves so that there is roughly the same number of them in each cubic centimeter of the room than there are for them all to remain in the top corner. Over time, they distribute themselves equally throughout the room as the entropy of the system rises. A cornerstone of the second law of thermodynamics is that entropy works in one direction only. En route to thermal equilibrium, the entropy of a system always increases or remains constant.

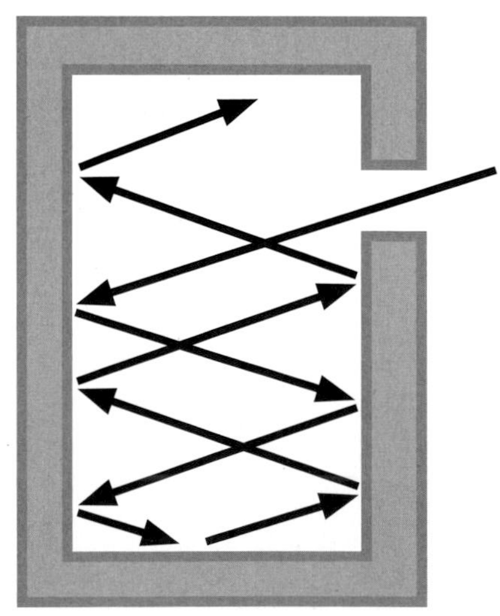

A cavity with a small hole will trap most of the radiation that enters through the hole, making it a good approximation of an ideal black body.

Planck reasoned that this principle should be evident in any theoretical black-body model.

The Wien–Planck Law

By the 1890s, experiments in Berlin came close to Kirchhoff's perfect black-body, using so-called cavity radiation. A small hole in a box kept at a constant temperature is a good approximation of a black body, as any radiation entering the box gets trapped inside, and the body's emissions are purely a result of its temperature.

The experimental results proved bothersome for Planck's colleague Wilhelm Wien, since the low-frequency emissions recorded did not fit his equations for radiation at all. Something had gone wrong. In 1899, Planck arrived at a revised equation—the Wien–Planck law—that attempted a better description of the spectrum of thermal radiation from a black body.

Ultraviolet catastrophe

A further challenge came a year later, when British physicists Lord Rayleigh and Sir James Jeans showed how classical physics predicts an absurd distribution of energy in black-body emission. The Rayleigh–Jeans Law predicted that, as the frequency of the radiation increased, the power it emitted would grow exponentially. This "ultraviolet catastrophe" was so radically at odds with experimental findings that the classical theory must have been seriously awry. If it were correct, a lethal dose of ultraviolet radiation would be emitted whenever a light bulb was turned on.

Planck was not too troubled by the Rayleigh–Jeans Law. He was more concerned about the Wien–Planck Law, which, even in its revised form, was not matching

No real-world object is a perfect black body, but the Sun, black velvet, and surfaces coated with lampblack, such as coal tar, come close.

the data—it could accurately describe the short-wavelength (high-frequency) spectrum of thermal emission from objects, but not the long-wavelength (low-frequency emissions). This is the point at which Planck broke with his conservatism and resorted to Ludwig Boltzmann's probabilistic approach to arrive at a new expression for his radiation law.

Boltzmann had formulated a new way to look at entropy by regarding a system as a large collection of independent atoms and molecules. While the second

> Science cannot solve the ultimate mystery of nature. And that is because, in the last analysis, we ourselves are a part of the mystery that we are trying to solve.
>
> **Max Planck**

law of thermodynamics remained valid, Boltzmann's reading gave it a probabilistic, rather than an absolute, truth. Thus, we observe entropy simply because it is overwhelmingly more likely than the alternative. A plate breaks but does not remake itself, but there is no absolute law preventing a plate from putting itself back together—it is just exceedingly unlikely to happen.

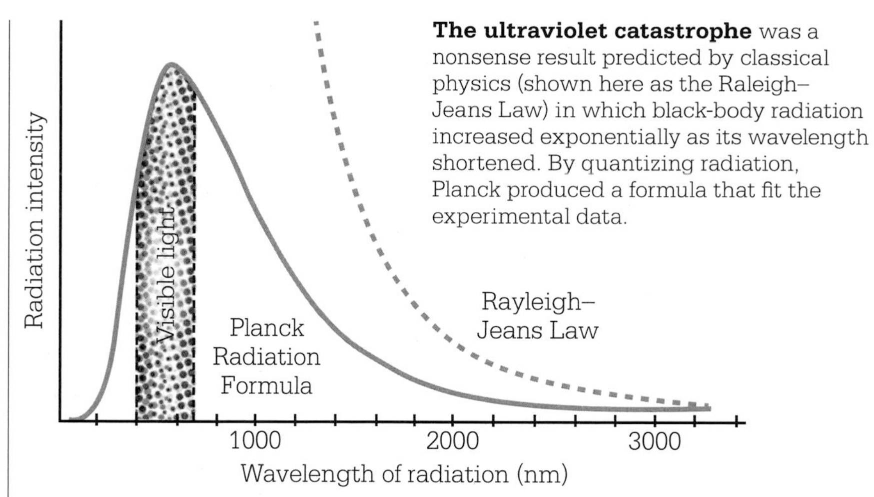

The ultraviolet catastrophe was a nonsense result predicted by classical physics (shown here as the Raleigh–Jeans Law) in which black-body radiation increased exponentially as its wavelength shortened. By quantizing radiation, Planck produced a formula that fit the experimental data.

Quantum of action

Planck used Boltzmann's statistical interpretation of entropy to arrive at a new expression for the radiation law. Imagining thermal radiation as being produced by individual "oscillators," he needed to count the ways in which a given energy could be distributed between them.

To do this, he divided the total energy into a finite number of discrete energy chunks—a process called "quantization." Planck was a gifted cellist and pianist and might have imagined these "quanta" in the same way that a fixed number of harmonics is available to the vibrating string of an instrument. The resulting equation was simple, and it fit the experimental data. Introducing "quanta" of energy reduced the number of states of energy available to the system, and in doing this (although it wasn't his goal), Planck solved the ultraviolet catastrophe. He thought of his quanta as a mathematical necessity—as a "trick"—rather than something that was real. But when Albert Einstein used the concept to explain the photoelectric effect in 1905, he insisted that quanta were a real property of light.

As with many of the pioneers of quantum mechanics, Planck spent the rest of his life struggling to come to terms with the consequences of his own work. While he was never in any doubt about the revolutionary impact of what he had done, he was—according to historian James Franck—"a revolutionary against his own will." He found the consequences of his equations not to his taste since they often gave descriptions of physical reality that clashed with our everyday experience of the world. But for better or worse, after Max Planck, the world of physics has never been the same. ■

Max Planck

Born in Kiel in northern Germany in 1858, Planck was an able pupil at school and graduated early, at 17. He chose to study physics at the University of Munich, where he soon became a pioneer of quantum physics. He received the Nobel Prize in Physics in 1918 for his discovery of energy quanta, although he never was able to satisfactorily describe the phenomena as a physical reality.

Planck's personal life was beset by tragedy. His first wife died in 1909, and his eldest son was killed in World War I. Both of his twin daughters died giving birth to their children. During World War II, an Allied bomb destroyed his house in Berlin and his papers, and in the closing stages of the war, his remaining son was caught up in the plot to assassinate Hitler and was executed. Planck himself died soon after the war.

Key works

1900 *Entropy and Temperature of Radiant Heat*
1901 *On the Law of Distribution of Energy in the Normal Spectrum*

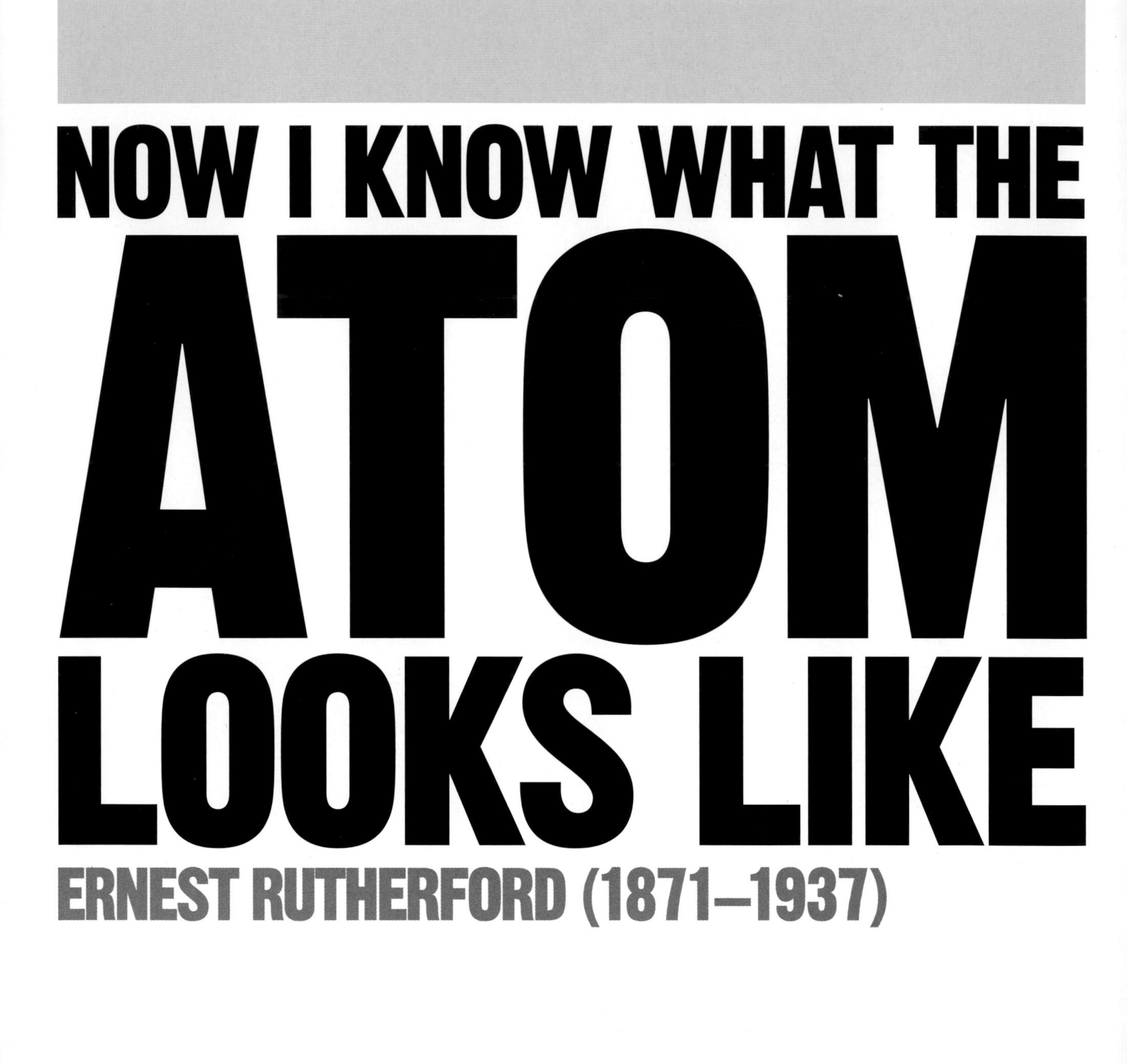

NOW I KNOW WHAT THE ATOM LOOKS LIKE

ERNEST RUTHERFORD (1871–1937)

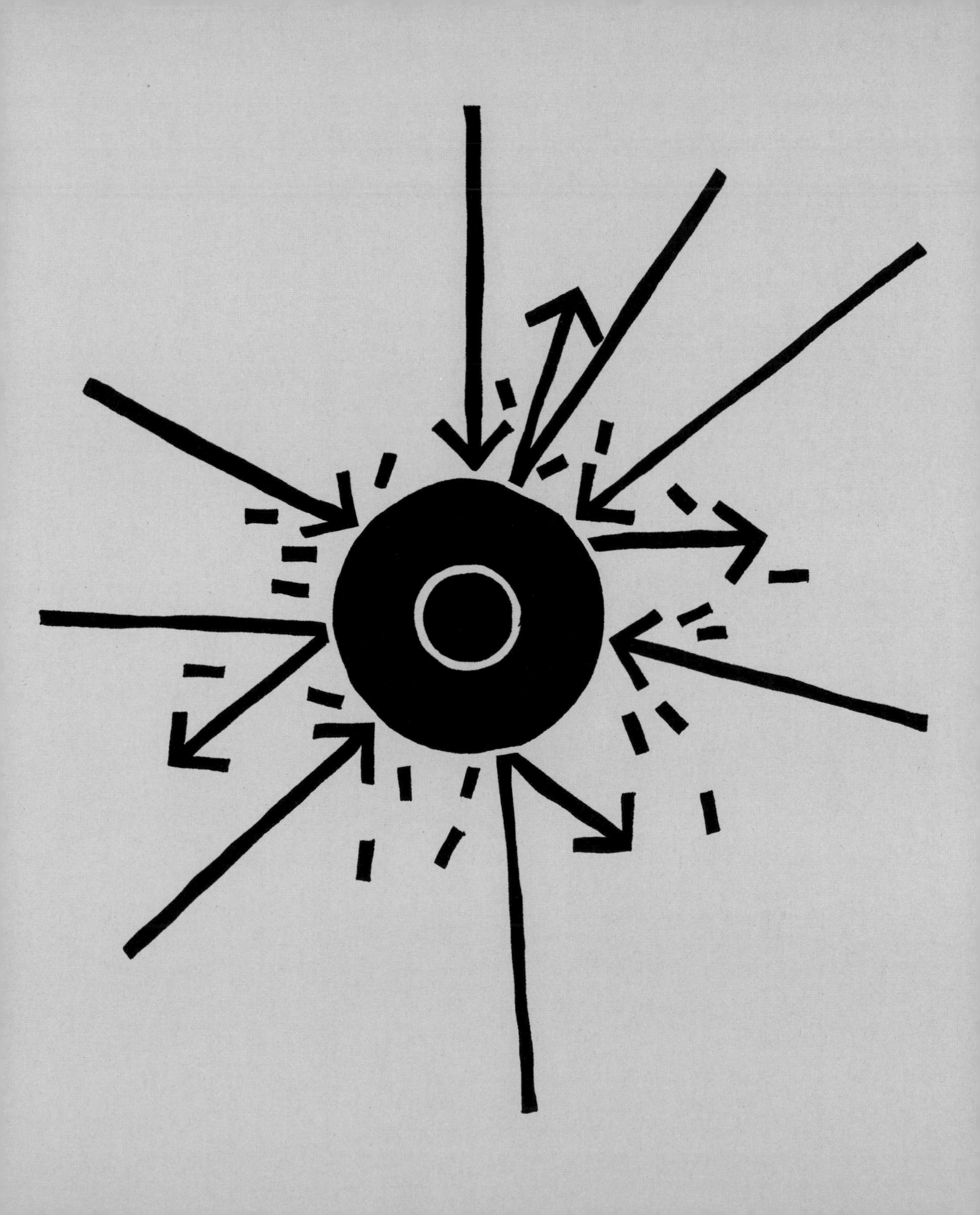

IN CONTEXT

BRANCH
Physics

BEFORE
c.400 BCE Greek philosopher Democritus envisages atoms as solid, indestructible building blocks of matter.

1805 John Dalton's atomic theory of matter marries chemical processes to physical reality and allows him to calculate atomic weights.

1896 Nuclear radiation is discovered by Henri Becquerel, and is used to reveal the internal structure of the atom.

AFTER
1938 Otto Hahn, Fritz Strassman, and Lise Meitner split the atomic nucleus.

2014 Firing increasingly energetic particles at the nucleus continues to reveal a slew of new subatomic particles and antiparticles.

The discovery at the turn of the 20th century that the basic constituent of matter—the atom—could be broken into smaller fragments was a watershed moment for physics. This astonishing breakthrough revolutionized ideas about how matter is constructed and the forces that hold it and the universe together. It revealed an entirely new world at the subatomic level—one that required a new physics to describe its interactions—and a slew of tiny particles that filled this infinitesimally small domain.

Atomic theories have a long history. The Greek philosopher Democritus developed the ideas of earlier thinkers that everything is composed of atoms. The Greek word *átomos*, which is credited to Democritus, means indivisible and referred to the basic units of matter. Democritus thought that the materials must reflect the atoms they are made of—so atoms of iron are solid and strong, while those of water are smooth and slippery.

At the turn of the 19th century, English natural philosopher John Dalton proposed a new atomic theory based on his "law of multiple proportions," which explained how elements (simple, uncombined substances) always combine in simple, whole-number ratios. Dalton saw that this meant that a chemical reaction between two substances is no more than the fusing of individual small components, repeated countless times. This was the first modern atomic theory.

A stable science

A self-congratulatory mood was detectable in physics at the end of the 19th century. Certain eminent physicists made grandstanding, declarations to the effect that the subject was all but finished—that the principal discoveries had all been made and the program going forward was one of improving the accuracy of known quantities "to the sixth decimal place." However, many research physicists of the time knew better. It was already clear that they were facing an entirely new and strange set of phenomena that defied explanation.

In 1896, Henri Becquerel, following a lead from Wilhelm Röntgen's discovery of mysterious "X-rays" the previous year, had

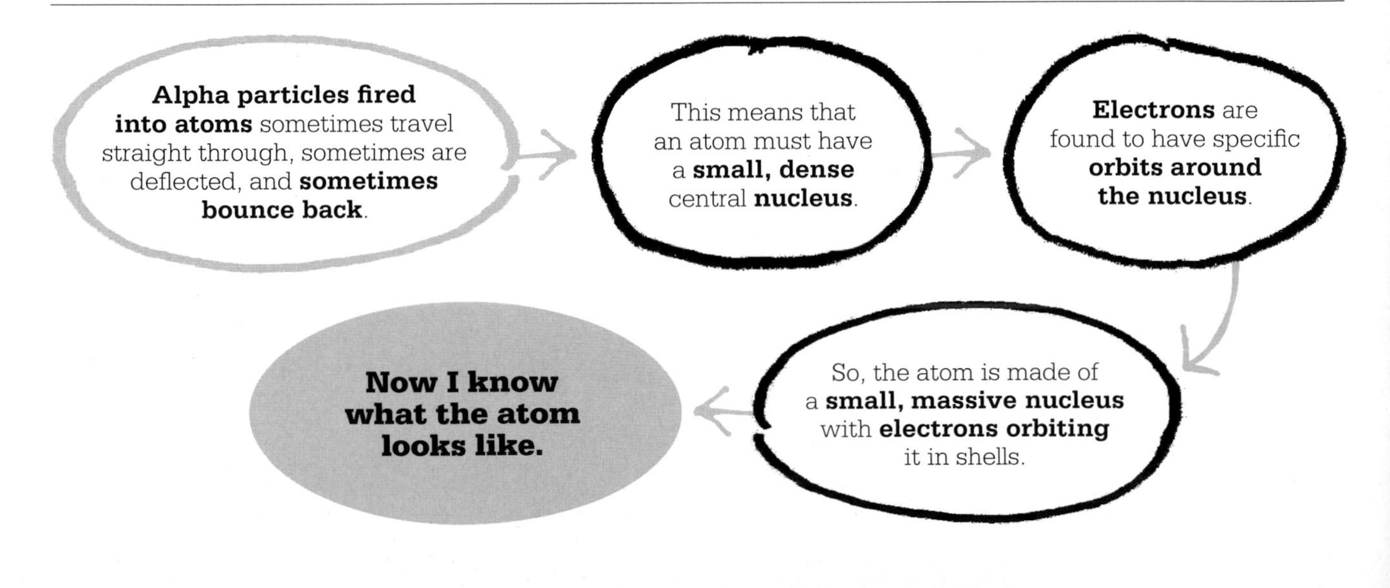

See also: John Dalton 112–13 ▪ August Kekulé 160–65 ▪ Wilhelm Röntgen 186–87 ▪ Marie Curie 190–95 ▪ Max Planck 202–05 ▪ Albert Einstein 214–21 ▪ Linus Pauling 254–59 ▪ Murray Gell-Mann 302–07

J. J. Thomson is pictured here at work in his Cambridge laboratory. Thomson's "plum pudding" model of the atom was the first to include the newly discovered electron.

found an unexplained radiation. What were these new radiations and where were they coming from? Becquerel correctly surmised that this radiation was emanating from within uranium salts. When Pierre and Marie Curie studied the decay of radium, they discovered a constant and seemingly inexhaustible source of energy inside radioactive elements. If this were the case, it would break several fundamental laws of physics. Whatever these radiations were, it was clear that there were large gaps in current models.

Discovery of the electron

The following year, the British physicist Joseph John (J. J.) Thomson caused a sensation when he demonstrated that he could break lumps out of atoms. While investigating the "rays" emanating from high-voltage cathodes (negatively charged electrodes), he found that this particular kind of radiation was made of individual "corpuscles," since it created momentary, pointlike sparkles of light on hitting a phosphorescent screen; it was negatively charged, since a beam could be deflected by an electric field; and it was exceedingly light, weighing less than a thousandth of the lightest atom, hydrogen. Moreover, the weight of the corpuscle was the same, no matter which element was used as a source. Thomson had discovered the electron. These results were totally unanticipated theoretically. If an atom contains charged particles, why shouldn't the opposing particles have equal mass? Previous atomic theories held that atoms were solid lumps. As befit their status as the most basic constituent of matter, they were entire, whole, and perfect. But when viewed in the light of Thomson's discovery, they clearly were divisible. Put together, these new radiations raised the suspicion that science had failed to understand the vital components of matter and energy. »

The plum-pudding model

Thomson's discovery of the electron earned him the Nobel Prize for Physics in 1906. He was enough of a theoretician, however, to see that a radical new model of the atom was needed to adequately incorporate his findings. His answer, produced in 1904, was the "plum-pudding" model. Atoms have no overall electric charge and, since the mass of this new electron was small, Thomson postulated that a larger positively charged sphere contained most of the atom's mass, and the electrons were embedded in it like plums in the dough of a Christmas pudding. With no evidence to suggest otherwise, it was sensible to assume that the point charges, like the plums in a pudding, were arbitrarily distributed across the atom.

> All science is either physics or stamp collecting.
> **Ernest Rutherford**

Rutherford revolution

However, the positively charged parts of the atom steadfastly refused to reveal themselves, and the hunt was on to locate the missing member of the atomic pair. The quest resulted in a discovery that would produce a very different visualization of the internal structure of the basic unit of all elements.

At the Physical Laboratories at the University of Manchester, Ernest Rutherford devised and directed an experiment to test Thomson's plum-pudding model. This charismatic New Zealander was a gifted experimentalist with a keen sense of which details to pursue. Rutherford had received the 1908 Nobel Prize in Physics for his "Theory of Atomic Disintegration."

The theory proposed that the radiations emanating from radioactive elements were the result of their atoms breaking apart. With the chemist Frederick Soddy, Rutherford had demonstrated that radioactivity involved one element spontaneously changing into another. Their work was to suggest new ways to probe the inside of the atom and see what was there.

Radioactivity

Although radioactivity was first encountered by Becquerel and the Curies, it was Rutherford who identified and named the three different types of what we would now call nuclear radiation. These are slow-moving, heavy, positively charged "alpha" particles; fast-moving, negatively charged "beta" particles; and highly energetic but uncharged "gamma" radiation (p.194). Rutherford classified these different forms of radiation by their penetrating power, from the least-penetrating alpha particles, which are blocked by thin paper, to gamma rays that require a thickness of lead to be stopped. He was the first to use alpha particles to explore the atomic realm. He was also the first to outline the notion of radioactive half-life and discover that "alpha particles" were helium nuclei—atoms stripped of their electrons.

Ernest Rutherford

Brought up in rural New Zealand, Ernest Rutherford was working in the fields when the letter from J. J. Thomson arrived informing him of a scholarship to Cambridge University. In 1895, he was made a research fellow at the Cavendish Laboratories, where he conducted experiments alongside Thomson that led to the discovery of the electron. In 1898, at 27 years old, Rutherford took up a professorial post at McGill University in Montreal, Canada. It was there that he carried out the work on radioactivity that won him the 1908 Nobel Prize in Physics.

Rutherford was an accomplished administrator, too, and during his lifetime he headed up the three top physics research laboratories. In 1907, he took the chair in physics at the University of Manchester where he discovered the atomic nucleus. In 1919, he returned to the Cavendish as director.

Key works

1902 *The Cause and Nature of Radioactivity, I & II*
1909 *The Nature of the α Particle from Radioactive Substances*

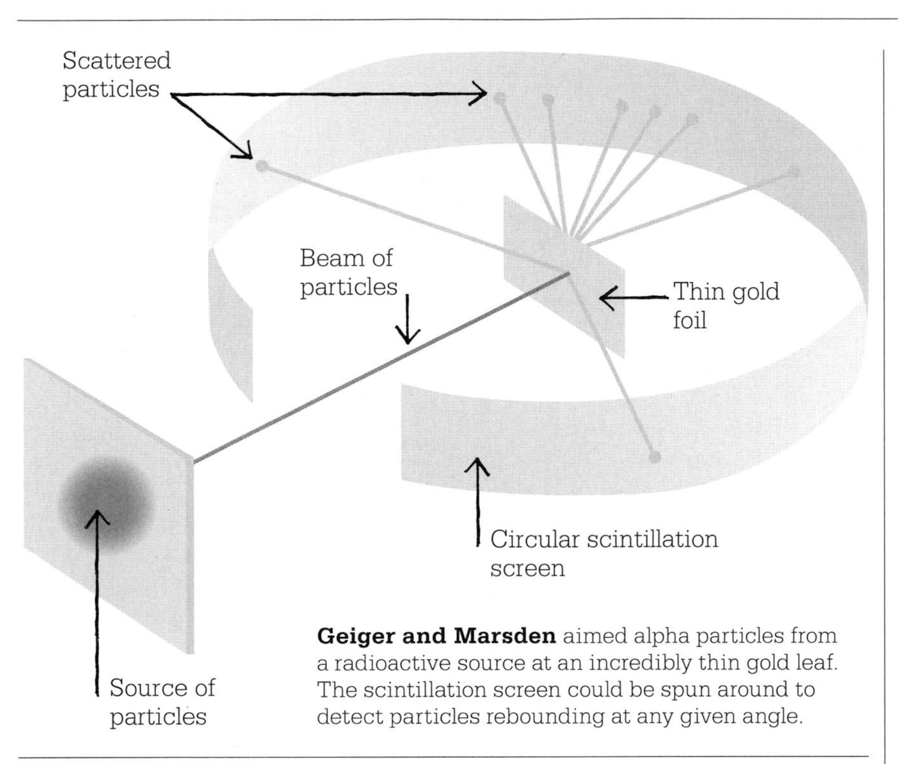

Geiger and Marsden aimed alpha particles from a radioactive source at an incredibly thin gold leaf. The scintillation screen could be spun around to detect particles rebounding at any given angle.

The gold foil experiment

In 1909, Rutherford set out to probe the structure of matter using alpha particles. The previous year, along with the German Hans Geiger, he had developed zinc sulphide "scintillation screens," which enabled individual collisions of alpha particles to be counted as brief bright flashes, or scintillations. With the help of undergraduate student Ernest Marsden, Geiger would use these screens to determine whether matter was infinitely divisible or whether atoms contained fundamental building blocks.

They fired a beam of alpha particles from a radium source at an extremely thin strip of gold leaf, just a thousand or so atoms thick. If, as the plum-pudding model held, gold atoms consisted of a diffuse cloud of positive charge with pointlike negative charges, then the massive, positively charged alpha particles would plough straight through the foil. Most of the particles would be deflected only slightly by interaction with the gold atoms and would be scattered across shallow angles.

Geiger and Marsden spent long hours sat in the darkened laboratory, peering down microscopes and counting the tiny flashes of light on the scintillation screens. Then, acting on a hunch, Rutherford instructed them to position screens that would catch any high-angle deflections as well as at the expected low-angle scintillations. With these new screens in place, they discovered that some of the alpha particles were being deflected by more than 90°, and others were rebounding off the foil back the way they came. Rutherford described the result as like firing a 15-inch shell at tissue paper and having it bounce back.

It was quite the most incredible event that has ever happened to me in my life. It was almost as incredible as if you fired a 15-inch shell at a piece of tissue paper and it came back and hit you.
Ernest Rutherford

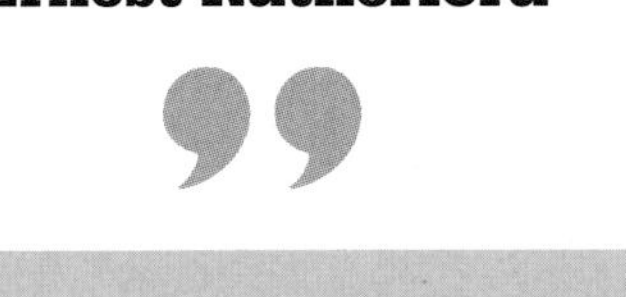

The nuclear atom

Halting heavy alpha particles in their tracks or deflecting them by high angles was possible only if the positive charge and mass of an atom were concentrated in small volume. In light of these results, in 1911, Rutherford published his conception of the structure of the atom. The "Rutherford Model" is a solar system in miniature, with electrons orbiting a small, dense, positively charged nucleus. The model's major innovation was the infinitesimally small nucleus, which forced the uncomfortable conclusion that the atom is not at all solid. Matter at an atomic scale is mostly space, governed by energy and force. This was a definitive break from the atomic theories of the previous century.

While Thomson's "plum-pudding" atom had been an instant hit, Rutherford's model was largely ignored by the scientific community. Its failings were all too plain to see. It was well established that accelerating electric charges emit energy as electromagnetic radiation. Thus, as electrons swoop around the nucleus—experiencing circular acceleration that keeps » them in their orbits—they ought to be continually emitting electromagnetic radiation. Steadily losing energy as they orbited, the electrons would spiral inexorably into the nucleus. According to Rutherford's model, atoms ought to be unstable, but clearly they are not.

A quantum atom

Danish physicist Niels Bohr saved the Rutherford model of the atom from languishing in obscurity by applying new ideas about quantization to matter. The quantum revolution had begun in 1900 when Max Planck had proposed the quantization of radiation, but the field was still in its infancy in 1913—it would have to wait until the 1920s for a formalized mathematical framework of quantum mechanics. At the time Bohr was working on this problem, quantum theory essentially consisted of no more than Einstein's notion that light comes in tiny "quanta" (discrete packets of energy) that we now call photons. Bohr sought to explain the precise pattern of absorption and emission of light from atoms. He suggested that each electron is confined to fixed orbits within atomic "shells," and that the energy levels of the orbits are "quantized"—that is, they can only take certain specific values.

In this orbital model, the energy of any individual electron is closely related to its proximity to the atom's nucleus. The closer an electron is to the nucleus, the less energy it has, but it can be excited into higher energy levels by absorbing electromagnetic radiation of a certain wavelength. Upon absorbing light, an electron leaps to a "higher," or outer, orbit. Upon attaining this higher state, the electron will promptly drop back into the lower-energy orbit, releasing a quantum of energy that precisely matches the energy gap between the two orbitals.

Bohr offered no explanation for what this meant or what it might look like—he simply stated that falling out of orbit into the nucleus was, for electrons, impossible. Bohr's was a purely theoretical model of the atom. However, it agreed with experiment and solved many associated problems in an elegant stroke. The way in which electrons would have to fill up empty shells in a strict order, getting progressively farther from the nucleus, matched the march of the properties of the elements seen across the periodic table as atomic number increases. Even

If your experiment needs statistics, you ought to have done a better experiment.
Ernest Rutherford

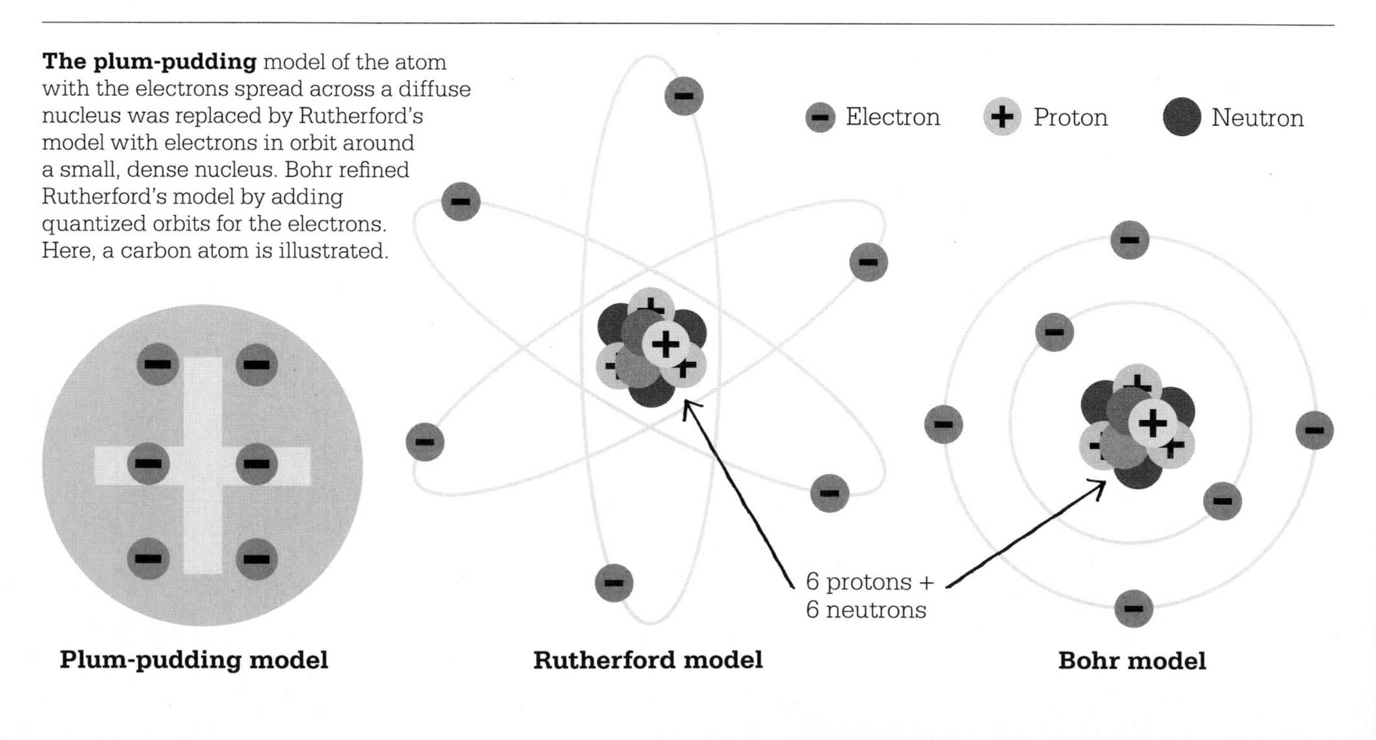

The plum-pudding model of the atom with the electrons spread across a diffuse nucleus was replaced by Rutherford's model with electrons in orbit around a small, dense nucleus. Bohr refined Rutherford's model by adding quantized orbits for the electrons. Here, a carbon atom is illustrated.

more convincing was the way in which the theoretical energy levels of the shells neatly fit actual "spectral series"—the frequencies of light absorbed and emitted by different atoms. A long sought after way to marry electromagnetism and matter had been realized.

Going inside the nucleus

Once this picture of the nuclear atom had been accepted, the next stage was to ask what lay inside the nucleus. In experiments reported in 1919, Rutherford found that his beams of alpha particles could generate hydrogen nuclei from many different elements. Hydrogen had long been recognized as the simplest of all the elements and thought of as a building block for all other elements, so Rutherford proposed that the hydrogen nucleus was in fact its own fundamental particle, the proton.

The next development in atomic structure was James Chadwick's 1932 discovery of the neutron, in which Rutherford once again had a hand. Rutherford had postulated the existence of the neutron in 1920 as a way to compensate for the repulsive effect of many point-sized positive charges crammed into a tiny nucleus. Like charges repel each other, so he theorized that there must be another particle that somehow dissipates the charge or binds the jostling protons tightly together. There was also extra mass in elements heavier than hydrogen, which could be accounted for by a third, neutral but equally massive subatomic particle.

However, the neutron proved difficult to detect and it took nearly a decade of searching to find it. Chadwick was working at the Cavendish Laboratory under the supervision of Rutherford. Guided by his mentor, he was studying a new kind of radiation that had been found by the German physicists Walther Bothe and Herbert Becker when they bombarded beryllium with alpha particles.

Chadwick duplicated the Germans' results and realized that this penetrating radiation was the neutron Rutherford had been looking for. A neutral particle, such as the neutron, is much more penetrating than a charged particle, such as a proton, because it feels no repulsion as it passes through matter. However, with mass slightly greater than a proton, it can easily knock protons out of the nucleus, something that otherwise only extremely energetic electromagnetic radiation can do.

Electron clouds

The discovery of the neutron completed the picture of the atom as a massive nucleus with electrons in orbit around it. New discoveries in quantum physics would refine our view of electrons in orbit around a nucleus. Modern models of the atom feature "clouds" of electrons, which represent only those areas in which we are most likely to find an electron, according to its quantum wavefunction (p.256).

The picture has been further complicated by the discovery that neutrons and protons are not fundamental particles, but are made of arrangements of smaller particles called quarks. Questions about the true structure of the atom are still actively being researched. ■

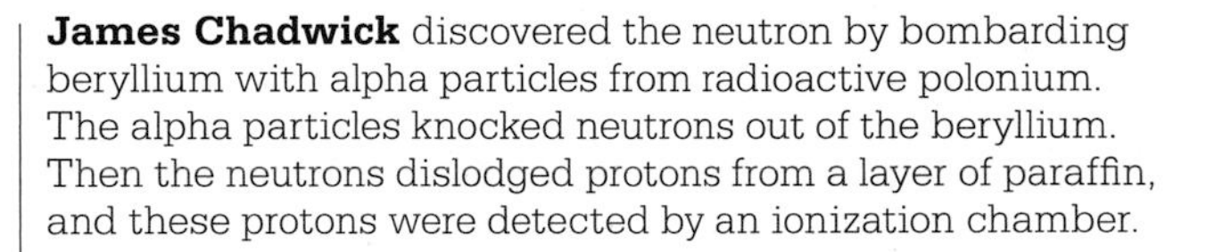

James Chadwick discovered the neutron by bombarding beryllium with alpha particles from radioactive polonium. The alpha particles knocked neutrons out of the beryllium. Then the neutrons dislodged protons from a layer of paraffin, and these protons were detected by an ionization chamber.

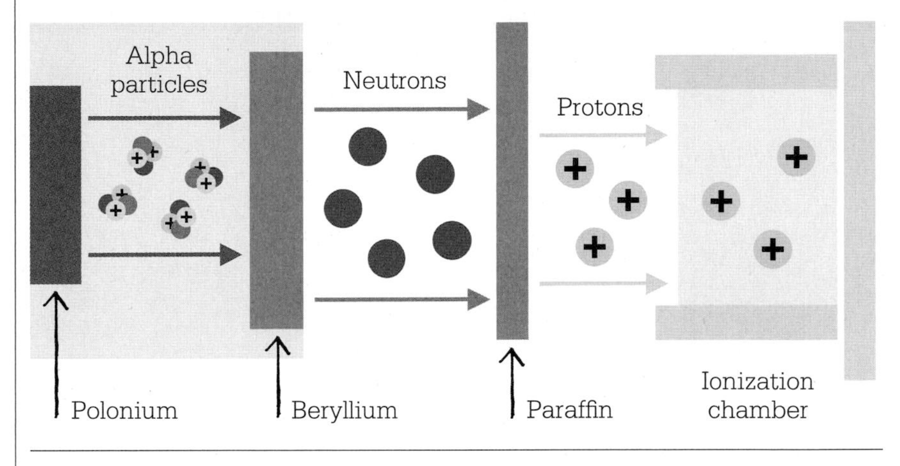

> The difficulties disappear if it be assumed that the radiation consists of particles of mass 1 and charge 0, or neutrons.
> **James Chadwick**

GRAVITY IS A DISTORTION IN THE SPACE-TIME CONTINUUM

ALBERT EINSTEIN (1879–1955)

IN CONTEXT

BRANCH
Physics

BEFORE
17th century Newtonian physics provides a description of gravity and motion, which is still adequate for most everyday calculations.

1900 Max Planck first argues that light can be considered to consist of individual packets, or "quanta," of energy.

AFTER
1917 Einstein uses general relativity to produce a model of the universe. Assuming that the universe is static, he introduces a factor called the cosmological constant to prevent its theoretical collapse.

1971 Time dilation due to general relativity is demonstrated by flying atomic clocks around the world in jet aircraft.

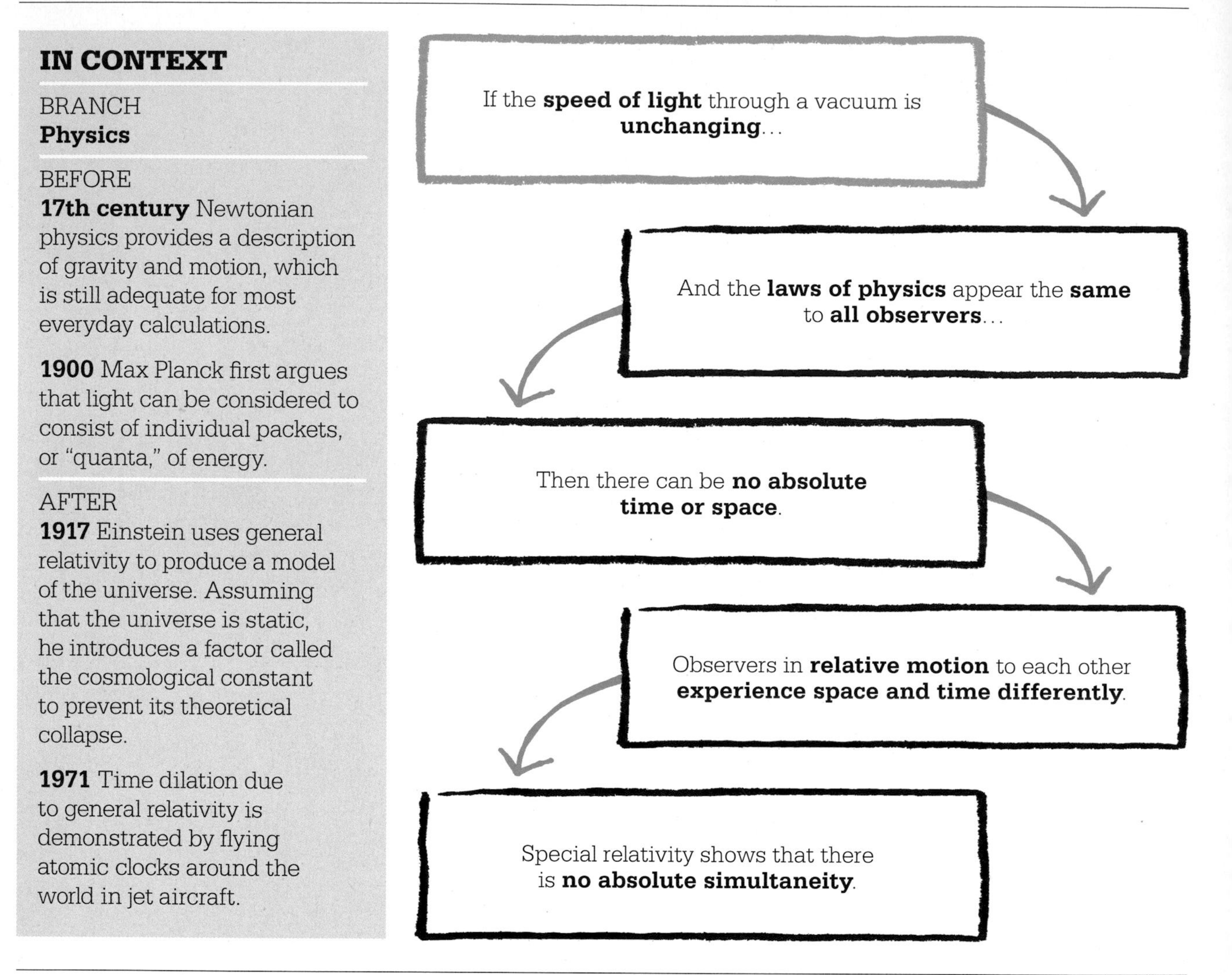

In the year 1905, the German scientific journal *Annalen der Physik* published four papers by a single author—a little-known 26-year-old physicist named Albert Einstein, then working at the Swiss patent office. Together, these papers would lay the foundations for much of modern physics.

Einstein resolved some fundamental problems that had appeared in the scientific understanding of the physical world toward the end of the 19th century. One of the papers of 1905 transformed understanding of the nature of light and energy. A second was an elegant proof that a long-observed physical effect called Brownian motion could demonstrate the existence of atoms. A third showed the presence of an ultimate speed limit to the universe, and considered the strange effects thereof, known as special relativity, while the fourth forever changed our understanding of the nature of matter, showing that it was interchangeable with energy. A decade later, Einstein followed up the implications of these latter papers with a theory of general relativity that presented a new and deeper understanding of gravity, space, and time.

Quantizing light

The first of Einstein's 1905 papers addressed a long-standing problem with the photoelectric effect. This phenomenon had been discovered by German physicist Heinrich Hertz in 1887. It involves metal electrodes producing a flow of electricity (that is, emitting electrons) when illuminated by certain wavelengths of radiation—typically ultraviolet light. The

See also: Christiaan Huygens 50–51 ▪ Isaac Newton 62–69 ▪ James Clerk Maxwell 180–85 ▪ Max Planck 202–05 ▪ Erwin Schrödinger 226–33 ▪ Edwin Hubble 236–41 ▪ Georges Lemaître 242–45

principle behind the emission is fairly easily described in modern terms (energy supplied by the radiation is absorbed by the outermost electrons in the metal's surface atoms, allowing them to break free). The puzzle was that the same materials stubbornly refused to emit electrons when illuminated by longer wavelengths, no matter how intense the light source.

This was a problem for the classical understanding of light, which assumed that intensity, above all, governed the amount of energy being delivered by a light beam. Einstein's paper, however, seized on the idea of "quantized light" recently developed by Max Planck. Einstein showed that if the beam of light is split into individual "light quanta" (what we would today call photons), then the energy carried by each quantum depends only on its wavelength—the shorter the wavelength, the higher the energy. If the photoelectric effect relies on interaction between an electron and a single photon, then it does not matter how many photons

> The grand aim of all science is to cover the greatest number of empirical facts by logical deduction from the smallest number of hypotheses or axioms.
> **Albert Einstein**

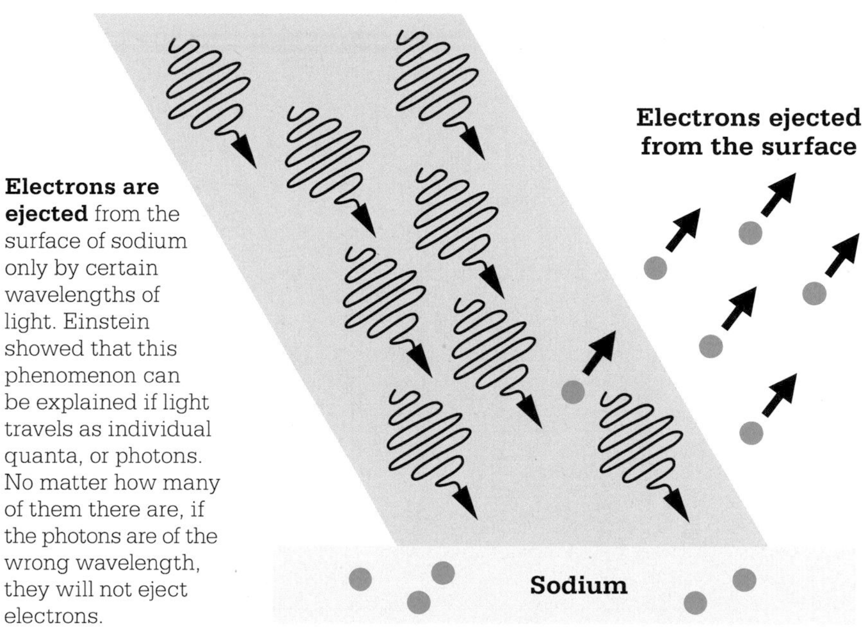

Electrons are ejected from the surface of sodium only by certain wavelengths of light. Einstein showed that this phenomenon can be explained if light travels as individual quanta, or photons. No matter how many of them there are, if the photons are of the wrong wavelength, they will not eject electrons.

bombard the surface (that is, how intense the light source is)—if none of them carries sufficient energy, the electrons will not break free.

Einstein's idea was rejected by leading figures of the day, including Planck, but his theory was shown to be correct by experiments conducted by American physicist Robert Millikan in 1919.

Special relativity

Einstein's greatest legacy was born in the third and fourth 1905 papers, which also involved an important reconceptualization of the true nature of light. Since the late 19th century, physicists had faced a crisis in their attempts to understand the speed of light. Its approximate value had been known and calculated with increasing accuracy since the 17th century, while James Clerk Maxwell's equations had demonstrated that visible light was just one manifestation of a wider spectrum of electromagnetic waves, all of which must move through the universe at a single speed.

Since light was understood to be a transverse wave, it was assumed that it propagated through a medium, just as water waves travel on the surface of a pond. The properties of this hypothetical substance, known as the "luminiferous ether," would give rise to the observed properties of electromagnetic waves, and since they could not alter from place to place, they would provide an absolute standard of rest.

One expected consequence of the fixed ether was that the speed of light from distant objects should vary depending on the relative motion of source and observer. For example, the speed of light from a distant star should vary » significantly depending on whether it was observed from one side of Earth's orbit, as our planet moved away from it at 20 miles/s (30 km/s), or on the opposite side, when the observer was moving toward it at a similar speed.

Measuring Earth's motion through the ether became an obsession for late 19th-century physicists. Such a measurement was the only way of confirming the existence of this mysterious substance, and yet the proof remained elusive. However precise the measuring equipment, light always seemed to move at the same speed. In 1887, US physicists Albert Michelson and Edward Morley devised an ingenious experiment to measure the so-called ether wind with high precision, but once again found no evidence for its existence. The negative result of the Michelson-Morley experiment shook the belief in the ether's existence, and similar results from attempts to repeat it over the following decades only intensified the sense of crisis.

Einstein's third 1905 paper, *On the Electrodynamics of Moving Bodies*, confronted the problem head on. Special relativity, as his theory became known, was developed from an acceptance of two simple postulates—that light moves through a vacuum with a fixed speed that is independent of the motion of the source, and that the laws of physics should appear the same to observers in all "inertial" frames of reference—that is, those not subject to external forces such as acceleration. Einstein was undoubtedly helped in accepting the first bold postulate by his previous acceptance of the quantum nature of light—conceptually, light quanta are often imagined as tiny self-contained packets of electromagnetic energy, able to travel through the vacuum of space with particle-like properties while still maintaining their wavelike characteristics.

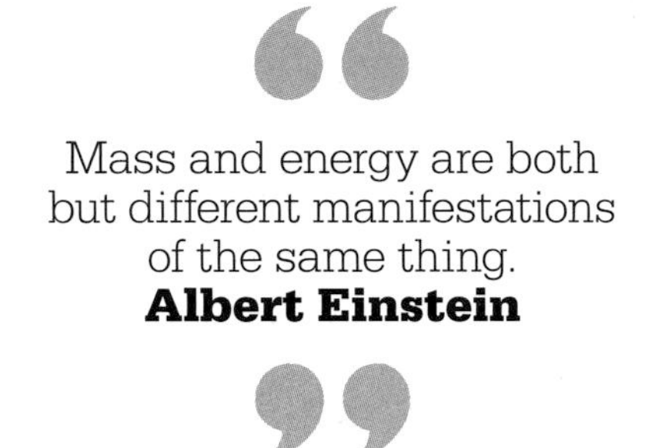

Mass and energy are both but different manifestations of the same thing.
Albert Einstein

Accepting these two postulates, Einstein considered the consequences for the rest of physics, and mechanics in particular. In order for the laws of physics to behave in the same way in all inertial reference frames, they would necessarily appear to be different when looking from one frame to another. Only relative motion mattered, and when the relative motion between two separate frames of reference approached the speed of light ("relativistic" speeds) strange things began to happen.

The Lorentz factor

Although Einstein's paper made no formal references to other scientific publications, it did mention the work of a handful of other contemporary scientists, for Einstein was certainly not the only person working toward an unorthodox solution to the ether crisis. Perhaps the most significant of these was Dutch physicist

Albert Einstein

Born in the southern German city of Ulm in 1879, Einstein had a somewhat bumpy secondary education, eventually training at Zurich Polytechnic to become a mathematics teacher. After failing to find teaching work, he took a job at the Swiss Patent Office in Bern, where he had plenty of spare time to develop the papers published in 1905. He attributed his success in this work to the fact that he had never lost his childlike sense of wonder.

Following the demonstration of general relativity, Einstein was propelled to global stardom. He continued to explore the implications of his earlier work, contributing to innovations in quantum theory. In 1933, fearing the rise of the Nazi party, Einstein elected not to return to Germany from a foreign tour, settling eventually at Princeton University in the United States.

Key works

1905 *On a Heuristic Viewpoint Concerning the Production and Transformation of Light*
1915 *The Field Equations of Gravitation*

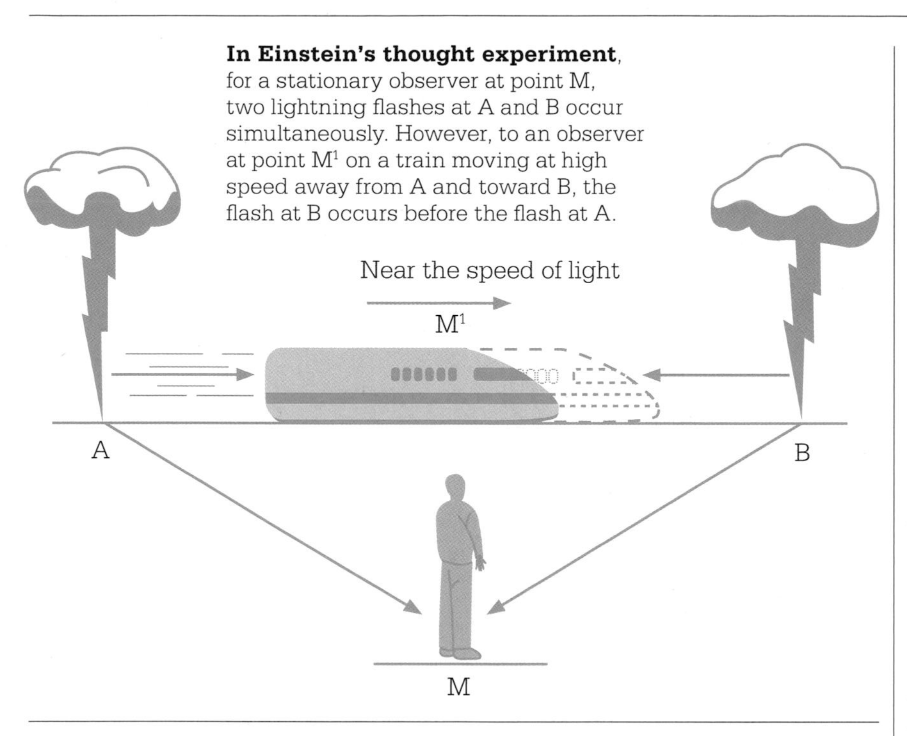

In Einstein's thought experiment, for a stationary observer at point M, two lightning flashes at A and B occur simultaneously. However, to an observer at point M[1] on a train moving at high speed away from A and toward B, the flash at B occurs before the flash at A.

Hendrik Lorentz, whose "Lorentz factor" lay at the heart of Einstein's description of physics close to the speed of light. It is defined mathematically as:

$$\frac{1}{\sqrt{1 - v^2 / c^2}}$$

Lorentz developed this equation to describe the changes in time and length measurements required in order to reconcile the Maxwell equations of electromagnetism with the principle of relativity. It was crucial to Einstein since it provided a term for transforming results as seen by one observer to show what they look like to another observer who is in motion relative to the first observer. In the term quoted above, v is the speed of one observer compared to the other, and c is the speed of light. In most situations, v will be very small compared to c, so v^2/c^2 will be close to zero, and the Lorentz factor close to 1, meaning that it makes almost no difference to calculations. Lorentz's work had been coolly received, largely because it could not be incorporated into standard ether theories. Einstein approached the problem from the other direction, showing that the Lorentz factor arose as an inevitable consequence of the principle of special relativity and reexamining the true meaning of measured time and distance intervals. An important result of this was the realization that events that appeared simultaneous for an observer in one reference frame were not necessarily so for someone in a different reference frame (a phenomenon known as the relativity of simultaneity). Einstein also showed how from the point of view of a distant observer, the length of moving objects in their direction of travel became compressed as they approached the speed of light, in accordance with a simple equation governed by the Lorentz factor. Even more strangely, time itself appears to run more slowly as measured from the observer's reference frame.

Illustrating relativity

Einstein illustrated special relativity by asking us to consider two frames of reference in motion relative to each other: a moving train and the embankment next to it. Two flashes of lightning, at points A and B, appear to occur simultaneously to an observer standing on the embankment at a midpoint between them, M. An observer on the train is at a position M[1] in a separate frame of reference. When the flashes occur, M[1] may be passing right by M. However, by the time the light has reached the observer on the train, the train has moved toward point B and away from point A. As Einstein puts it, the observer is "riding ahead of the beam of light coming from A." The observer on the train concludes that lightning strike B occurred before strike A. Einstein now insists that: "Unless we are told the reference-body to which the statement of time refers, there is no meaning in a statement of the time of an event." Both time and position are relative concepts.

Mass-energy equivalence

The last of Einstein's 1905 papers was called *Does the Inertia of a Body Depend on its Energy Content?* Its three brief pages expanded on an idea touched on in the previous paper—that the mass of a body is a measure of its energy. Here, Einstein demonstrated that if a body radiates away a certain amount of energy (E) in the form of electromagnetic radiation, its mass will diminish by an amount equivalent to E/c^2. This equation is easily rewritten to show that the energy of a stationary particle » within a particular reference frame is given by the equation $E = mc^2$. This principle of "mass-energy equivalence" was to become a keystone of 20th-century science, with applications that range from cosmology to nuclear physics.

Gravitation fields

Although Einstein's papers in that *annus mirabilis* seemed too obscure at first to make much impression beyond the rarefied world of physics, it propelled him to fame within that community.

Our **experience of gravity** is equivalent to that of being inside a **constantly accelerating** frame of reference.

The acceleration can be explained by a **distortion** in the **space-time manifold**.

If **objects with mass** distort space-time, this explains their gravitational attraction.

General relativity explains gravity as a distortion in the space-time manifold.

Over the next few years, many scientists reached the conclusion that special relativity offered a better description of the universe than the discredited ether theory, and devised experiments that demonstrated relativistic effects in action. Meanwhile, Einstein was already moving on to a new challenge, extending the principles that he had now established in order to consider "noninertial" situations—those involving acceleration and deceleration.

As early as 1907, Einstein had hit upon the idea that a situation of "free fall" under the influence of gravity is equal to an inertial situation—the equivalence principle. In 1911, he realized that a stationary frame of reference influenced by a gravitational field is equivalent to one undergoing constant acceleration. Einstein illustrated this idea by imagining a person standing in a sealed elevator in space. The elevator is being accelerated in one direction by a rocket. The person feels a force pushing up from the floor, and pushes back against the floor with equal and opposite force following Newton's Third Law. Einstein realized that the person in the elevator would feel exactly as they would if they were standing still in a gravitational field.

In an elevator under constant acceleration, a beam of light fired at an angle perpendicular to the acceleration would be deflected onto a curved path, and Einstein reasoned that the same would occur in a gravitational field. It was this effect of gravity on light—known as gravitational lensing—that would first demonstrate general relativity.

Einstein considered what this said about the nature of gravity. In particular, he predicted that

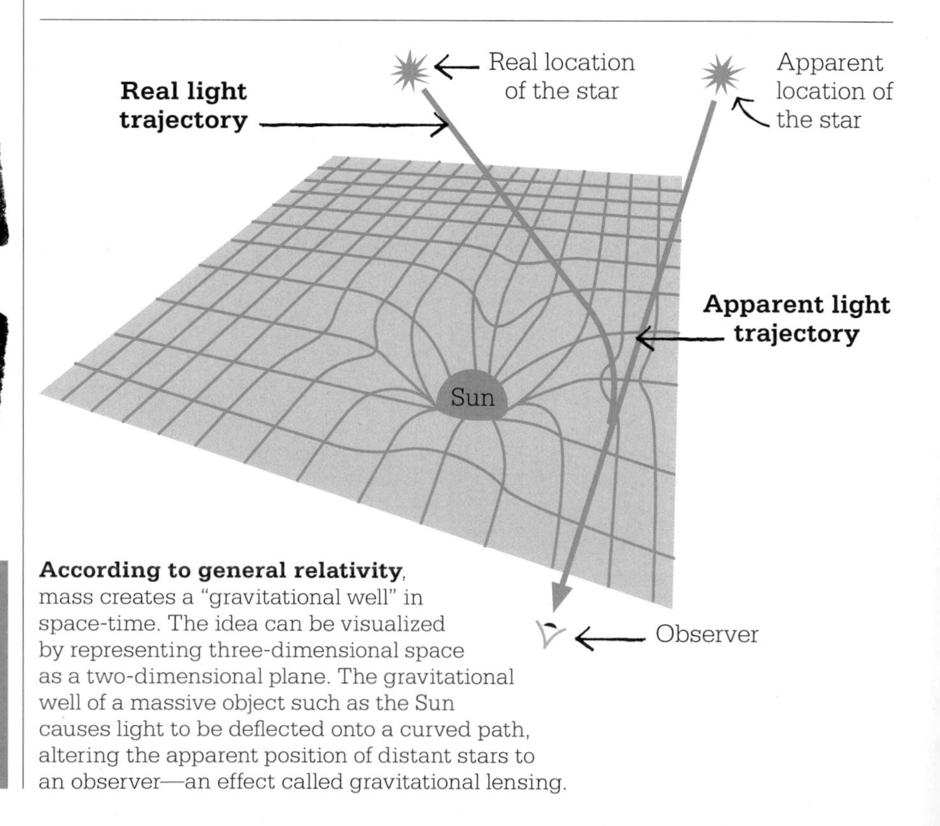

According to general relativity, mass creates a "gravitational well" in space-time. The idea can be visualized by representing three-dimensional space as a two-dimensional plane. The gravitational well of a massive object such as the Sun causes light to be deflected onto a curved path, altering the apparent position of distant stars to an observer—an effect called gravitational lensing.

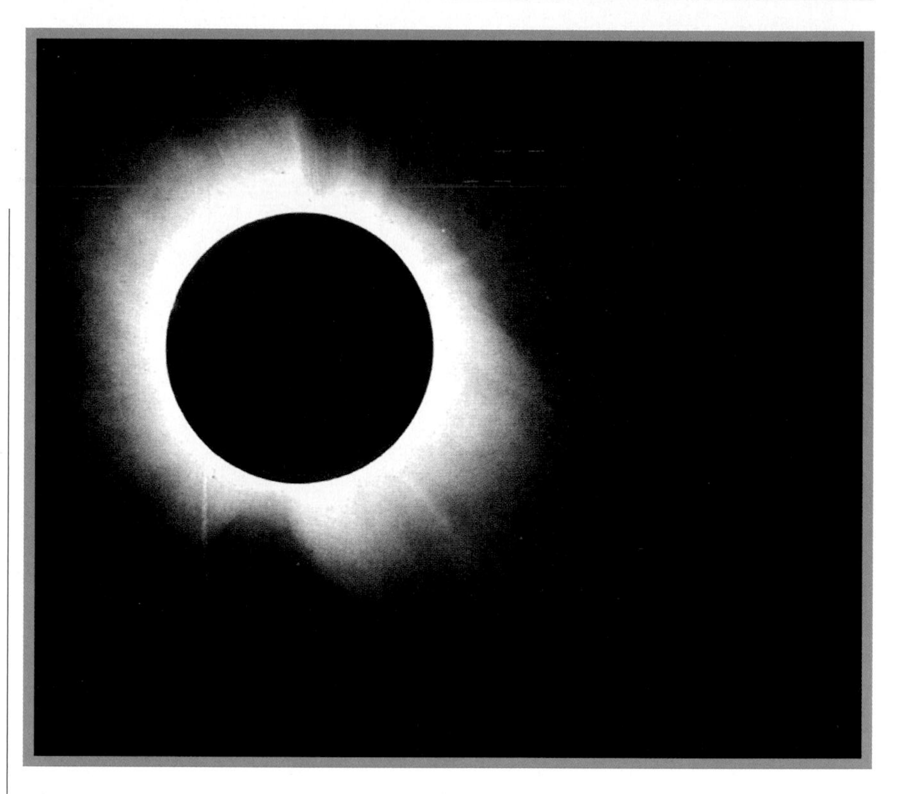

Arthur Eddington's photographs of a solar eclipse in 1919 provided the first evidence for general relativity. Stars around the Sun appeared out of place, just as Einstein had predicted.

relativistic effects such as time dilation should occur in strong gravitational fields. The closer a clock is to a source of gravitation, the more slowly it will tick. This effect remained purely theoretical for many years, but has now been confirmed using atomic clocks.

Space-time manifold

Meanwhile, also in 1907, Einstein's former tutor Hermann Minkowski had hit upon another important part of the puzzle. Considering the effective trade-offs between the dimensions of space and time involved in special relativity, he developed the idea of combining the three dimensions of space with one of time in a space-time manifold. In Minkowski's interpretation, relativistic effects could be described in geometrical terms by considering distortions in the way that observers in relative motion observe the manifold in a different frame of reference.

In 1915, Einstein published his complete theory of general relativity. In its finished form, it was nothing less than a new description of the nature of space, time, matter, and gravity. Adopting Minkowski's ideas, Einstein saw the "stuff of the universe" as a space-time manifold that could be distorted thanks to relativistic motion, but could also be warped by the presence of large masses such as stars and planets in a way that was experienced as gravity. The equations that described the link between mass, distortion, and gravity were fiendishly complex, but Einstein used an approximation to solve a long-standing mystery—the way in which Mercury's closest approach to the Sun (aphelion) precesses, or rotates, around the Sun much more quickly than predicted by Newtonian physics. General relativity solved the puzzle.

Gravitational lensing

Einstein published at a time when much of the world was swept up in World War I, and English-speaking scientists had other things on their minds. General relativity was a complex theory and might have languished in obscurity for many years had it not been for the interest of Arthur Eddington, a conscientious objector to the war, and, as it happened, Secretary of the Royal Astronomical Society.

Eddington became aware of Einstein's work thanks to letters from Dutch physicist Willem de Sitter, and soon became its chief advocate in Britain. In 1919, a few months after the end of the war, Eddington led an expedition to the island of Príncipe, off the west coast of Africa, in order to test the theory of general relativity and its prediction of gravitational lensing in the most spectacular circumstances. Einstein had predicted as early as 1911 that a total solar eclipse would allow the effects of gravitational lensing to be seen, in the form of apparently out-of-place stars around the eclipsed disk (a result of their light being deflected as it passed through the warped space-time around the Sun). Eddington's expedition delivered both impressive images of the solar eclipse and convincing proof of Einstein's theory. When published the following year, they proved to be a worldwide sensation, propelling Einstein to global fame and ensuring that our ideas about the nature of the universe would never be the same again. ■

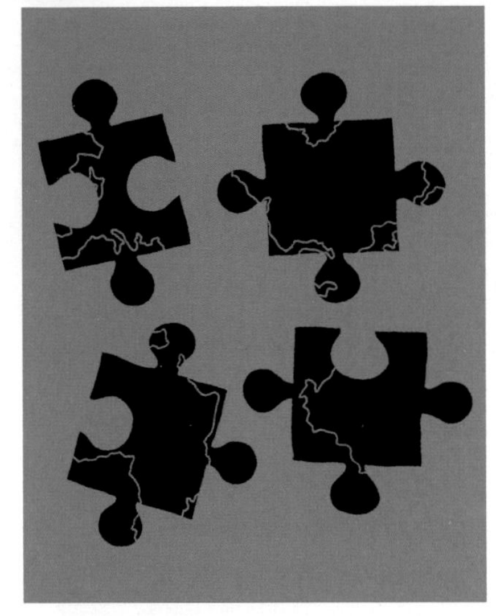

EARTH'S DRIFTING CONTINENTS ARE GIANT PIECES IN AN EVER-CHANGING JIGSAW

ALFRED WEGENER (1880–1930)

IN CONTEXT

BRANCH
Earth science

BEFORE
1858 Antonio Snider-Pellegrini makes a map of the Americas connected to Europe and Africa, to account for identical fossils found on opposite sides of the Atlantic Ocean.

1872 French geographer Élisée Reclus proposes that motion of the continents caused the formation of the oceans and mountain ranges.

1885 Eduard Suess suggests the southern continents were once linked by land bridges.

AFTER
1944 British geographer Arthur Holmes proposes convection currents in Earth's mantle as the mechanism that moves the crust at the surface.

1960 American geologist Harry Hess proposes that seafloor spreading pushes the continents apart.

In 1912, German meteorologist Alfred Wegener combined several strands of evidence to put forward a theory of continental drift, which suggested that Earth's continents were once connected but moved apart over millions of years. Scientists only accepted his theory once they had figured out what made such vast landmasses move.

Looking at the first maps of the New World and Africa, Francis Bacon had noted, in 1620, that the eastern coasts of the Americas are roughly parallel with the western coasts of Europe and Africa. This led scientists to speculate that these landmasses were once connected, challenging conventional notions of a solid, unchanging planet.

In 1858, Paris-based geographer Antonio Snider-Pellegrini showed that similar plant fossils had been found on either side of the Atlantic, dating back to the Carboniferous period, 359–299 million years ago.

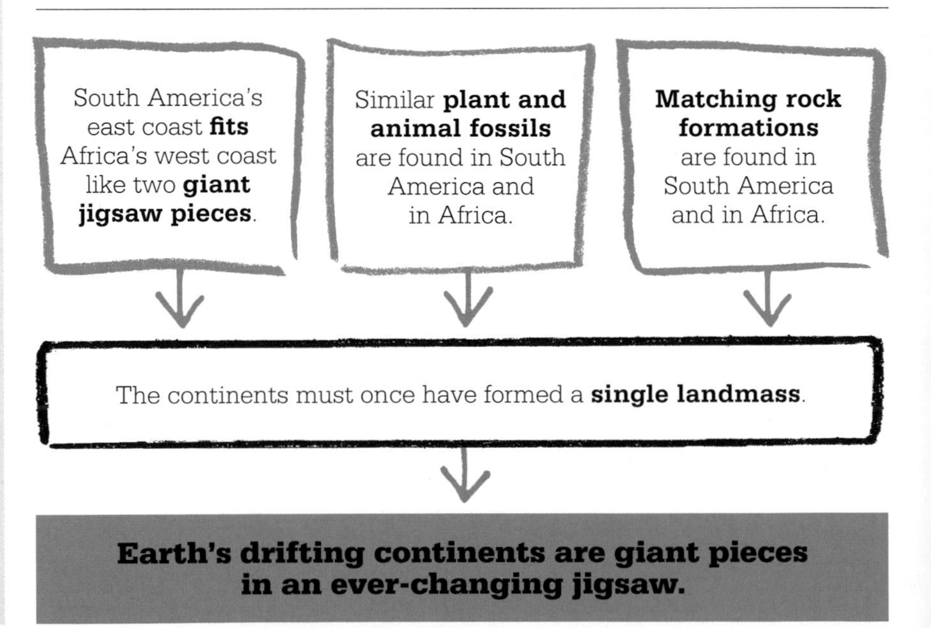

See also: Francis Bacon 45 ▪ Nicholas Steno 55 ▪ James Hutton 96–101 ▪ Louis Agassiz 128–29 ▪ Charles Darwin 142–49

He made maps showing how the American and African continents may once have fit together, and attributed their separation to the biblical Flood. When fossils of *Glossopteris* ferns were found in South America, India, and Africa, Austrian geologist Eduard Suess argued that they must have evolved on a single landmass. He suggested that the southern continents were once linked by land bridges across the sea, forming a supercontinent that he called Gondwanaland.

Wegener found more examples of similar organisms separated by oceans, but also similar mountain ranges and glacial deposits. Instead of earlier ideas that portions of a supercontinent had sunk beneath the waves, he thought perhaps it had split apart. Between 1912 and 1929, he expanded on this theory. His supercontinent—Pangaea—connected Suess's Gondwanaland to the northern continents of North America and Eurasia. Wegener dated the fragmentation of this single landmass to the end of the Mesozoic era, 150 million years ago, and pointed to Africa's Great Rift Valley as evidence of ongoing continental breakup.

Search for a mechanism

Wegener's theory was criticized by geophysicists for not explaining how continents move. In the 1950s, however, new geophysical techniques revealed a wealth of new data. Studies of Earth's past magnetic field indicated that the ancient continents lay in a different position relative to the poles. Sonar mapping of the seabed revealed signs of more recent ocean-floor formation. This was found to occur at mid-ocean ridges, as molten rock erupts through cracks in Earth's crust and spreads away from the ridges as new rock erupts.

In 1960, Harry Hess realized that seafloor spreading provided the mechanism for continental drift, and presented his theory of plate tectonics. Earth's crust is made up of giant plates that continually shift as convection currents in the mantle below bring new rock to the surface, and it is the formation and destruction of ocean crust that leads to the displacement of continents. This theory not only vindicated Wegener but is now the bedrock of modern geology. ■

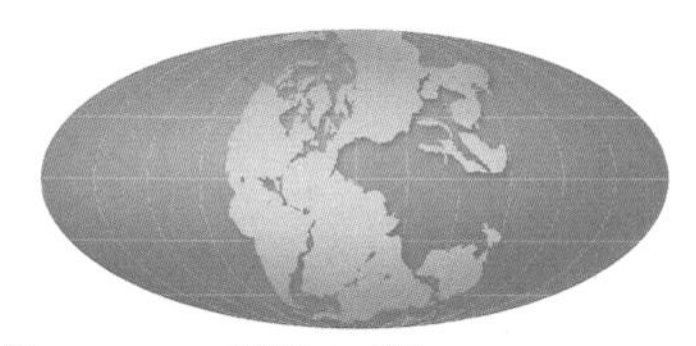

Pangaea, 200 million years ago

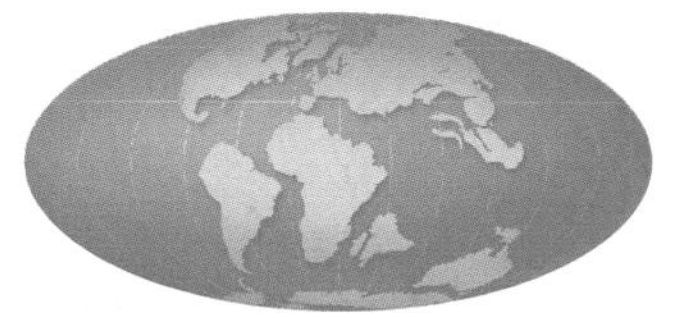

75 million years ago

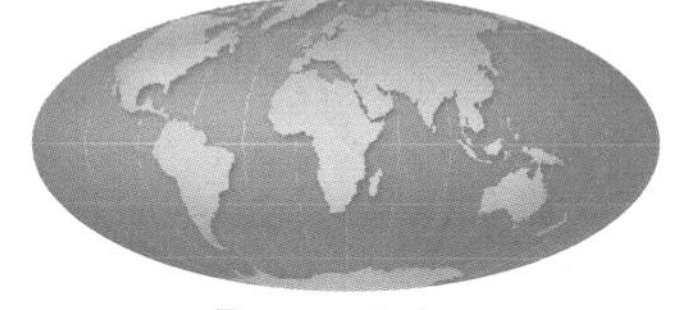

Present day

Wegener's supercontinent is just one in a long series. Geologists think the continents may be converging again, to form another supercontinent 250 million years from now.

Alfred Wegener

Born in Berlin, Alfred Lothar Wegener obtained a doctorate in astronomy from the University of Berlin in 1904, but soon became more interested in earth science. Between 1906 and 1930, he made four trips to Greenland as part of his pioneering meteorological studies of Arctic air masses. He used weather balloons to track air circulation and took samples from deep within the ice for evidence of past climates.

In between these expeditions, Wegener developed his theory of continental drift in 1912, and published it in a book in 1915. He produced revised and expanded editions in 1920, 1922, and 1929, but was frustrated by the lack of recognition for his work.

In 1930, Wegener led a fourth expedition to Greenland, hoping to collect evidence in support of the drift theory. On November 1, his 50th birthday, he set out across the ice to get badly needed supplies, but he died before reaching the main camp.

Key work

1915 *The Origin of Continents and Oceans*

CHROMOSOMES PLAY A ROLE IN HEREDITY

THOMAS HUNT MORGAN (1866–1945)

IN CONTEXT

BRANCH
Biology

BEFORE
1866 Gregor Mendel describes laws of inheritance, concluding that inherited characteristics are controlled by discrete particles, later called genes.

1900 Dutch botanist Hugo de Vries reaffirms Mendel's laws.

1902 Theodor Boveri and Walter Sutton independently conclude that chromosomes are involved in inheritance.

AFTER
1913 Morgan's student Alfred Sturtevant constructs the first genetic "map," of the fruit fly.

1930 Barbara McClintock discovers that genes can shift positions on chromosomes.

1953 James Watson and Francis Crick's double-helix model of DNA explains how genetic information is passed on during reproduction.

When cells divide, their **chromosomes** split and replicate in ways that **parallel the emergence of inherited characteristics**.

This suggests that **genes** controlling these characteristics **occur on the chromosomes**.

Some characteristics depend on **the sex of the organism**, so must be controlled by sex-determining chromosomes.

Chromosomes play a role in heredity.

During the 19th century, biologists observing cells divide under a microscope noticed the appearance of pairs of tiny threads in every cell's nucleus. These threads could be stained by dyes for observation, and came to be called chromosomes, meaning "colored bodies." The biologists soon began to wonder whether chromosomes had something to do with heredity.

In 1910, experiments conducted by American geneticist Thomas Hunt Morgan would confirm the roles of genes and chromosomes in inheritance, explaining evolution at a molecular level.

Particles of inheritance

By the early 20th century, scientists had traced the chromosome's precise movements at cell division, and noticed that the number of chromosomes varied between species, but that the number in the body cells of the same species were usually the same. In 1902, German biologist Theodor Boveri, having studied the fertilization of a sea

See also: Gregor Mendel 166–71 ▪ Barbara McClintock 271 ▪ James Watson and Francis Crick 276–83 ▪ Michael Syvanen 318–19

urchin, concluded that an organism's chromosomes had to be present in a full set for an embryo to develop properly. Later that same year, an American student named Walter Sutton concluded from his work on grasshoppers that chromosomes might even mirror the theoretical "particles of inheritance" proposed in 1866 by Gregor Mendel.

Mendel had done exhaustive experiments in the breeding of pea plants and, in 1866, suggested that their inherited characteristics were determined by discrete particles. Four decades later, to test the link between chromosomes and Mendel's theory, Morgan embarked on research that would combine breeding experiments with modern microscopy, in what came to be known as the "Fly Room" at Columbia University, New York.

From peas to fruit flies

Fruit flies (*Drosophila*) are gnat-sized insects that can be bred in small glass bottles and can produce the next generation—with a great many offspring—in just 10 days. This made the fruit fly ideal for studying inheritance. Morgan's team isolated and crossbred flies with particular characteristics, and then analyzed the proportions of variations in the offspring—just as Mendel had done with his peas.

Morgan finally corroborated Mendel's results after he spotted a male with eyes that were white rather than the normal red. Mating a white-eyed male with a red-eyed female produced only red-eyed offspring, which suggested that red was a dominant trait and white was recessive. When those offspring were crossbred, one in four of the next generation was white-eyed, and always male. The "white gene" must be linked to sex. When other traits linked to sex appeared, Morgan concluded that all these traits must be inherited jointly and the genes responsible for them must all be carried on the chromosome that determines sex. The females had a pair of X chromosomes, while males had an X and a Y. During reproduction, the offspring inherits an X from the mother, and an X or a Y from the father. The "white gene" is carried by the X. The Y chromosome has no corresponding gene.

Further work led Morgan to the notion that specific genes were not only located on specific chromosomes, but occupied particular positions on them. This opened up the idea that scientists could "map" an organism's genes. ■

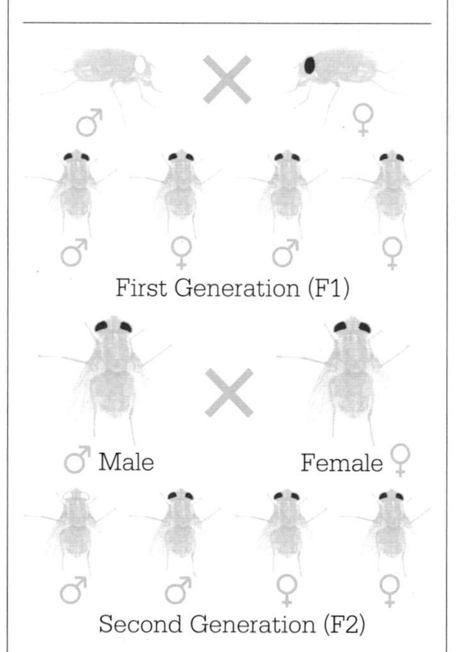

Crossbreeding fruit flies over two generations shows how the white-eyed trait is passed only to some males, through the sex chromosomes.

Thomas Hunt Morgan

Born in Kentucky, US, Thomas Hunt Morgan trained as a zoologist before going on to study the development of embryos. After moving to Columbia University in New York in 1904, he began to focus on the mechanism of inheritance. Initially sceptical of Mendel's conclusions, and even of Darwin's, he focused his efforts on the breeding of fruit flies to test his ideas about genetics. His success with fruit flies would lead many researchers to use them in genetics experiments.

Morgan's observation of stable, inherited mutations in fruit flies eventually led him to realize that Darwin was right, and in 1915, he published a work explaining how heredity functioned according to Mendel's laws. Morgan continued his research at the California Institute of Technology (Caltech) and, in 1933, he was awarded the Nobel Prize in Genetics.

Key works

1910 *Sex-limited Inheritance in* Drosophila
1915 *The Mechanism of Mendelian Heredity*
1926 *The Theory of the Gene*

PARTICLES HAVE WAVELIKE PROPERTIES

ERWIN SCHRÖDINGER (1887–1961)

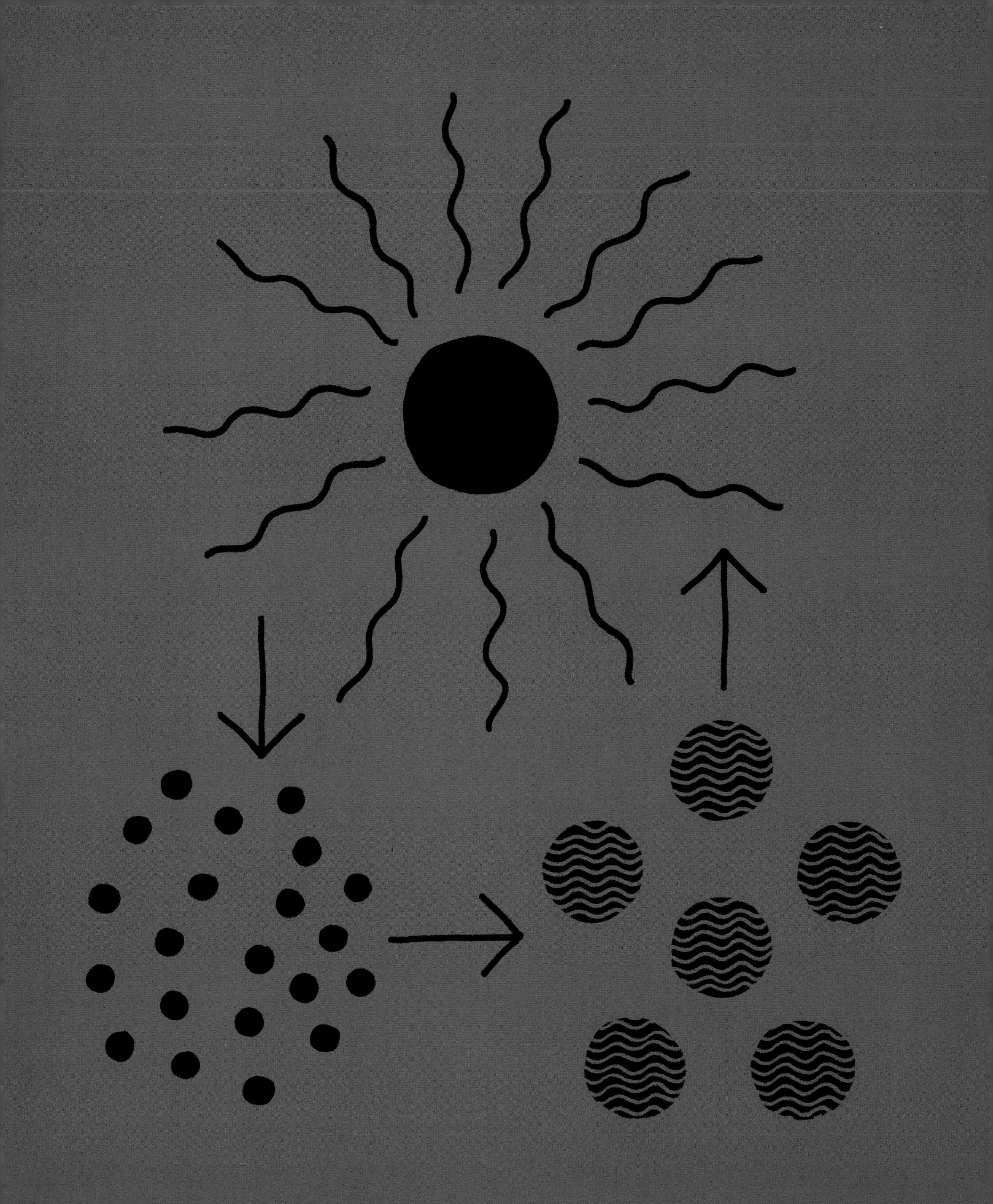

IN CONTEXT

BRANCH
Physics

BEFORE
1900 A crisis in the understanding of light inspires Max Planck to find a theoretical solution that involves treating light as quantized packets of energy.

1905 Albert Einstein demonstrates the reality of Planck's quantized light through his explanation of the photoelectric effect.

1913 Niels Bohr's model of the atom uses the idea that electrons shifting between energy levels within an atom emit or absorb individual quanta of light (photons).

AFTER
1930s Schrödinger's work, along with that of Paul Dirac and Werner Heisenberg, forms the foundation of modern particle physics.

Erwin Schrödinger was a key figure in the advancement of quantum physics—the science that explains the tiniest levels of subatomic matter. His star contribution was a famous equation that showed how particles moved in waves. It formed the basis of today's quantum mechanics and revolutionized the way we perceive the world. But this revolution did not happen suddenly. The process of discovery was a long one, with many pioneers along the way.

Quantum theory was originally limited to the understanding of light. In 1900, as part of an attempt to solve a troubling problem in theoretical physics known as the "ultraviolet catastrophe," the German physicist Max Planck proposed treating light as though it came in discrete packets, or quanta, of energy. Albert Einstein then took the next step and argued that light quanta were indeed a real physical phenomenon.

Danish physicist Niels Bohr knew that Einstein's idea was saying something fundamental about the nature of light and atoms, and in 1913 used it to solve an old problem—the precise wavelengths of light emitted when certain elements were heated. By modeling the structure of the atom with electrons orbiting in discrete "shells" whose distance from the nucleus determined their energy, Bohr could explain the emission spectra (distribution of light wavelengths) of atoms in terms of photons of energy given off as electrons jumped between orbits. However, Bohr's model lacked a theoretical explanation, and could only predict the emissions from hydrogen, the simplest atom.

Wavelike atoms?

Einstein's idea had breathed new life into the old theory of light as streams of particles, even though light had also been proved, through Thomas Young's double-slit experiment, to behave as a wave. The puzzle of how light could possibly be both particle and wave received a new twist in 1924 from

1927 saw a gathering of greats at the Solvay Conference of physics in Brussels. Among others are: **1**. Schrödinger, **2**. Pauli, **3**. Heisenberg, **4**. Dirac, **5**. de Broglie, **6**. Born, **7**. Bohr, **8**. Planck, **9**. Curie, **10**. Lorentz, **11**. Einstein.

See also: Thomas Young 110–11 ▪ Albert Einstein 214–21 ▪ Werner Heisenberg 234–35 ▪ Paul Dirac 246–47
Richard Feynman 272–73 ▪ Hugh Everett III 284–85

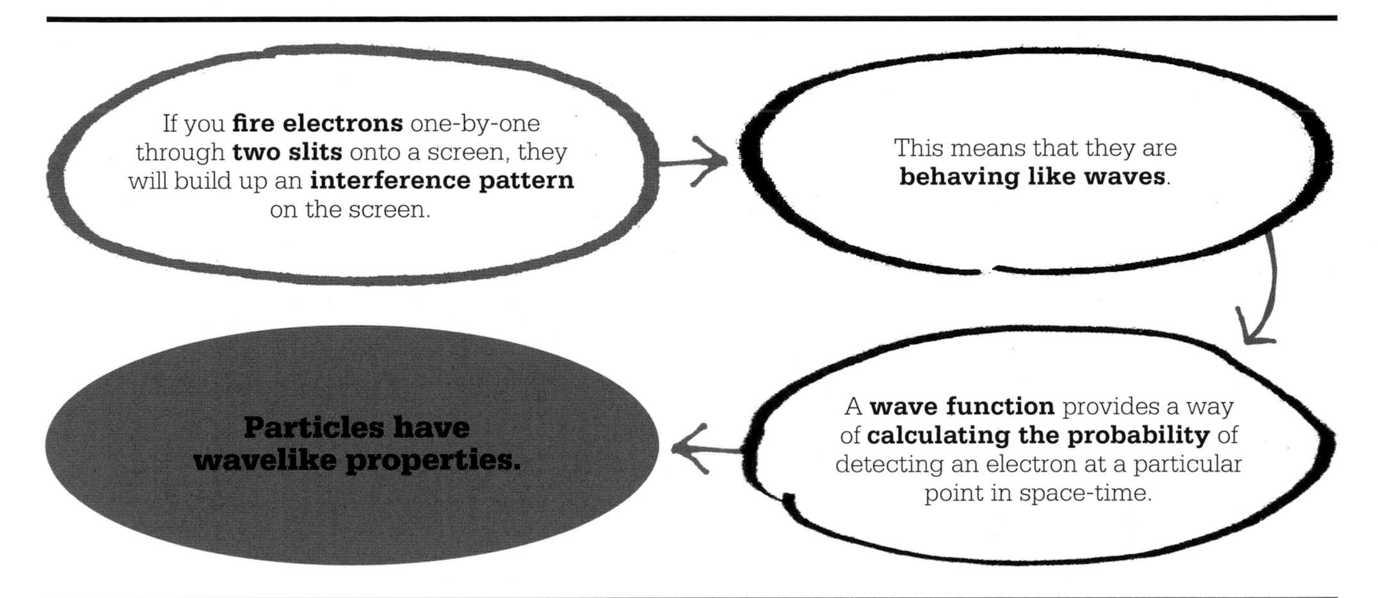

a French PhD student, Louis de Broglie, whose suggestion led the quantum revolution into a dramatic new phase. Not only did de Broglie demonstrate with a simple equation how, in the subatomic world, particles could equally be waves, he also showed how any object, of whatever mass, could behave as a wave to some extent. In other words, if light waves had particle-like properties, then particles of matter—such as electrons—must have wavelike properties.

Planck had calculated the energy of a light photon with the simple equation $E = hv$, where E is the energy of the electromagnetic quanta, v is the wavelength of the radiation involved, and h is a constant, today known as the Planck constant. De Broglie showed that a light photon also has momentum, something normally only associated with particles with mass and given by multiplying the particle's mass with its speed. De Broglie showed that a light photon had a momentum of h divided by its wavelength. However, since he was dealing with particles whose energy and mass might be affected by motion at speeds close to that of light, de Broglie incorporated the Lorentz factor (p.219) into his equation. This produced a more sophisticated version that took into account the effects of relativity.

De Broglie's idea was radical and daring, but it soon had influential supporters, including Einstein. The hypothesis was also relatively easy to test. By 1927, scientists in two separate laboratories had conducted experiments to show that electrons diffracted and interfered with each other in exactly the same way as photons of light. De Broglie's hypothesis was proved.

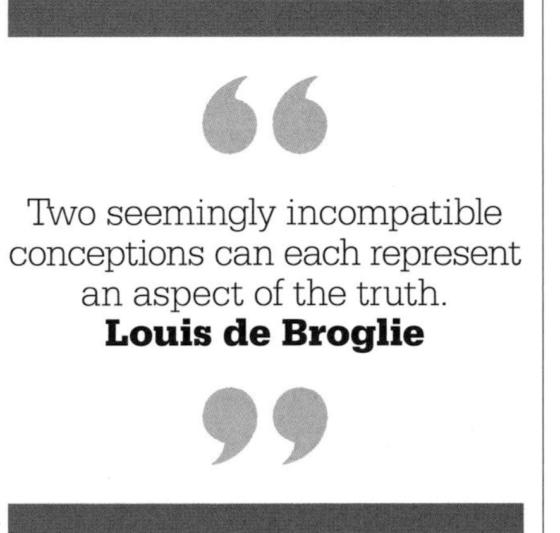

Growing significance

In the meantime, a number of theoretical physicists were sufficiently intrigued by de Broglie's hypothesis to investigate it further. In particular, they wanted to know how the properties of such matter waves could give rise to the pattern of specific energy levels among the electron orbitals of the hydrogen atom proposed by Bohr's model of the atom. De Broglie himself had suggested that the pattern arose because the circumference of each orbital must accommodate a whole number of wavelengths of the matter wave. Since the electron's energy level depends on its distance from the atom's positively charged nucleus, this meant that » only certain distances, and certain energy levels, would be stable. However, de Broglie's solution relied on treating the matter wave as a one-dimensional wave trapped in orbit around the nucleus—a full description would need to describe the wave in three dimensions.

The wave equation

In 1925, three German physicists, Werner Heisenberg, Max Born, and Pascual Jordan, tried to explain the quantum jumps that occurred in Bohr's model of the atom with a method called matrix mechanics, in which the properties of an atom were treated as a mathematical system that could change over time. However, the method could not explain what was actually happening inside the atom, and its obscure mathematical language did not make it very popular.

A year later, an Austrian physicist working in Zurich, Erwin Schrödinger, hit upon a better approach. He took de Broglie's wave-particle duality a step further and began to consider whether there was a mathematical equation of wave motion that would describe how a subatomic particle might move. To formulate his wave equation, he began with the laws governing energy and momentum in ordinary mechanics, then amended them to include the Planck constant and de Broglie's law connecting the momentum of a particle to its wavelength.

When he applied the resulting equation to the hydrogen atom, it predicted exactly the specific energy levels for the atom that had been observed in experiments. The equation was a success. But one awkward issue remained, because no one, not even Schrödinger, knew exactly what the wave equation really described. Schrödinger tried to interpret it as the density of electric charge, but this was not entirely successful. It was Max Born who eventually suggested what it really was—it was a probability amplitude. In other words, it expressed the likelihood of a measurement finding the electron in that particular place. Unlike matrix mechanics, the Schrödinger wave equation or "wave function" was embraced by physicists, although it threw open a whole range of wider questions about its proper interpretation.

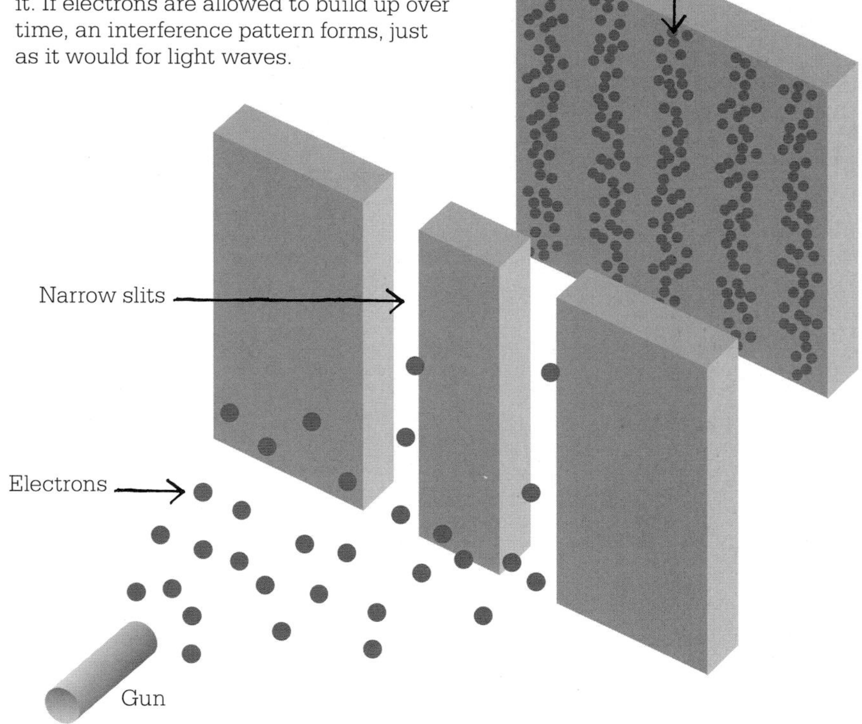

A classic illustration of wave-particle duality involves firing electrons from a "gun" through a barrier with two slits in it. If electrons are allowed to build up over time, an interference pattern forms, just as it would for light waves.

Pauli's exclusion principle

Another important piece of the puzzle fell into place in 1925 courtesy of another Austrian, Wolfgang Pauli. In order to describe why the electrons within an atom did not all automatically fall directly into the lowest possible energy state, Pauli developed the exclusion principle. Reasoning that a particle's overall quantum state could be defined by a certain number of properties, each with a fixed number of possible discrete values, his principle stated that it was impossible for two particles within the same system to have the same quantum state simultaneously.

In order to explain the pattern of electron shells that was apparent from the periodic table, Pauli calculated that electrons must be described by four distinct quantum numbers. Three of these—the principal, azimuthal, and magnetic quantum numbers—define the electron's precise place within the available orbital shells and subshells, with the values of the latter pair limited by the value of the principal number. The fourth number, with two possible values, was needed to explain why two electrons can exist in each subshell with slightly different energy levels. Together, the numbers neatly explained the existence of atomic orbitals that accept 2, 6, 10, and 14 electrons respectively.

Today, the fourth quantum number is known as spin; it is a particle's intrinsic angular momentum (which is created by its rotation as it orbits), and has positive or negative values that are either whole- or half-integer numbers. A few years later, Pauli would demonstrate that values of spin split all particles into two major groups—fermions such as electrons (with half-integer spins), which obey a set of rules known as Fermi–Dirac statistics (pp.246–47), and bosons such as photons (with zero or whole-number spin), which obey different rules known as Bose–Einstein statistics. Only fermions obey the exclusion principle, and this has important implications for the understanding of everything from collapsing stars to the elementary particles that make up the universe.

Schrödinger's success

Combined with Pauli's exclusion principle, Schrödinger's wave equation allowed a new and deeper understanding of the orbitals, shells, and subshells within an atom. Rather than imagining them as classical orbits—well-defined paths on which the electrons circle the nucleus—the wave equation shows that they are actually clouds of probability—doughnut-shaped and lobe-shaped regions in which a particular electron with certain quantum numbers is likely to be found (p.256).

Another major success for Schrödinger's approach was that it offered an explanation for radioactive alpha decay—in which a fully formed alpha particle (consisting of two protons and two neutrons) escapes from an atomic nucleus. According to classical physics, in order to remain intact, the nucleus had to be surrounded by a potential well steep enough to prevent particles escaping from it. (A potential well is a region in space where the potential energy is lower than its surroundings, meaning that it traps particles.) If the well was not sufficiently steep, the nucleus would disintegrate completely. How, then, could the intermittent emissions seen in alpha decay happen while allowing the remaining nucleus to survive intact? The wave equations overcame the problem because they allowed the energy of the alpha particle within the nucleus to vary. Most of the time, its energy would be low enough to keep it trapped, but occasionally it would rise high enough to overcome the »

Schrödinger's equation, in its most general form, shows the development of a quantum system over time. It requires the use of complex numbers.

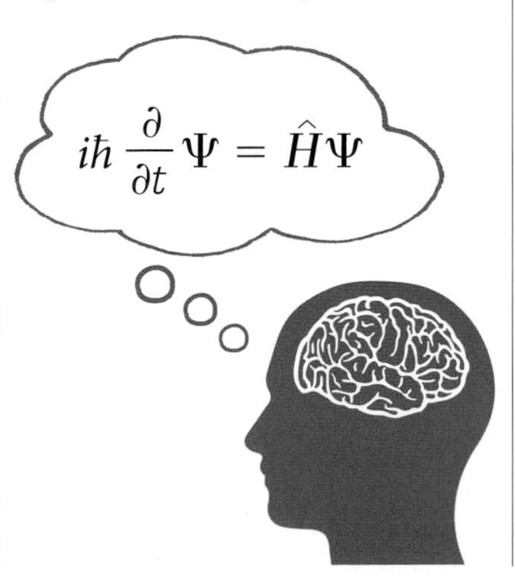

Erwin Schrödinger

Born in Vienna, Austria, in 1887, Erwin Schrödinger studied physics at the University of Vienna, attaining an assistant's post there before serving in World War I. After the war, he moved first to Germany, and then to the University of Zurich, Switzerland, where he did his most important work, immersing himself in the emerging field of quantum physics. In 1927, he returned to Germany, and succeeded Max Planck at the Humboldt University of Berlin.

Schrödinger was a vocal opponent of the Nazis, and left Germany for a post at Oxford University in 1934. It was there that he learned he had been awarded the 1933 Nobel Prize in Physics, with Paul Dirac, for the quantum wave equation. By 1936, he was back in Austria, but had to flee again following Germany's annexation of the country. He settled in Ireland for the rest of his career before retiring to Austria in the 1950s.

Key works

1920 *Color Measurement*
1926 *Quantization as an Eigenvalue Problem*

wall and escape (an effect now known as quantum tunneling). The probability predictions of the wave equation matched the unpredictable nature of the radioactive decay.

Uncertainty principle

The great debate that shaped the development of quantum physics during the middle years of the 20th century (and remains essentially unresolved today) surrounded what the wave function actually meant for reality. In an echo of the Planck/Einstein debate two decades previously, de Broglie saw his and Schrödinger's equations as mere mathematical tools for describing movement: for de Broglie, the electron was still essentially a particle—just one that had a wave property governing its motion and location. For Schrödinger, however, the wave equation was far more fundamental—it described the way in which the properties of the electron were physically "smeared out" across space. Opposition to Schrödinger's approach inspired Werner Heisenberg to develop another of the century's great ideas—the uncertainty principle (pp.234–35). This was a realization that the wave function meant that a particle can never be "localized" to a point in space and at the same time have a defined wavelength. The more accurately a particle's position was pinned down, for example, the harder its momentum was to measure. Thus, particles defined by a quantum wave function existed in a general state of uncertainty.

God knows I am no friend of probability theory, I have hated it from the first moment when our dear friend Max Born gave it birth.
Erwin Schrödinger

Dane Niels Bohr (left) collaborated with Werner Heisenberg, to formulate the Copenhagen interpretation of Schrödinger's wave function.

The road to Copenhagen

Measuring the properties of a quantum system always revealed the particle to be in one location, rather than in its wavelike smear. On the scale of classical physics and everyday life, most situations involved definite measurements and definite outcomes, rather than myriad overlapping possibilities. The challenge of reconciling quantum uncertainty with reality is called the measurement problem, and various approaches to it have been put forward, known as interpretations.

The most famous of these is the Copenhagen interpretation, devised by Niels Bohr and Werner Heisenberg in 1927. This states simply that it is the very interaction between the quantum system and a large-scale, external observer or apparatus (subject to the classical laws of physics) that causes the wave function to "collapse" and a definite outcome to arise. This interpretation is perhaps the most widely (though not universally) accepted, and appears to be borne out by experiments such as electron diffraction and the double-slit experiment for light waves. It is possible to devise an experiment that reveals the wavelike aspects of light or electrons, but impossible to record the properties of individual particles in the same apparatus.

However, while the Copenhagen interpretation seems reasonable when dealing with small-scale systems such as particles, its implication that nothing is determined until it is measured troubled many physicists. Einstein famously commented that "God does not throw dice," while Schrödinger devised a thought experiment to illustrate what he viewed as a ridiculous situation.

Schrödinger's cat

Taken to its logical conclusion, the Copenhagen interpretation resulted in a seemingly absurd paradox. Schrödinger imagined a cat sealed in a box that contains a vial of poison linked to a radioactive source. If the source decays and emits a particle of radiation, a mechanism will release a hammer that breaks the vial of poison. According to the Copenhagen interpretation, the radioactive source remains in its wave function form (as a so-called superposition of two possible outcomes) until it is observed. But if that is the case, the same has to be said of the cat.

New interpretations

Dissatisfaction with apparent paradoxes such as Schrödinger's cat has spurred scientists to develop various alternative interpretations of quantum mechanics. One of the best known is the "Many Worlds Interpretation" put forward in 1956 by American physicist Hugh Everett III. This resolved the paradox by suggesting that during any quantum event, the universe splits into mutually unobservable alternate histories for each of the possible outcomes. In other words, Schrödinger's cat would both live and die.

The "Consistent Histories" approach addresses the problem in a rather less radical way, using complex mathematics to generalize the Copenhagen interpretation. This avoids the issues around the collapse of the wave function, but instead allows probabilities to be assigned to various scenarios, or "histories," on both a quantum and classical scale. The approach accepts that only one of these histories eventually conforms to reality, but does not allow prediction of which outcome that will be—instead it simply describes how quantum physics can give rise to the universe we see without wave function collapse.

The ensemble, or statistical, approach is a minimalist mathematical interpretation that was favored by Einstein. The de Broglie–Bohm theory, which developed from de Broglie's initial reaction to the wave equation, is an attempt at a strictly causal, rather than probabilistic, explanation, and postulates the existence of a hidden "implicate" order to the universe. The transactional approach involves waves traveling both forward and backward in time.

Perhaps the most intriguing possibility of all, however, is one that verges on the theological. Working in the 1930s, Hungarian-born mathematician John von Neumann concluded that the measurement problem implied that the entire universe is subject to an all-encompassing wave equation known as the universal wave function, and that it is constantly collapsing as we measure its various aspects. Von Neumann's colleague and countryman Eugene Wigner took the theory and expanded it to suggest that it was not simply interaction with large-scale systems (as in the Copenhagen interpretation) that caused the wave function to collapse—it was the presence of intelligent consciousness itself. ■

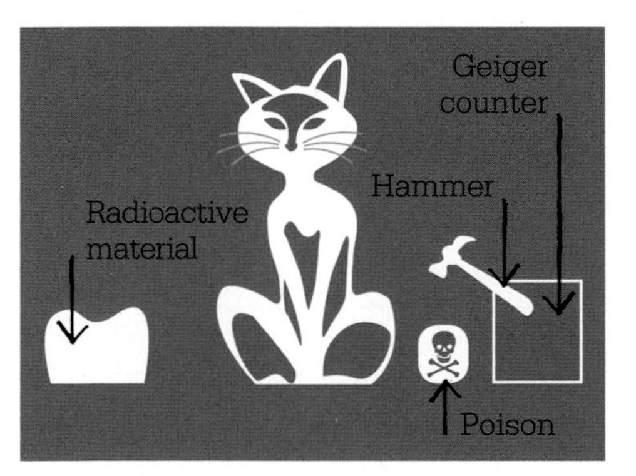

A cat inside a sealed box remains alive as long as a radioactive source in the box does not decay.

If the source decays, it releases poison and the cat dies.

We must measure the system to find out whether the source has decayed. Until then, we must think of the cat as both dead and alive.

Schrödinger's thought experiment produces a situation in which, according to a strict reading of the Copenhagen interpretation, a cat is both alive and dead at the same time.

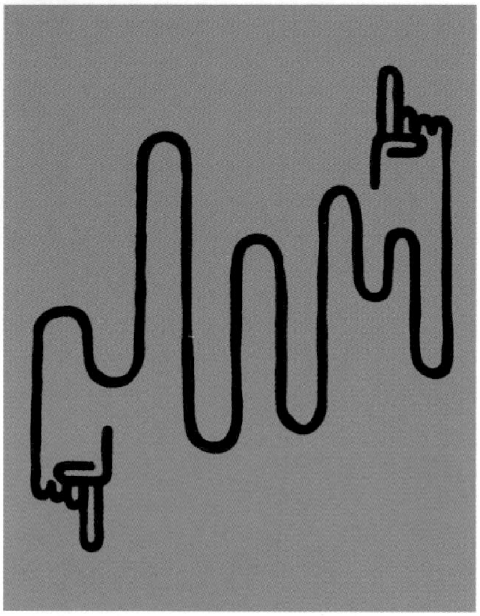

UNCERTAINTY IS INEVITABLE

WERNER HEISENBERG (1901–1976)

IN CONTEXT

BRANCH
Physics

BEFORE
1913 Niels Bohr uses the concept of quantized light to explain the specific energy levels associated with electrons inside atoms.

1924 Louis de Broglie proposes that just as light can exhibit particle-like properties so, on the smallest scale, particles might sometimes show wavelike behavior.

AFTER
1927 Heisenberg and Bohr put forward the highly influential Copenhagen interpretation of the way that quantum-level events affect the large-scale (macroscopic) world.

1929 Heisenberg and Wolfgang Pauli work on the development of quantum field theory, whose foundations have been laid by Paul Dirac.

Following Louis de Broglie's suggestion in 1924 that on the smallest scales of matter, subatomic particles could display wavelike properties (pp.226–33), a number of physicists turned their attention to the question of understanding how the complex properties of an atom could arise from the interaction of "matter waves" associated with its constituent particles. In 1925, German scientists Werner Heisenberg, Max Born, and Pascual Jordan used "matrix mechanics" to model the hydrogen atom's development over time. This approach was later supplanted by Erwin Schrödinger's wave function. Working with Danish physicist Niels Bohr, Heisenberg built on Schrödinger's work to develop the "Copenhagen interpretation" of the way that quantum systems, governed by the laws of probability, interact with the large-scale world. One key element of this is the "uncertainty principle," which limits the accuracy to which we can determine properties in quantum systems.

The uncertainty principle arose as a mathematical consequence of matrix mechanics. Heisenberg realized that his mathematical method would not allow certain pairs of properties to be determined simultaneously with absolute

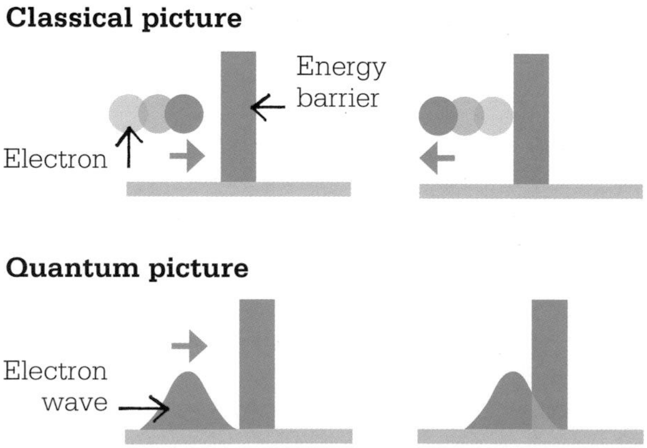

Quantum tuneling is explained by Heisenberg's principle. There is a nonzero chance that an electron can pass through a barrier even if it appears to have too little energy to do so.

See also: Albert Einstein 214–21 ▪ Erwin Schrödinger 226–33 ▪ Paul Dirac 246–47 ▪ Richard Feynman 272–73 ▪ Hugh Everett III 284–85

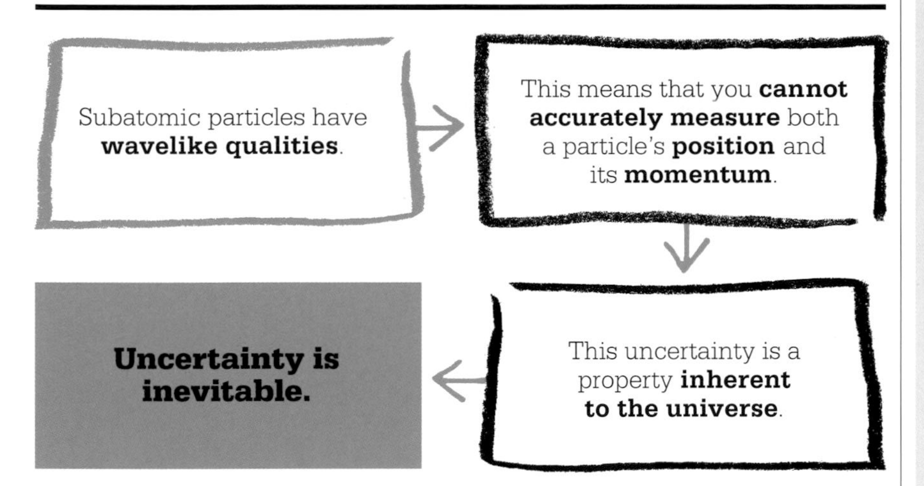

precision. For example, the more accurately one measures a particle's position, the less accurately one can determine its momentum, and vice versa. Heisenberg found that for these two properties in particular, the relationship could be written as:

$$\Delta x \Delta p \geq \hbar/2$$

where Δx is the uncertainty of position, Δp the uncertainty of momentum, and $\hbar$ is a modified version of Planck's constant (p.202).

An uncertain universe

The uncertainty principle is often described as a consequence of quantum-scale measurements—for example, it is sometimes said that determining a subatomic particle's position involves the application of a force of some sort that means its kinetic energy and momentum are less well defined. This explanation, put forward at first by Heisenberg himself, led various scientists including Einstein to spend time devising thought experiments that might obtain a simultaneous and accurate measurement of position and momentum by some form of "trickery." However, the truth is far stranger—it turns out that uncertainty is an inherent feature of quantum systems.

A helpful way of thinking about the issue is to consider the matter waves associated with the particles: in this situation, the particle's momentum affects its overall energy and therefore its wavelength—but the more tightly we pin down the particle's position, the less information we have about its wave function, and therefore about its wavelength. Conversely, accurately measuring the wavelength requires us to consider a broader region of space, and therefore sacrifices information about the particle's precise location. Such ideas might seem strangely at odds with those we experience in the large-scale world, but they have nevertheless been proved real by many experiments, and form an important foundation of modern physics. The uncertainty principle explains seemingly strange real-life phenomena such as quantum tunneling, in which a particle can "tunnel" through a barrier even if its energy suggests that it should not be able to. ■

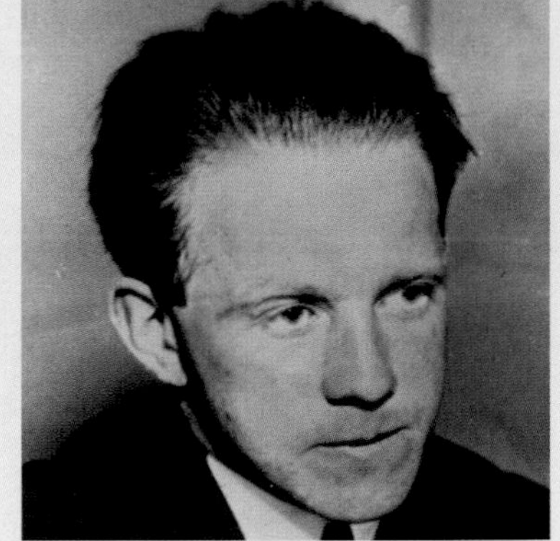

Werner Heisenberg

Born in the southern German town of Würzburg in 1901, Werner Heisenberg studied mathematics and physics at the universities of Munich and Göttingen, where he studied under Max Born and met his future collaborator Niels Bohr for the first time.

He is best known for his work on the Copenhagen interpretation and the uncertainty principle, but Heisenberg also made important contributions to quantum field theory and developed his own theory of antimatter. Awarded the Nobel Prize in Physics in 1932, he became one of its youngest recipients, and his stature enabled him to speak out against the Nazis after they seized power the following year. However, he chose to stay in Germany and led the country's nuclear energy program during World War II.

Key works

1927 *Quantum Theoretical Re-interpretation of Kinematic and Mechanical Relations*
1930 *The Physical Principles of the Quantum Theory*
1958 *Physics and Philosophy*

THE UNIVERSE IS BIG… AND GETTING BIGGER

EDWIN HUBBLE (1889–1953)

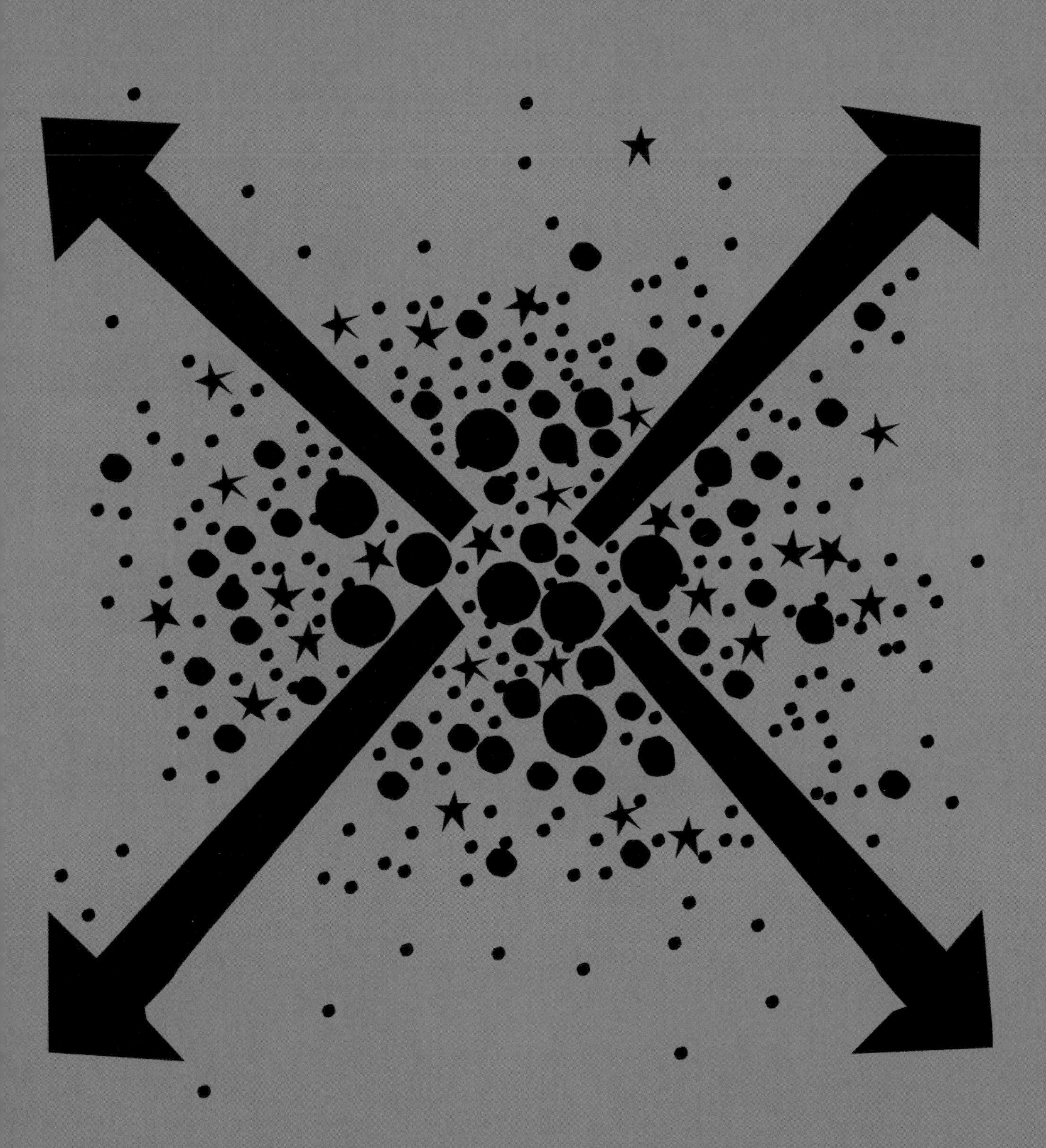

IN CONTEXT

BRANCH
Cosmology

BEFORE
1543 Nicolaus Copernicus concludes that Earth is not the center of the universe.

17th century The changing view of stars offered by Earth's orbit around the Sun gives rise to the parallax method for measuring stellar distances.

19th century Improvements to telescopes pave the way for the study of starlight and the rise of astrophysics.

AFTER
1927 Georges Lemaître proposes that the universe can be traced back to a single point of origin.

1990s Astronomers discover that the expansion of the universe is accelerating, driven by a force known as dark energy.

By the early 20th century, ideas about the scale of the universe divided astronomers into two schools of thought—those who believed that the Milky Way galaxy was, generally speaking, its entire extent, and those who thought that the Milky Way could be just one galaxy among countless others. Edwin Hubble was to solve the puzzle, and show that the universe is much larger than anyone imagined.

Key to the debate was the nature of "spiral nebulae." Today, a nebula is the term used for an interstellar cloud of dust and gas, but at the time of this debate, it was the name used for any amorphous cloud of light, including objects that were later found to be galaxies beyond the Milky Way.

> There is a simple relation between the brightness of the variables and their periods.
> **Henrietta Leavitt**

As telescopes improved dramatically during the 19th century, some of the objects catalogued as nebulae began to reveal distinctive spiral features. At the same time, the development of spectroscopy (the study of the interaction between matter and radiated energy) suggested that these spirals were in fact made up of countless individual stars, blending seamlessly together.

The distribution of these nebulae was interesting too—unlike other objects that clustered together in the plane of the Milky Way, they were more common in the dark skies away from the plane. As a result, some astronomers adopted an idea from the German philosopher Immanuel Kant, who in 1755 suggested that nebulae were "island universes"—systems similar to the Milky Way but vastly more distant, and only visible where the distribution of material in our galaxy permits clear views into what we now call intergalactic space. Those who continued to believe that the universe was far

Edwin Hubble

Born in Marshfield, Missouri, in 1889, Edwin Powell Hubble had a fiercely competitive nature that manifested itself in his youth as a gifted athlete. Despite his interest in astronomy, he followed his father's wishes and studied law, but at 25 years old, after his father's death, he resolved to follow his early passion. His studies were interrupted by service in World War I, but after his return to the United States he began to work at the Mount Wilson Observatory. There he did his most important work, publishing his study on "extragalactic nebulae" in 1924–25, and his proof of cosmic expansion in 1929. In later years, he campaigned for astronomy to be recognized by the Nobel Prize Committee. The rules were only changed after his death in 1953 and so he was never awarded the prize himself.

Key works

1925 *Cepheid Variables in Spiral Nebulae*
1929 *A Relation Between Distance and Radial Velocity among Extra-galactic Nebulae*

See also: Nicolaus Copernicus 34–39 ▪ Christian Doppler 127 ▪ Georges Lemaître 242–45

Henrietta Leavitt received little recognition in her lifetime, but her discoveries relating to Cepheid variable stars were the key that allowed astronomers to measure the distance from Earth to faraway galaxies.

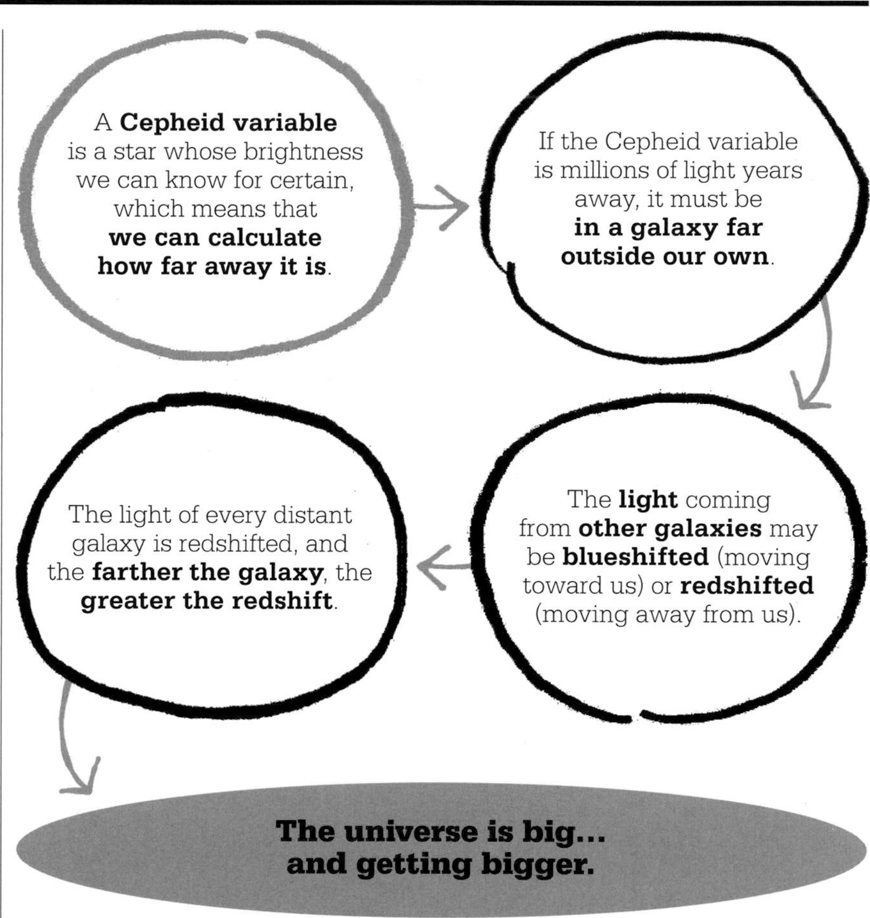

more limited in extent argued that the spirals might be suns or solar systems in the process of formation, in orbit around the Milky Way.

Stars with a pulse

The answers to this long-standing puzzle came in several stages, but perhaps the most important was the establishment of an accurate means of measuring the distance to stars. The breakthrough came with the work of Henrietta Swan Leavitt, one of the team of female astronomers at Harvard University who were analyzing the properties of starlight.

Leavitt was intrigued by the behavior of variable stars. These were stars whose brightness appeared to fluctuate, or pulse, because they periodically expanded and contracted as they neared the end of their lives. She began to study photographic plates of the Magellanic Clouds, two small patches of light visible from the southern sky that look like isolated "clumps" of the Milky Way. Each of the clouds, she found, contained huge numbers of variable stars, and by comparing them across many different plates, she not only saw that their light was varying in a regular cycle, she could also figure out the period of the cycle.

By concentrating on these small, faint, isolated star clouds, Leavitt could safely assume that the stars within them were all at more or less the same distance from Earth. Though she could not know the distance itself, this was still enough to assume that differences in the "apparent magnitude" (observed brightness) of the stars were an indication of differences in their "absolute magnitude" (actual brightness). Publishing her first results in 1908, Leavitt noted in passing that some stars seemed to show a relationship between their variability period and their absolute magnitude, but it took another four years for her to figure out what this relationship was. It turned out that, for a certain type of variable star known as a Cepheid variable, stars with greater luminosity have longer variability periods.

Leavitt's "period-luminosity" law would prove the key to unlocking the scale of the »

> We are reaching into space, farther and farther, until, with the faintest nebulae that can be detected…we arrive at the frontier of the known universe.
> **Edwin Hubble**

universe—if you could figure out the star's absolute magnitude from its variability period, then the star's distance from Earth could be calculated from its apparent magnitude. The first step in figuring this out was to calibrate the scale, which was done in 1913 by Swedish astronomer Ejnar Hertzsprung. He figured out the distances to 13 relatively nearby Cepheids using the parallax method (p.39). Cepheids were immensely bright—thousands of times more luminous than our Sun (in modern terminology they are "yellow supergiants"). In theory, then, they were an ideal "standard candle"—stars whose brightness could be used to measure huge cosmic distances. But despite the best efforts of astronomers, Cepheids within the spiral nebulae remained stubbornly elusive.

The Great Debate

In 1920, the Smithsonian Museum in Washington DC hosted a debate between the two rival cosmological schools, hoping to settle the issue of the scale of the universe once and for all. Respected Princeton astronomer Harlow Shapley spoke for the "small universe" side. He had been the first to use Leavitt's work on Cepheids to measure the distance to globular clusters (dense star clusters in orbit around the Milky Way), and discovered that they were typically several thousand light years away. In 1918, he had used RR Lyrae stars (fainter stars that behave like Cepheids) to estimate the size of the Milky Way and show that the Sun was nowhere near its center. His arguments appealed to public scepticism toward notions of an enormous universe with many galaxies, but also cited specific evidence (later to be proved inaccurate), such as reports that over many years some astronomers had actually observed the spiral nebulae rotating. For this to be true without parts of the nebula exceeding the speed of light, they must be relatively small.

The "island universe" supporters were represented by Heber D. Curtis of the University of Pittsburgh's Allegheny Observatory. He based his arguments on comparisons between the rates of bright "nova" explosions in distant spirals and in our own Milky Way. Novae are very bright star explosions that can serve as distance indicators.

Curtis also cited the evidence of another, crucial factor—the high redshift exhibited by many spiral nebulae. This phenomenon had been discovered by Vesto Slipher of the Flagstaff Observatory, Arizona, in 1912—apparent through distinctive shifts in the pattern of a nebula's spectral lines toward the red end of the spectrum. Slipher, Curtis, and many others believed that they were caused by the

By measuring the light from Cepheid variable stars in the Andromeda nebula, Hubble established that Andromeda was 2.5 million light years way—and was a galaxy in its own right.

Doppler effect (a change in the wavelength of light due to relative motion between source and observer), and therefore indicated that the nebulae were moving away from us at very high speeds—far too fast for the Milky Way's gravity to keep hold of them.

Measuring the universe

By 1922–23, Edwin Hubble and Milton Humason of California's Mount Wilson Observatory were in a position to end the mystery once and for all. Using the observatory's new 100 in (2.5 m) Hooker Telescope (the largest in the world at that time), they set out to identify Cepheid variables shining within the spiral nebulae, and this time they were successful in finding Cepheids in many of the largest and brightest nebulae.

Hubble then plotted their periods of variability and therefore their absolute magnitude. From this, a simple comparison to a star's apparent magnitude revealed its distance, producing figures that were typically millions of light years. This proved conclusively that the spiral nebulae were really huge, independent star systems, far beyond the Milky Way and rivaling it in size. Spiral nebulae are now correctly called spiral galaxies.

> Equipped with his five senses, man explores the universe around him and calls the adventure science.
> **Edwin Hubble**

As if this revolution in the way we see the universe were not enough, Hubble then went on to look at how galaxy distances related to the redshifts already discovered by Slipher—and here he found a remarkable relationship. By plotting the distances for more than 40 galaxies against their redshifts, he showed a roughly linear pattern: the farther away a galaxy is, on average, the greater its redshift and therefore the faster it is receding from Earth. Hubble immediately realized that this could not be because our galaxy is uniquely unpopular, but must be the result of a general cosmic expansion—in other words, space itself is expanding and carrying every single galaxy with it. The wider the separation between two galaxies, the faster the space between them will expand. The rate of expansion of space soon became known as the "Hubble Constant." It was conclusively measured in 2001 by the space telescope bearing Hubble's name.

Long before then, Hubble's discovery of the expanding universe had given rise to one of the most famous ideas in the history of science—the Big Bang theory (pp.242–45). ■

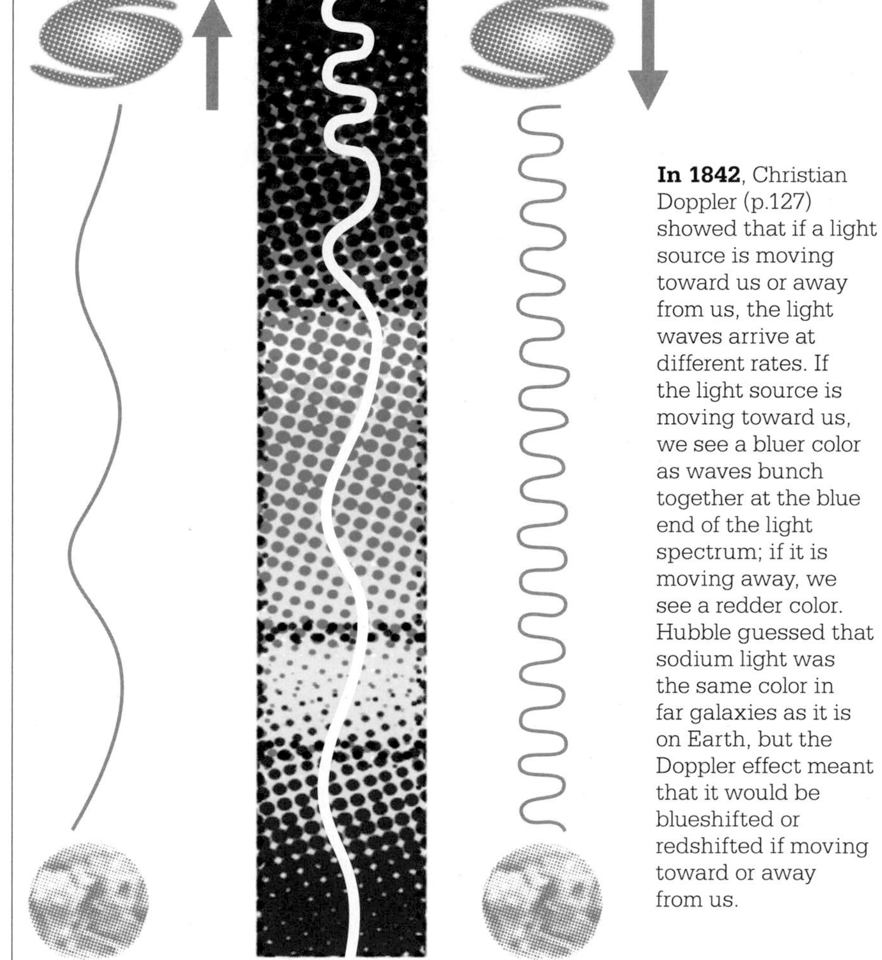

In 1842, Christian Doppler (p.127) showed that if a light source is moving toward us or away from us, the light waves arrive at different rates. If the light source is moving toward us, we see a bluer color as waves bunch together at the blue end of the light spectrum; if it is moving away, we see a redder color. Hubble guessed that sodium light was the same color in far galaxies as it is on Earth, but the Doppler effect meant that it would be blueshifted or redshifted if moving toward or away from us.

THE RADIUS OF SPACE BEGAN AT ZERO

GEORGES LEMAÎTRE (1894–1966)

IN CONTEXT

BRANCH
Astronomy

BEFORE
1912 US astronomer Vesto Slipher discovers the high redshifts of spiral nebulae, suggesting they are moving away from Earth at high speeds.

1923 Edwin Hubble confirms that the spiral nebulae are distant, independent galaxies.

AFTER
1980 US physicist Alan Guth proposes a brief period of dramatic inflation in the early universe to produce the conditions we see today.

1992 The COBE (Cosmic Background Explorer) satellite detects tiny ripples in the cosmic microwave background radiation (CMBR)—hints of the first structure that emerged in the early universe.

The idea that the universe began with a Big Bang, expanding from a tiny, superdense, and extremely hot point in space, is the basis of modern cosmology, and one that is often said to have originated with Edwin Hubble's 1929 discovery of cosmic expansion. But the precursors of the theory predate Hubble's breakthrough by several years, and first sprang from interpretations of Albert Einstein's theory of general relativity as it applied to the universe as a whole.

When formulating his theory, Einstein drew on the available evidence of the time to assume that the universe was static—

See also: Isaac Newton 62–69 ▪ Albert Einstein 214–21 ▪ Edwin Hubble 236–41 ▪ Fred Hoyle 270

Since the Big Bang 13.8 billion years ago, the expansion of the universe has been through different phases. There was an initial period of rapid expansion known as inflation. After that, expansion slowed, then started to speed up once more.

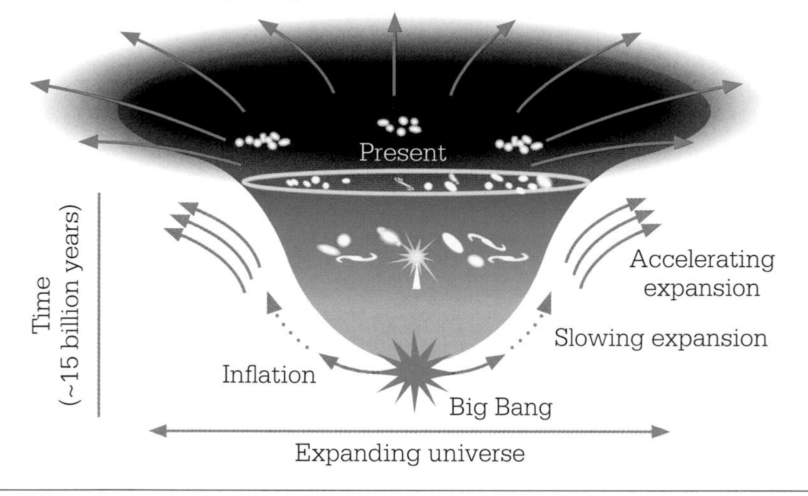

neither expanding nor contracting. General relativity indicated that the universe should collapse under its own gravity, so Einstein fudged his own equations by adding a term known as the cosmological constant. Einstein's constant mathematically counteracted the gravitational contraction to produce the presumed static universe.

The first stages of the expansion consisted of a rapid expansion determined by the mass of the initial atom, almost equal to the present mass of the universe.
Georges Lemaître

Famously, Einstein later called the constant his greatest mistake, but even at the time he proposed it there were some who found it unsatisfactory. The Dutch physicist Willem de Sitter and Russian mathematician Alexander Friedmann independently suggested a solution to general relativity in which the universe was expanding, and, in 1927, Belgian astronomer and priest Georges Lemaître reached the same conclusion, two years ahead of Hubble's observational proof.

Beginning in fire

In an address to the British Association in 1931, Lemaître took the idea of cosmic expansion to its logical conclusion, suggesting that the universe had sprung from a single point that he called the "primeval atom." The response to this radical idea was mixed.

The astronomical establishment of the time was attached to the idea of an eternal universe »

Georges Lemaître

Born in Charleroi, Belgium, in 1894, Lemaître studied civil engineering at the Catholic University of Louvain and served in World War I before returning to academia, where he studied physics and mathematics as well as theology. From 1923, he studied astronomy in Britain and the United States. On his return to Louvain in 1925 as a lecturer, Lemaître began to develop his theory of an expanding universe as an explanation for the redshifts of the extragalactic nebulae.

First published in 1927, in a little-read Belgian journal, Lemaître's ideas took off after he published an English translation with Arthur Eddington. He lived until 1966—long enough to see proof that his ideas were correct with the discovery of the cosmic microwave background radiation (CMBR).

Key works

1927 *A Homogeneous Universe of Constant Mass and Growing Radius Accounting for the Radial Velocity of Extragalactic Nebulae*
1931 *The Evolution of the Universe: Discussion*

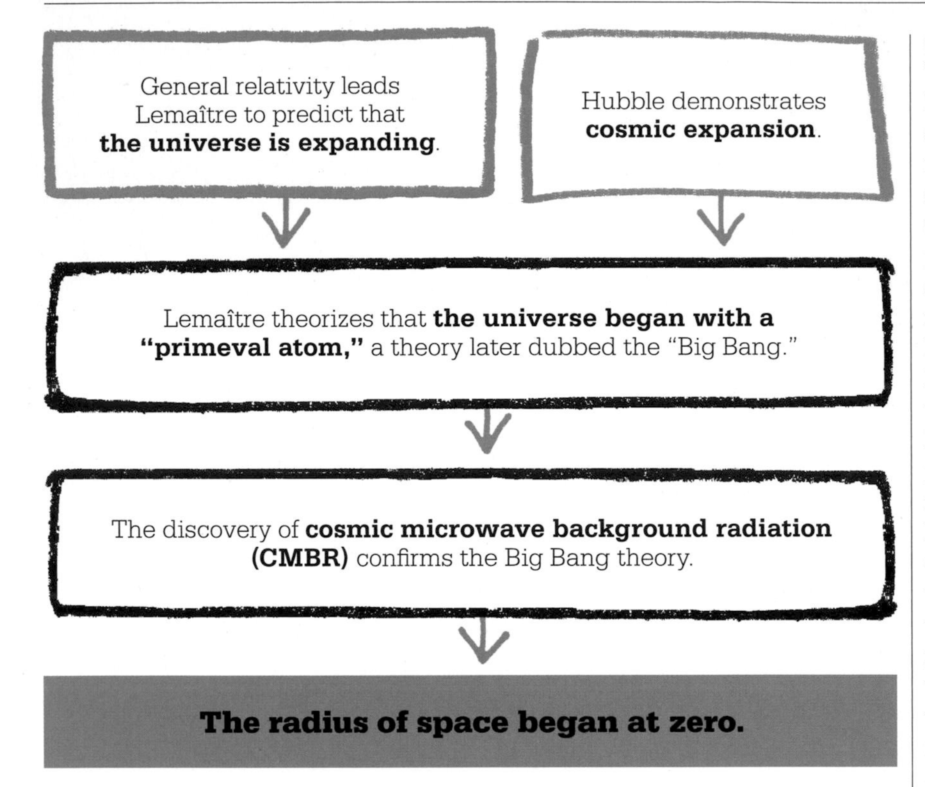

without end or beginning, and the prospect of a distinct point of origin (especially when proposed by a Catholic priest) was seen as introducing an unnecessary religious element into cosmology.

However, Hubble's observations were undeniable, and some kind of model was needed to explain the expanding universe. Numerous theories were put forward in the 1930s, but by the late 1940s, just two remained in play—Lemaître's primeval atom, and the rival "steady state" model, in which matter was continuously created as the universe expanded. British astronomer Fred Hoyle was the champion of the steady state idea. In 1949, Hoyle scornfully referred to the rival theory as a "Big Bang." The name stuck.

Making the elements

By the time Hoyle had inadvertently named the theory, a persuasive piece of evidence in favor of Lemaître's hypothesis had been published, tipping the balance away from a steady state universe. This was a 1948 paper written by Ralph Alpher and George Gamow of the Johns Hopkins University in the US. It was called *The Origin of Chemical Elements*, and described in detail how subatomic particles and lightweight chemical elements could have been produced from the raw energy of the Big Bang, in accordance with Einstein's equation $E = mc^2$. But this theory, later known as Big Bang nucleosynthesis, explained a process that could form only the four lightest elements—hydrogen, helium, lithium, and beryllium. Only later was it discovered that the heavier elements of the universe are the product of stellar nucleosynthesis (a process that takes place inside stars). Ironically, the evidence showing how stellar nucleosynthesis worked was to be developed by Fred Hoyle.

Nevertheless, there was still no direct observational evidence to determine the truth of either the Big Bang or a steady state universe. Early attempts to test the theories were made in the 1950s using a basic radio telescope known as the Cambridge Interferometer. These tests relied on a simple principle: if the steady state theory was true, then the

Tiny variations have been found in the cosmic microwave background radiation—the different colors in this image show temperature differences of less than 400 millionths of a Kelvin.

Arno Penzias and Robert Wilson detected the background radiation by accident. At first, they thought the interference had been caused by bird droppings on their radio antenna.

universe must be essentially uniform in both time and space; but if it originated 10–20 billion years ago, as the Big Bang theory suggested, and evolved throughout its history, then distant reaches of the universe, whose radiation had taken billions of years to reach Earth, should appear substantially different. (This cosmic time machine effect, whereby we see more distant celestial objects as they were in the distant past, is known as "lookback time.") By measuring the number of distant galaxies emitting radiation above a certain brightness, it should be possible to distinguish between the two scenarios.

The first of the Cambridge experiments delivered a result that seemed to support the Big Bang. However, problems were discovered with the radio detectors, so the results had to be disregarded. Later results proved more equivocal.

Traces of the Big Bang

Fortunately, the question soon resolved itself by other means. As early as 1948, Alpher and his colleague Robert Herman had predicted that the Big Bang would have left a residual heating effect throughout the universe. According to the theory, when the universe was about 380,000 years old, it had cooled enough to become transparent, allowing light photons to travel freely through space for the first time. The photons that existed at this time had been propagating through space ever since, growing longer and redder as space expanded. In 1964, Robert Dicke and his colleagues at Princeton University set out to build a radio telescope that could detect this faint signal, which they thought would take the form of low-energy radio waves. However, they were ultimately beaten to the prize by Arno Penzias and Robert Wilson, two engineers working at the nearby Bell Telephone Laboratories. Penzias and Wilson had built a radio telescope for satellite communication, but found themselves plagued by an unwanted background signal that they could not eliminate. Coming from all over the sky, it corresponded to microwave emission from a body at a temperature of 3.5K—just 6°F (3.5°C) above absolute zero. When Bell Labs contacted Dicke to ask for help with their problem, Dicke realized that they had found the remnants of the Big Bang—now known as the cosmic microwave background radiation (CMBR).

The discovery that the CMBR permeates the universe—a phenomenon for which the steady state theory had no explanation—decided the case in favor of the Big Bang. Subsequent measurements have shown that the CMBR's true average temperature is about 2.73K, and high-precision satellite measurements have revealed minute variations in the signal that allow us to study conditions in the universe back to 380,000 years after the Big Bang.

Later developments

Despite being proved correct in principle, the Big Bang theory has undergone many transformations since the 1960s to match it to our growing understanding of the universe. Among the most significant are the introduction of dark matter and dark energy to the story, and the addition of a violent growth spurt in the instant after creation, known as Inflation. The events that triggered the Big Bang remain beyond our reach but measurements of the rate of cosmic expansion, aided by instruments such as the Hubble Space Telescope, now allow us to pin down the epoch of cosmic creation with great accuracy—the universe came into existence 13.798 billion years ago, give or take 0.037 billion years. Various theories exist about the future of the universe, but many think that it is set to continue expanding until it reaches a state of thermodynamic equilibrium, or "heat death," in which matter has disintegrated into cold subatomic particles, in around 10^{100} years' time. ■

EVERY PARTICLE OF MATTER HAS AN ANTIMATTER COUNTERPART

PAUL DIRAC (1902–1984)

IN CONTEXT

BRANCH
Physics

BEFORE
1925 Werner Heisenberg, Max Born, and Pascual Jordan develop matrix mechanics to describe the wavelike behavior of particles.

1926 Erwin Schrödinger develops a wave function describing the change in an electron over time.

AFTER
1932 The existence of the positron, the antiparticle to the electron, is confirmed by Carl Anderson.

1940s Richard Feynman, Sin-Itiro Tomonaga, and Julian Schwinger develop quantum electrodynamics—a mathematical way to describe the interaction between light and matter, which fully unites quantum theory with special relativity.

Dirac corrects Schrödinger's **wave equation** to take into account **relativistic effects**.

↓

Dirac's **new equation** predicts the existence of **antimatter**.

↓

Antimatter is subsequently **discovered**, confirming Dirac's prediction.

↓

Every particle of matter has an antimatter counterpart.

English physicist Paul Dirac contributed a huge amount to the theoretical framework of quantum physics in the 1920s, but is probably best known today for predicting the existence of antiparticles through mathematics.

Dirac was a postgraduate student at Cambridge University when he read Werner Heisenberg's groundbreaking paper on matrix mechanics, which described how particles jump from one quantum state to another. Dirac was one of the few people capable of grasping the paper's difficult mathematics, and noticed parallels between Heisenberg's equations and parts of the classical (pre-quantum) theory of particle motion known as Hamiltonian mechanics. This allowed Dirac to develop a method by which classical systems could be understood on a quantum level.

One early result of this work was a derivation of the idea of quantum spin. Dirac formulated a set of rules now known as "Fermi-Dirac statistics" (since they were also independently found by Enrico Fermi). Dirac named particles such as electrons that have a half-integer spin value "fermions," after Fermi. The rules describe how large

See also: James Clerk Maxwell 180–85 ▪ Albert Einstein 214–21 ▪ Erwin Schrödinger 226–33 ▪ Werner Heisenberg 234–35 ▪ Richard Feynman 272–73

numbers of fermions interact with one another. In 1926, Dirac's PhD supervisor Ralph Fowler used his statistics to calculate the behavior of a collapsing stellar core and explain the origin of superdense white dwarf stars.

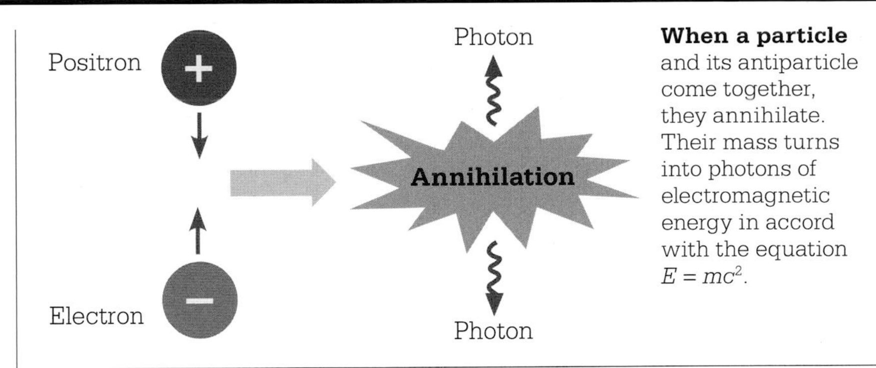

When a particle and its antiparticle come together, they annihilate. Their mass turns into photons of electromagnetic energy in accord with the equation $E = mc^2$.

Quantum field theory

While much of schoolbook physics focuses on the properties and dynamics of individual particles and bodies under the influence of forces, a deeper understanding can be gained by developing field theories. These describe the way that forces make their influence felt across space. The importance of fields as independent entities was first recognized in the mid-19th century by James Clerk Maxwell while he was developing his theory of electromagnetic radiation. Einstein's general relativity is another example of a field theory.

Dirac's new interpretation of the quantum world was a quantum field theory. In 1928, it allowed him to produce a relativistic version of Schrödinger's wave equation for the electron (that is, one that could take into account the effects of particles moving close to the speed of light, and therefore model the quantum world more accurately than Schrödinger's nonrelativistic equation). The so-called Dirac equation also predicted the existence of particles with identical properties to particles of matter but with opposite electric charge. They were dubbed "antimatter" (a term that had been bandied around in wilder speculations since the late 19th century).

The antielectron particle, or positron, was experimentally confirmed by US physicist Carl Anderson in 1932, detected first in cosmic rays (high-energy particles showered into Earth's atmosphere from deep space), and then in certain types of radioactive decay. Since then, antimatter has become a subject for intense physical research, and also beloved of science-fiction writers (particularly for its habit of "annihilating" with a burst of energy on contact with normal matter). Perhaps more importantly, however, Dirac's quantum field theory laid the foundations for the theory of quantum electrodynamics brought to fruition by a later generation of physicists. ■

Paul Dirac

Paul Dirac was a mathematical genius who made several key contributions to quantum physics, sharing the Nobel Prize in Physics with Erwin Schrödinger in 1933. Born in Bristol, England, to a Swiss father and an English mother, he earned degrees in electrical engineering and mathematics at the city's university, before continuing his studies at Cambridge, where he pursued his fascination with general relativity and quantum theory. After his groundbreaking advances of the mid-1920s, he continued his work at Göttingen and Copenhagen before returning to Cambridge, having been appointed the Lucasian Chair in Mathematics. Much of his later career was focused on quantum electrodynamics. He also pursued the idea of unifying quantum theory with general relativity, but this endeavor met with limited success.

Key works

1930 *Principles of Quantum Mechanics*
1966 *Lectures on Quantum Field Theory*

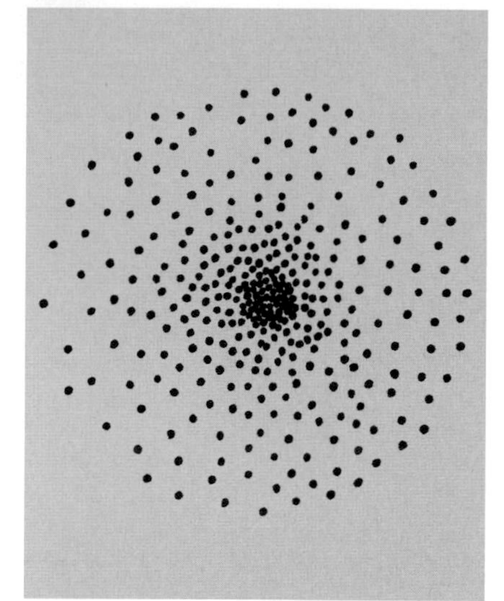

THERE IS AN UPPER LIMIT BEYOND WHICH A COLLAPSING STELLAR CORE BECOMES UNSTABLE

SUBRAHMANYAN CHANDRASEKHAR (1910–1995)

IN CONTEXT

BRANCH
Astrophysics

BEFORE
19th century White dwarf stars are discovered when astronomers identify a star that has far more mass than its tiny size would suggest.

AFTER
1934 Fritz Zwicky and Walter Baade propose that explosions known as supernovae mark the deaths of massive stars, and the collapse of their cores form neutron stars.

1967 British astronomers Jocelyn Bell and Anthony Hewish detect rapidly pulsing radio signals from an object now known as a "pulsar"—a rapidly rotating neutron star.

1971 X-ray emissions from a source known as Cygnus X-1 are found to originate from hot material spiraling into what is probably a black hole—the first such object to be confirmed.

The development of quantum physics in the 1920s had implications for astronomy, where it was applied to the understanding of superdense stars known as white dwarfs. These are the burned-out cores of sunlike stars that have exhausted their nuclear fuel and collapsed, under their own gravity, to objects about the size of Earth. In 1926, physicists Ralph Fowler and Paul Dirac explained that collapse stops at this size due to the "degenerate electron pressure" that arises whenever electrons are packed together so tightly that the Pauli exclusion principle (p.230)—that no two particles can occupy the same quantum state—comes into play.

Forming a black hole

In 1930, Indian astrophysicist Subrahmanyan Chandrasekhar figured out that there was an upper limit to the mass of a stellar core beyond which gravity would overcome the degenerate electron pressure. The stellar core would collapse to a single point in space known as a singularity—forming a black hole. This "Chandrasekhar Limit" for a collapsing stellar core is now known to be 1.44 solar masses (or 1.44 times the mass of the Sun). However, there is a middle stage between white dwarf and black hole—a city-sized neutron star stabilized by another quantum effect called "neutron degeneracy pressure." Black holes are created only when the neutron star's core exceeds an upper limit somewhere between 1.5 and 3 solar masses. ■

> The black holes of nature are the most perfect macroscopic objects in the Universe.
> **Subrahmanyan Chandrasekhar**

See also: John Michell 88–89 ▪ Albert Einstein 214–21 ▪ Paul Dirac 246–47 ▪ Fritz Zwicky 250–51 ▪ Stephen Hawking 314

LIFE ITSELF IS A PROCESS OF OBTAINING KNOWLEDGE

KONRAD LORENZ (1903–1989)

IN CONTEXT

BRANCH
Biology

BEFORE
1872 Charles Darwin describes inherited behavior in *The Expression of the Emotions in Man and Animals.*

1873 Douglas Spalding makes a distinction between innate (genetic) and learned behavior in birds.

1890s Russian physiologist Ivan Pavlov demonstrates that dogs can be conditioned to salivate in a simple form of learning.

AFTER
1976 British zoologist Richard Dawkins publishes *The Selfish Gene*, in which he emphasizes the role of genes in driving behavior.

2000s New research reveals growing evidence of the importance of teaching among many species of animal, from insects to killer whales.

Among the first to conduct scientific experiments on the behavior of animals was 19th-century English biologist Douglas Spalding, who studied birds. The prevailing view was that complex behavior in birds was learned, but Spalding thought that some behavior was innate: it was inherited and essentially "hardwired"—such as the tendency of a hen to incubate her eggs.

Modern ethology—the study of animal behavior—accepts that behavior includes both learned and innate components: innate behavior is stereotypical and, because it is inherited, it can evolve by natural selection, whereas learned behavior can be modified by experience.

Imprinting geese

In the 1930s, Austrian biologist Konrad Lorenz focused on a form of learned behavior in birds that he called "imprinting." He studied the way that greylag geese imprint on, or follow, the first eligible moving stimulus they see—usually their mother—within a critical period after hatching. The mother's example triggers an instinctive behavior known as a "fixed action pattern" in her offspring.

These cranes and geese, hatched and raised by Christian Moullec, imprinted on him and follow him everywhere. Taking to the air in his microlight, he teaches them their migratory routes.

Lorenz demonstrated this with goslings, which adopted him as their mother and followed him everywhere. They would even imprint on inanimate objects, and followed a model train in circles on its track. Together with Dutch biologist Nikolaas Tinbergen, Lorenz was awarded the Nobel Prize in Physiology in 1973. ■

See also: Charles Darwin 142–49 ▪ Gregor Mendel 166–71 ▪ Thomas Hunt Morgan 224–25

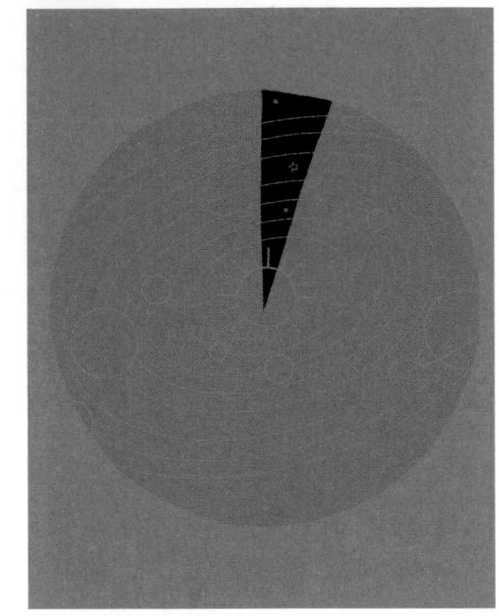

95 PERCENT OF THE UNIVERSE IS MISSING

FRITZ ZWICKY (1898–1974)

IN CONTEXT

BRANCH
Physics and cosmology

BEFORE
1923 Edwin Hubble confirms the true nature of galaxies as independent star systems millions of light years beyond the Milky Way.

1929 Hubble establishes that the universe is expanding, and that galaxies move away from us more rapidly the farther away they are (the so-called Hubble Flow).

AFTER
1950s American astronomer George Abell compiles the first detailed catalogue of galaxy clusters. Subsequent studies of galaxy clusters have repeatedly confirmed the existence of dark matter.

1950s–present Various models of the Big Bang predict that it should have generated much more matter than that which is currently visible.

The idea that the universe might be dominated by something other than detectable luminous matter was first proposed by Swiss astronomer Fritz Zwicky. In 1922–23, Edwin Hubble had realized that "nebulae" were in fact distant galaxies.

A decade later, Zwicky set out to measure the overall mass of the Coma cluster of galaxies. He used a mathematical model called the Virial theorem, which allowed him to calculate the overall mass from the relative velocities of individual cluster galaxies. To Zwicky's

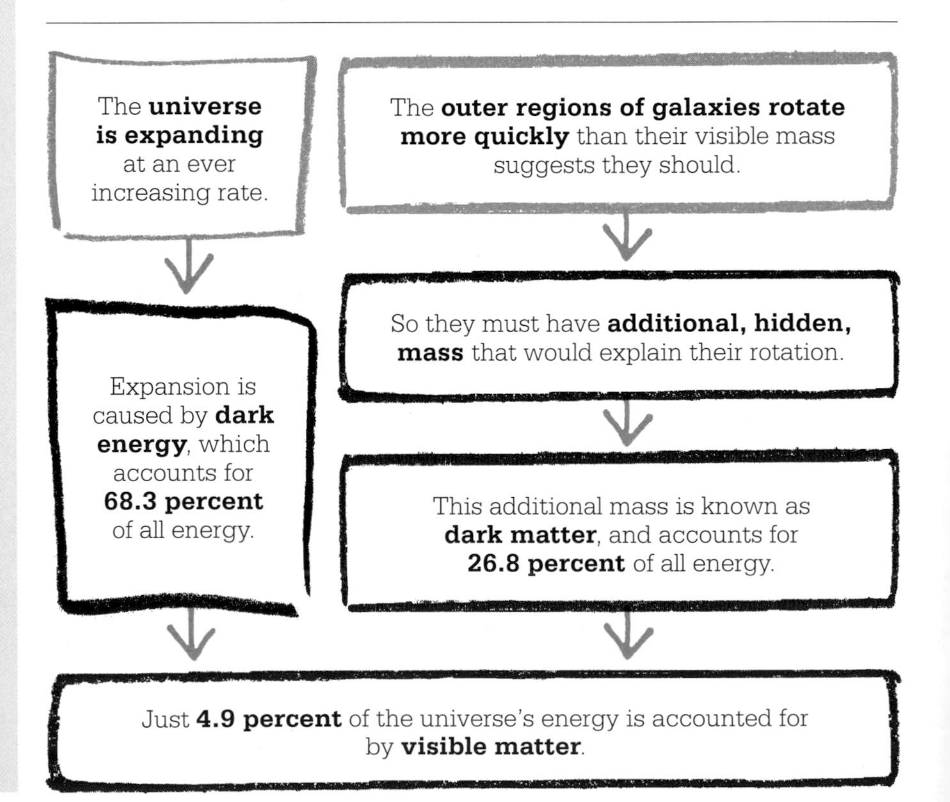

See also: Edwin Hubble 236–41 ▪ Georges Lemaître 242–45
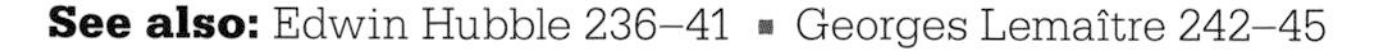

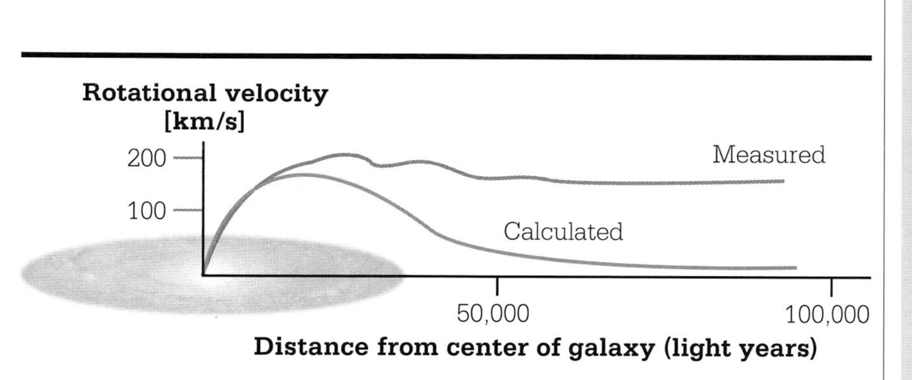

If our galaxy's mass distribution matched that of its visible matter, then stars in the galaxy's outer disk would move more slowly at greater distances from the massive center. Vera Rubin's research found that beyond a certain distance the stars tend to move at a uniform speed regardless of their distance from the hub, revealing dark matter in the galaxy's outer halo.

surprise, his results suggested that the cluster contained about 400 times more mass than that suggested by the combined light of its stars. Zwicky called this staggering amount of unseen matter "dark matter."

Zwicky's conclusion was largely overlooked at the time, but by the 1950s, new technology had opened up new means of detecting nonluminous material. It was clear that large amounts of matter are too cool to glow in visible light but still radiate in infrared and radio wavelengths. As scientists began to understand the visible and invisible structure of our galaxy and others, the amount of "missing mass" fell substantially.

The invisible is real

The reality of dark matter was finally recognized in the 1970s, after US astronomer Vera Rubin mapped the velocity of stars orbiting in the Milky Way and measured the distribution of its mass. She showed that large amounts of mass are distributed beyond the galaxy's visible confines, in a region known as the galactic halo. Today it is widely accepted that dark matter constitutes around 84.5 percent of the mass in the universe. Any hopes that it might actually be normal matter in hard-to-detect forms, such as black holes or rogue planets, have not been borne out by research. It is now thought that dark matter comprises so-called Weakly Interacting Massive Particles (WIMPs). The properties of these hypothetical subatomic particles are still unknown—they are not only dark and transparent, but they do not interact with normal matter or radiation except through gravity.

Since the late 1990s, it has become clear that even dark matter is dwarfed by "dark energy." This phenomenon is the force accelerating the expansion of the universe (pp.236–41), and its nature is still unknown—it may be an integral feature of space-time itself, or a fifth fundamental force known as "quintessence." Dark energy is thought to account for 68.3 percent of all the energy in the universe, with the energy of dark matter amounting to 26.8 percent, and normal matter a mere 4.9 percent. ■

Fritz Zwicky

Born in Varna, Bulgaria, in 1898, Fritz Zwicky was raised by his Swiss grandparents and showed an early talent for physics. In 1925, he left for the US to work at the California Institute of Technology (Caltech), where he spent the rest of his career.

Aside from his work on dark matter, Zwicky is also known for his research into massive exploding stars. He and Walter Baade were the first to show the existence of neutron stars intermediate in size between white dwarfs and black holes, and coined the term "supernovae" for the enormous stellar explosions in which these massive stellar remnants are born. By showing that one class of supernovae always reach the same peak brightness during their explosions, they also provided a means of measuring the distance to far-off galaxies independently of Hubble's Law, paving the way for the later discovery of dark energy.

Key works

1934 *On Supernovae* (with Walter Baade)
1957 *Morphological Astronomy*

A UNIVERSAL COMPUTING MACHINE

ALAN TURING (1912–1954)

IN CONTEXT

BRANCH
Computer science

BEFORE

1906 US electrical engineer Lee De Forest invents the triode valve, the mainstay of early electronic computers.

1928 German mathematician David Hilbert formulates the "decision problem," asking if algorithms can deal with all kinds of input.

AFTER

1943 Valve-based Colossus computers, using some of Turing's code-breaking ideas, begin work at Bletchley Park.

1945 US-based mathematician John von Neumann describes the basic logical structure, or architecture, of the modern stored-program computer.

1946 The first general-purpose electronic programmable computer, ENIAC, based partly on Turing's concepts, is unveiled.

Computing the answers to many number problems can be **reduced** to a series of mathematical steps, or **algorithm**.

A **Turing machine** can, with the right instructions, compute the solution to any **solvable algorithm**.

Varied tasks can be solved using different sets of instructions in a **programmable device**.

This is a universal computing machine.

Imagine sorting 1,000 random numbers, for example 520, 74, 2395, 4, 999..., into ascending order. Some kind of automatic procedure could help. For instance:
A Compare the first pair of numbers.
B If the second number is lower, swap the numbers, go back to A. If it is the same or higher, go to C.
C Make the second number of the last pair the first of a new pair. If there is a next number, make it the second number of the pair, go to B. If there is no next number, **finish**.

This set of instructions is a sequence known as an algorithm. It begins with a starting condition or state; receives data or input; executes itself a finite number of times; and yields a finished result, or output. The idea is familiar to any computer programmer today. It was first formalized in 1936, when British mathematician and logician Alan Turing conceived of machines now known as Turing machines to perform such procedures. His work was initially

See also: Donald Michie 286–91 ▪ Yuri Manin 317

theoretical—an exercise in logic. He was interested in reducing a numbers task to its simplest, most basic, automatic form.

The a-machine

To help envisage the situation, Turing conceived a hypothetical machine. The "a-machine" ("a" for automatic) was a long paper tape divided into squares, with one number, letter, or symbol in each square, and a read/print tape head. With instructions in the form of a table of rules, the tape head reads the symbol of the square it sees, and alters it by erasing and printing another, or leaves it alone, as per the rules. It then moves to one square either to the left or right, and repeats the procedure. Each time there is a different overall configuration of the machine, with a new sequence of symbols.

The whole process can be compared to the number-sorting algorithm above. This algorithm is constructed for one particular task. Similarly, Turing envisaged a range of machines, each with a set of instructions or rules for a particular undertaking. He added, "We have only to regard the rules as being capable of being taken out and exchanged for others and we have something very akin to a universal computing machine."

Now known as the Universal Turing Machine (UTM), this device had an infinite store (memory) containing both instructions and data. The UTM could therefore simulate any Turing machine. What Turing called changing the rules would now be called programming. In this way, Turing first introduced the concept of the programmable computer, adaptable for many tasks, with input, processing of information, and output. ■

A computer would deserve to be called intelligent if it could deceive a human into believing that it was human.
Alan Turing

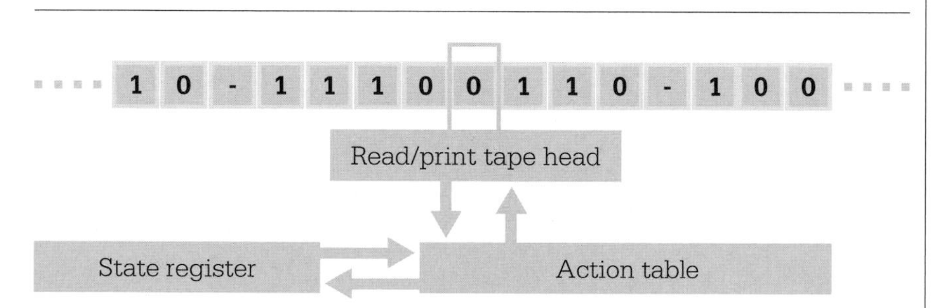

A Turing machine is a mathematical model of a computer. The head reads a number on the infinitely long tape, writes a new number on it, and moves left or right according to rules contained in the action table. The state register keeps track of the changes and feeds this input back into the action table.

Alan Turing

Born in London in 1912, Turing showed a prodigious talent for mathematics at school. He earned a first class degree in mathematics from Kings College, Cambridge, in 1934, and worked on probability theory. From 1936 to 1938, he studied at Princeton University in the US, where he proposed his theories about a generalized computing machine.

During World War II, Turing designed and helped build a fully functioning computer known as the "Bombe" to crack German codes made by the so-called Enigma machine. Turing was also interested in quantum theory, and shapes and patterns in biology. In 1945, he moved to the National Physics Laboratory in London, then to Manchester University to work on computer projects. In 1952, he was tried for homosexual acts (then illegal), and two years later died from cyanide poisoning—it seems likely this was by suicide rather than by accident.
In 2013, Turing was granted a posthumous pardon.

Key work

1939 *Report on the Applications of Probability to Cryptopgraphy*

THE NATURE OF THE CHEMICAL BOND

LINUS PAULING (1901–1994)

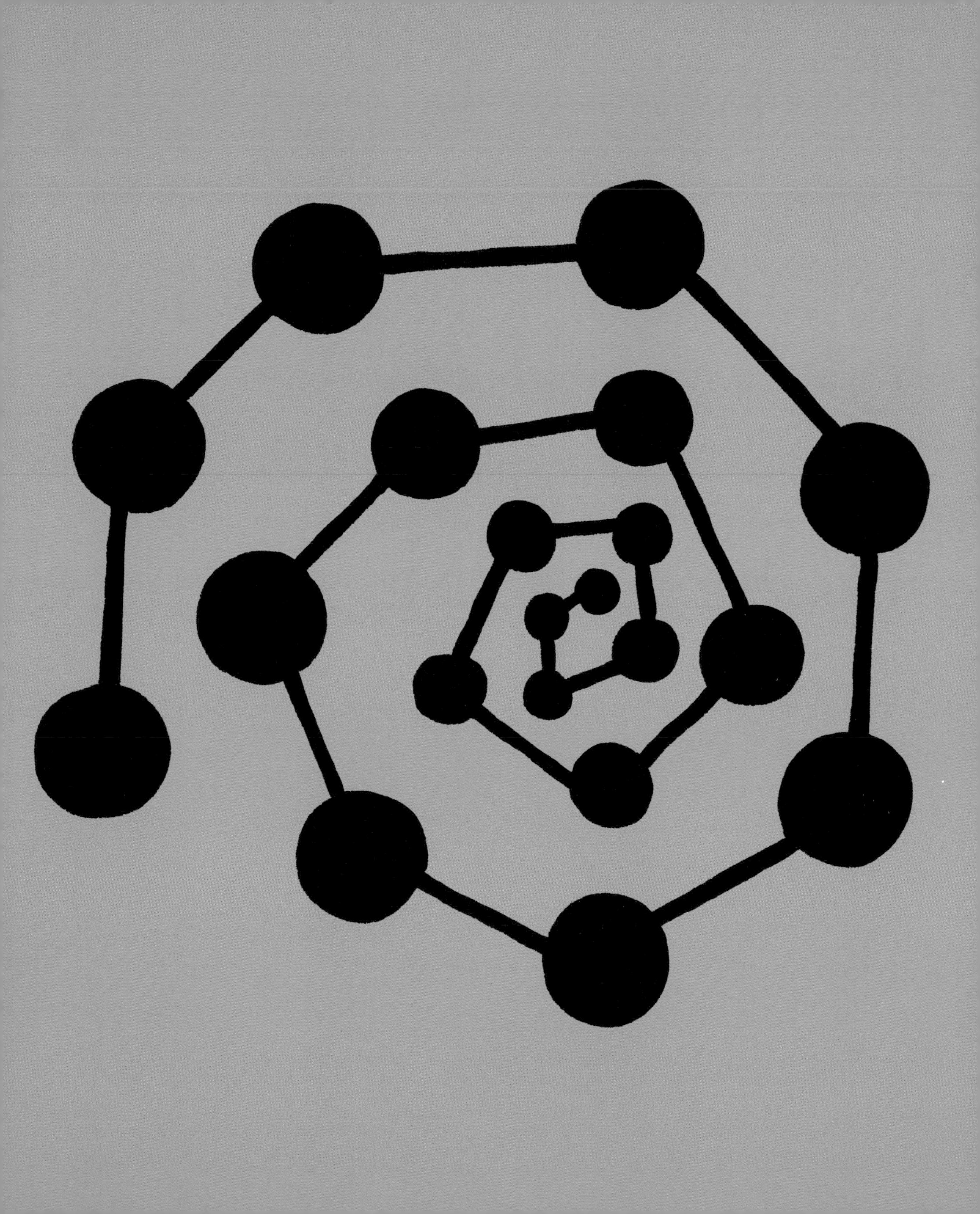

IN CONTEXT

BRANCH
Chemistry

BEFORE
1800 Alessandro Volta lists metals in decreasing order of electropositivity.

1852 British chemist Edward Frankland states that atoms have definite combining power, which determines the formulae of compounds.

1858 August Kekulé shows that carbon has a valency of four—it forms four bonds with other atoms.

1916 American physical chemist Gilbert Lewis shows that a covalent bond is a pair of electrons shared by two atoms in a molecule.

AFTER
1938 British mathematician Charles Coulson calculates an accurate molecular orbital wave function for hydrogen.

In the late 1920s and early 1930s, in a series of landmark papers, American chemist Linus Pauling figured out a quantum-mechanical explanation of the nature of chemical bonds. Pauling had studied quantum mechanics in Europe with the German physicist Arnold Sommerfeld in Munich, with Niels Bohr in Copenhagen, and with Erwin Schrödinger in Zurich. He had already decided that he wanted to research the bonding within molecules, and realized that quantum mechanics gave him the right tools to do so.

Hybridization of orbitals

When he returned to the US, Pauling published about 50 papers, and, in 1929, he laid down a set of five rules for interpreting the X-ray diffraction patterns of complicated crystals, now known as Pauling's rules. At the same time, he was turning his attention to the bonding between atoms in covalent molecules (molecules in which atoms are bonded by sharing two electrons with each other), especially of organic compounds—those based on carbon.

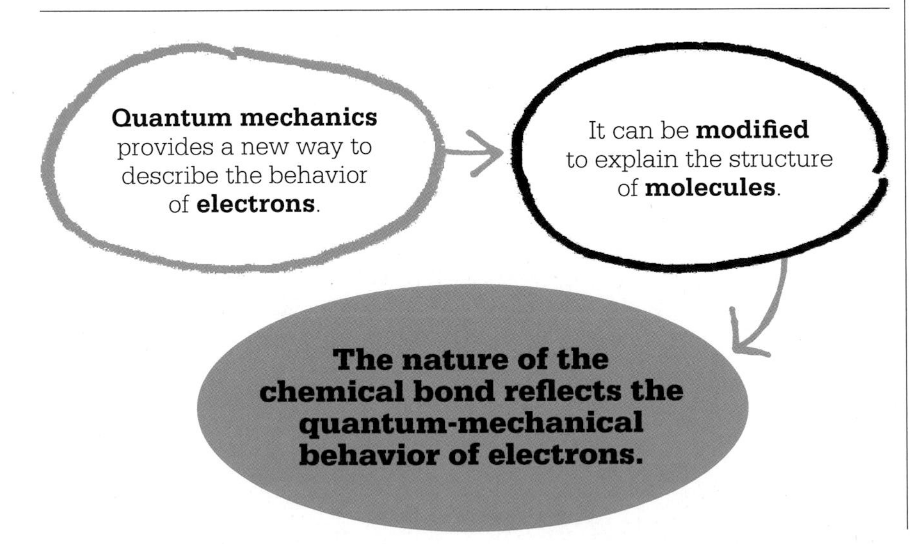

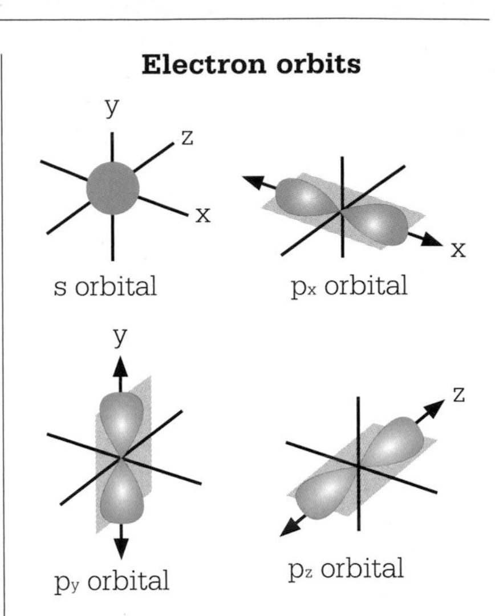

Electrons orbit an atomic nucleus in various ways—in shells around the center (s) or lobes along one axis (p).

A carbon atom has six electrons in total. The European pioneers of quantum mechanics designated the first two as "1s-electrons": these have a spherical orbital or shell around the carbon nucleus—like a balloon inflated around a golf ball in the center. Outside the 1s shell is another shell containing two "2s-electrons." The 2s shell is like another, bigger balloon outside the first. Lastly, there are "p-orbitals," which have big lobes sticking out either side of the nucleus. The p_x orbital lies on the x-axis, the p_y on the y axis, and the p_z orbital on the z-axis. The last two electrons of the carbon atom occupy two of these orbitals—perhaps one in p_x and one in p_y.

The new quantum-mechanical picture of electrons treated their orbits as "clouds" of probability densities. It was no longer quite right to think of the electrons as points moving around their orbits; rather, their existence was smeared across the orbits. This new nonlocal picture of reality allowed for some radical new ideas for chemical

See also: August Kekulé 160–65 ▪ Max Planck 202–05 ▪ Erwin Schrödinger 226–33 ▪ Harry Kroto 320–21

bonding. Bonds could either be strong “sigma” bonds, in which orbitals overlap head-on, or weaker, more diffuse “pi” bonds, in which orbitals are parallel to each other.

Pauling came up with the idea that in a molecule, as opposed to a bare atom, carbon’s atomic orbitals could combine, or “hybridize,” to give stronger bonds to other atoms. He showed that the s and p orbitals could hybridize to form four sp^3 hybrids, which would all be equivalent, and would project from the nucleus toward the corners of a tetrahedron, with inter-bond angles of 109.5°. Each sp^3 orbital can form a sigma bond with another atom. This is consistent with the fact that all the hydrogen atoms in methane (CH_4), and all the chlorine atoms in carbon tetrachloride (CCl_4), behave the same way. As the structures of various carbon compounds were studied, the four closest neighboring atoms were often found in a tetrahedral arrangement. The crystal structure of diamond was among the first structures to be resolved by X-ray crystallography, in 1914.

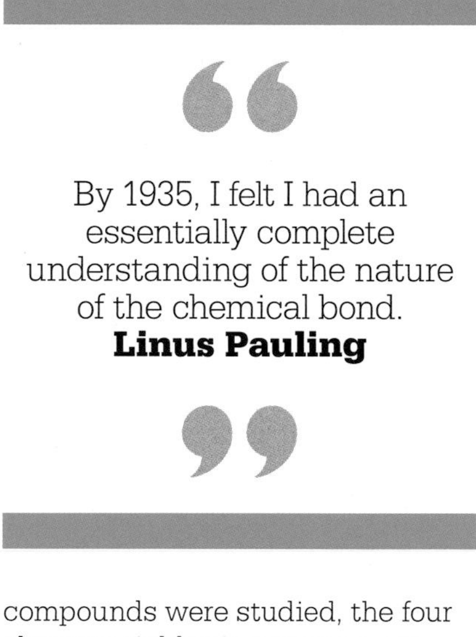

> By 1935, I felt I had an essentially complete understanding of the nature of the chemical bond.
> **Linus Pauling**

Diamond is pure carbon, and in the crystal each carbon atom is bonded to four others by sigma bonds at the corners of a tetrahedron. This structure explains diamond’s hardness.

Another possible way for carbon atoms to bond to other atoms is for an s-orbital to mix with two p-orbitals to form three sp^2 hybrids. These stick out from the nucleus in one plane, with angles of 120° between them. This is consistent with the geometry of molecules such as ethylene, which has the double-bond structure $H_2C{=}CH_2$. Here, a sigma bond is formed between the carbon atoms by one of the sp^2 hybrids, and a pi bond by the fourth, unhybridized orbital.

Lastly, an s-orbital can mix with one p-orbital to form two sp hybrids, whose lobes stick out in a straight line, 180° apart. This is consistent »

Methane

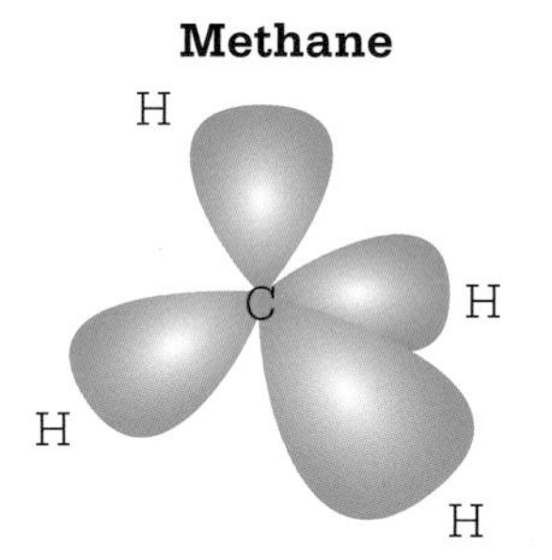

Four electrons in the carbon atom hybridize to form four sp^3 orbitals.

Ethylene

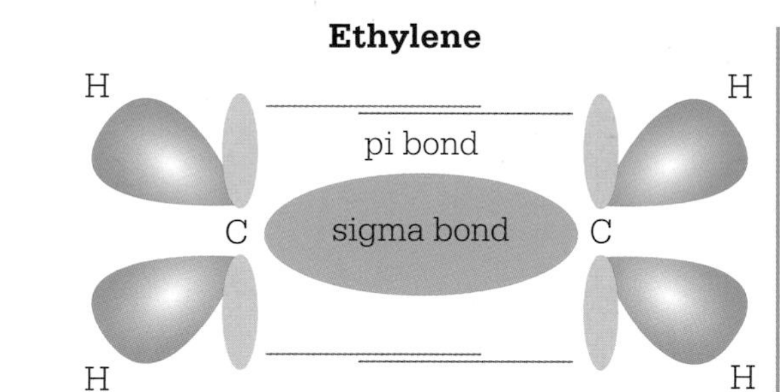

Three electrons in the carbon atoms hybridize to form three sp^2 orbitals. The remaining unhybridized orbitals form a second pi bond between the carbon atoms.

Carbon dioxide

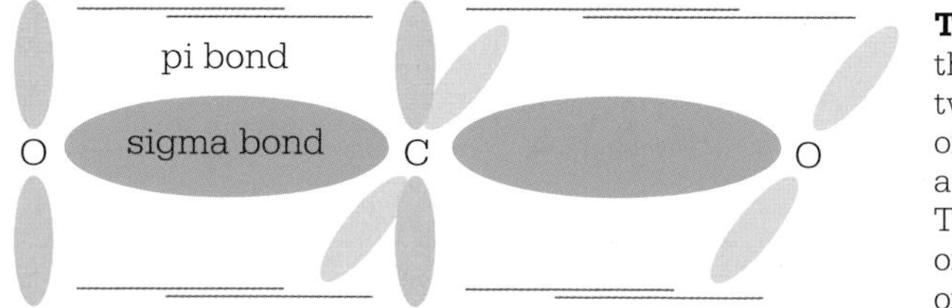

Two electrons in the carbon atom form two sp orbitals, each of which bonds with an oxygen atom. The remaining two orbitals bond to the oxygen in a pi bond.

Diamond

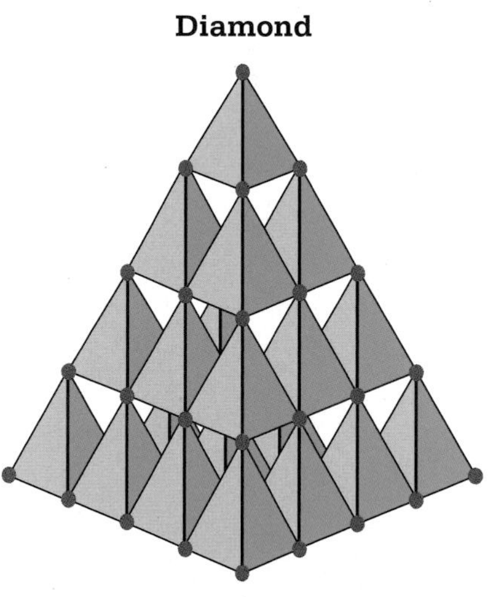

Each carbon atom in a diamond is bonded by sp^3 hybrids to four other atoms to form the corner of a tetrahedron. The result is an infinite lattice held together by covalent carbon–carbon bonds, which are immensely strong.

with the structure of carbon dioxide (CO_2), where the sp hybrids each form a sigma bond with the oxygen, and a second pi bond is formed by the remaining two unhybridized orbitals.

A new structure of benzene

The structure of benzene, C_6H_6, had worried August Kekulé when he first proposed that it was a ring, more than 60 years earlier. He eventually suggested that the carbon atoms must be connected with alternate single and double bonds, and that the molecule oscillated between the two equivalent structures (p.164).

Pauling's alternative solution was elegant. He said that the carbon atoms were all sp^2 hybridized, so that the bonds between them and the hydrogen atoms all lie in the same xy plane and form an angle of 120° with each other. Each carbon atom has one remaining electron in a p_z orbital. These electrons combine to form a bond connecting all six carbon atoms. This is a pi bond, and, in it, the electrons remain above and below the ring, and away from the carbon nuclei (see right).

Ionic bonding

Methane and ethylene are gases at room temperature. Benzene and many other organic compounds based on carbon are liquids. They have small, lightweight molecules that can easily move around in the gas or liquid state. Salts such as calcium carbonate and potassium nitrate, by contrast, are almost invariably solids, and melt only at high temperatures. And yet a unit of sodium chloride (NaCl) has a molecular weight of 62, while benzene has a molecular weight of 78. The difference in their behavior is explained not by their weight, but by their structure. Benzene is held together in single molecules by covalent bonds between the atoms; that is, each bond comprises one pair of electrons shared between two specific atoms.

Sodium chloride has quite different properties. The silvery metal sodium burns energetically in the greenish gas chlorine to produce the white solid sodium chloride. The sodium atom has a stable complete shell of electrons around the nucleus, plus one spare electron outside that. The chlorine atom is one electron short of a stable complete shell. When they react, an electron is transferred from the sodium atom to the chlorine atom, and both acquire stable complete shells of electrons, but now the sodium has become a sodium ion Na^+, and chlorine has become the chloride ion Cl^- (see above). They have no spare electrons to form covalent bonds, but the ions are now charged: the sodium atom has lost a negatively charged electron so now has a

Ionic bonding

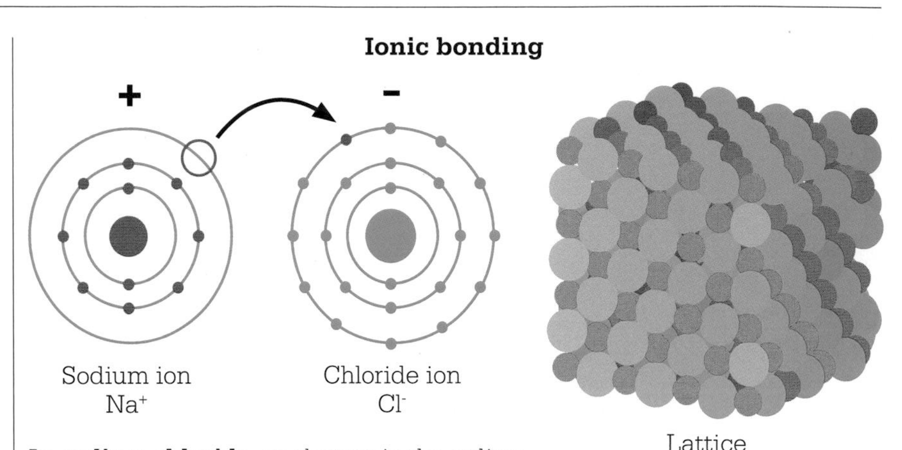

In sodium chloride, an electron in the sodium atom moves into a chlorine atom to form two charged, stable ions. The ions are held together by electrostatic attraction to form a stable lattice.

Benzene ring

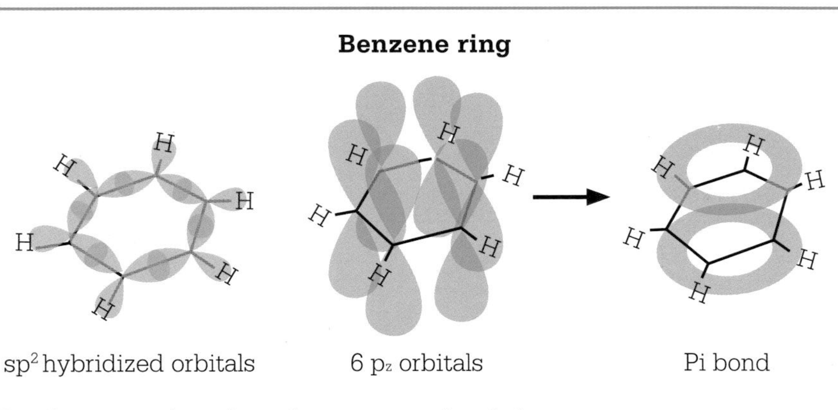

In a benzene ring, the carbon atoms are bonded to each other and a hydrogen atom by sp^2 hybridized orbitals. The rings are bonded to each other by a nonlocalized pi-bond formed from the six p_z orbitals.

There is no area of the world that should not be investigated by scientists. There will always remain some questions that have not been answered. In general, these are the questions that have not yet been posed.
Linus Pauling

positive overall charge; the chlorine atom has gained an electron and has a negative charge. The ions are held together by electrostatic attraction, plus to minus—a strong bond.

Sodium chloride was the first compound to be analyzed by X-ray crystallography. It was found that in reality there is no such thing as a molecule of NaCl. The structure comprises an infinite array of alternating sodium and chloride ions. Each sodium ion is surrounded by six chloride ions, and each chloride is surrounded by six sodiums. Many other salts have similar structures: infinite lattices of one type of ion with different ions filling all the gaps.

Electronegativity

Pauling explained ionic bonding in compounds such as sodium chloride, which is purely ionic, and also compounds in which the bonding is neither purely ionic nor purely covalent but somewhere in between. This work led him to develop the concept of electronegativity, which to some extent echoed the list of metals in decreasing order of electropositivity first introduced by Alessandro Volta in 1800. Pauling discovered that the covalent bond formed between atoms of two different elements (e.g. C–O) is stronger than might be expected from the average of the strengths of C–C bonds and O–O bonds. He thought that there must be some electrical factor that strengthened the bond, and set out to calculate values for this factor. The scale is now known as the Pauling scale.

The electronegativity of an element (strictly speaking in a particular compound) is a measure of how strongly an atom of the element attracts electrons toward itself. The most electronegative element is fluorine; the least electronegative (or the most electropositive) of the well-known elements is cesium. In the compound cesium fluoride, each fluorine atom pulls an electron entirely away from a cesium atom, resulting in an ionic compound Cs^+F^-.

In a covalent compound such as water (H_2O), there are no ions, but oxygen is much more electronegative than hydrogen, and the result is that the water molecule is polar, with a small negative charge on the oxygen atom and a small positive charge on the hydrogen atoms. The charges make the water molecules stick together strongly. This explains why water has so much surface tension and such a high boiling point.

Pauling first proposed a scale of electronegativity in 1932, and he and others developed it further in subsequent years. For his work elucidating the nature of the chemical bond, he won the Nobel Prize in Chemistry in 1954. ■

Linus Pauling

Linus Carl Pauling was born in Portland, Oregon, US. He first heard about quantum mechanics while still in Oregon, and won a scholarship to study the subject under some of the world experts in Europe in 1926. He returned to become assistant professor at California Institute of Technology, where he remained for most of his life.

Pauling took great interest in biological molecules, and he discovered that sickle-cell anemia is a molecular disease. He was also a peace campaigner, and was awarded the Nobel Peace Prize in 1963 for attempting to mediate between the US and Vietnam.

In later life, his reputation was damaged as a result of his enthusiasm for alternative medicine. He championed the use of high-dose vitamin C as a defense against the common cold, a treatment that has subsequently been shown to be ineffective.

Key work

1939 *The Nature of the Chemical Bond and the Structure of Molecules and Crystals*

AN AWESOME POWER IS LOCKED INSIDE THE NUCLEUS OF AN ATOM

J. ROBERT OPPENHEIMER (1904–1967)

IN CONTEXT

BRANCH
Physics

BEFORE
1905 Albert Einstein's famous mass-energy equivalence equation $E = mc^2$ describes how tiny masses "store" large amounts of energy.

1932 John Cockcroft and Ernest Walton's experiments splitting lithium nuclei with protons hint at the enormous energy locked inside the nucleus.

1939 Leó Szilárd spots that a single fission event of uranium-235 releases three neutrons and suggests that a chain reaction is possible.

AFTER
1954 The USSR's Obninsk Nuclear Power Plant goes into operation. It is the first nuclear power station to generate electricity for a country's national grid.

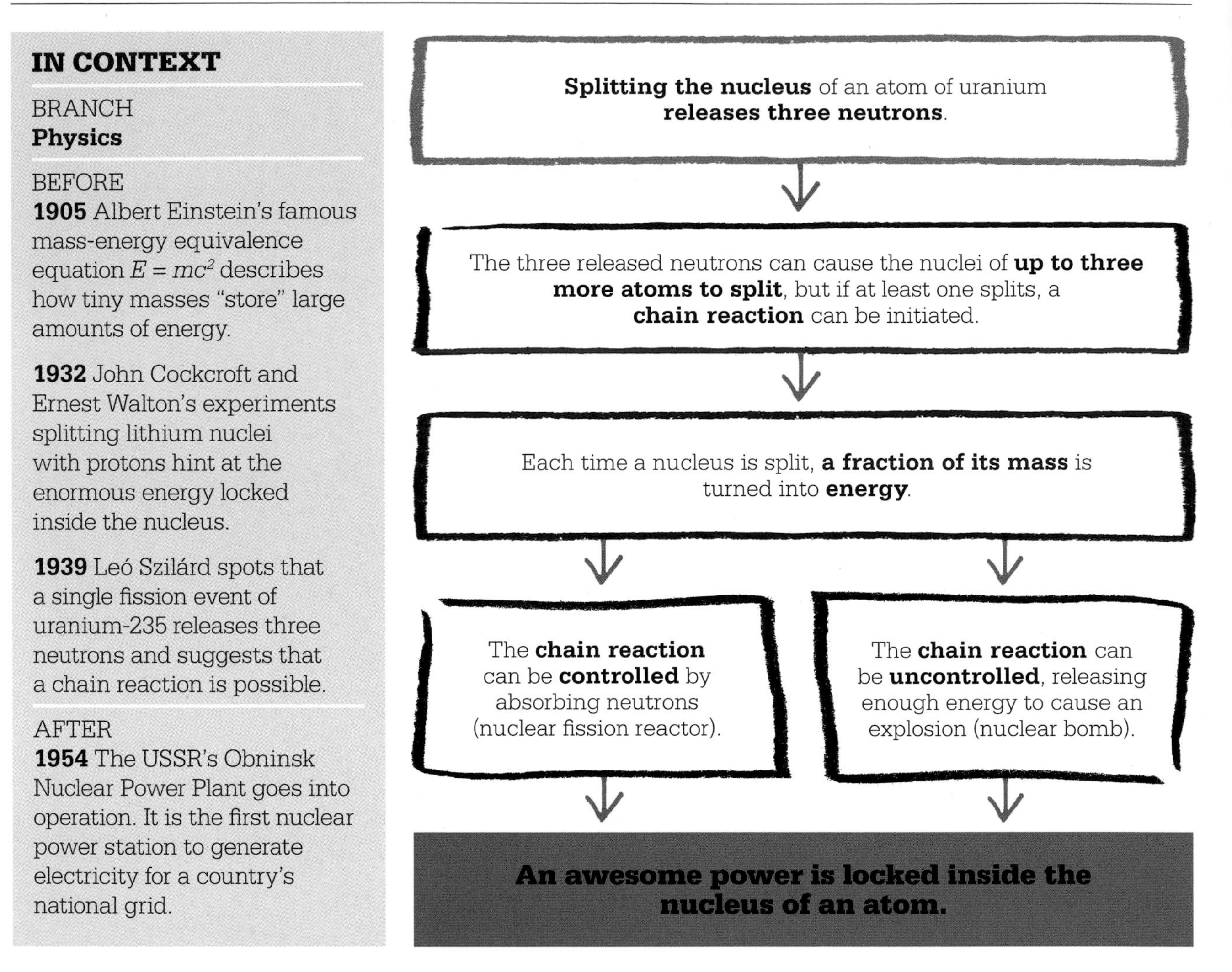

In 1938, the world stood at the threshold of the atomic age. One man would step forward to lead the scientific drive that would usher in this new era. For J. Robert Oppenheimer, this decision would ultimately destroy him. He was the administrator of the largest scientific project the world had seen—the Manhattan Project—but came to deeply regret his part in it.

Drive to the center

Oppenheimer's varied professional life had been characterized by a ruthless drive to "be where it's at" and this compulsion took the newly graduated Harvard man to Europe, the center of a blossoming of theoretical physics. At Göttingen University, Germany, in 1926, he produced the Born–Oppenheimer approximation with Max Born, used to explain, as Oppenheimer put it, "why molecules are molecules." This method extended quantum mechanics beyond single atoms to describe the energy of chemical compounds. It was an ambitious mathematical exercise as a dizzying range of possibilities for each electron in a molelcule had to be computed. Oppenheimer's work in Germany has proved crucial to calculating energy in modern chemistry, but the final breakthrough that would lead to the atomic bomb came after he had returned to the US.

Fission and black holes

The chain reaction that led to the building of the atomic bomb began in mid-December 1938, when German chemists Otto Hahn and Fritz Strassmann "split the atom" in their Berlin laboratory. They had been firing neutrons at uranium,

See also: Marie Curie 190–95 ▪ Ernest Rutherford 206–13 ▪ Albert Einstein 214–21

We knew the world would not be the same. A few people laughed. A few people cried. Most people were silent. I remembered the line from the Hindu scripture: "Now I am become Death, the destroyer of worlds."
J. Robert Oppenheimer

but instead of creating heavier elements by neutron absorption, or lighter elements by emission of one or more nucleons (protons or neutrons), the pair found that the lighter element barium was released, which had 100 fewer nucleons than the uranium nucleus. No nuclear process understood at the time could account for the loss of 100 nucleons.

Perplexed, Hahn sent a letter to colleagues Lise Meitner and Otto Frisch in Copenhagen. Within the month, Meitner and Frisch had figured out the basic mechanism of nuclear fission, recognizing how uranium was split into barium and krypton, the missing nucleons were converted into energy, and a chain reaction could follow. In 1939, Danish physicist Niels Bohr took the news to the US. His account, along with the publication of the Meitner–Frisch paper in the journal *Nature*, set the East Coast scientific community ablaze with excitement. Conversations between Bohr and John Archibald Wheeler at Princeton after the annual Theoretical Physics Conference, led to the Bohr–Wheeler theory of nuclear fission.

All the atoms of the same element have nuclei with the same number of protons in them, but the number of neutrons can vary, making different isotopes of the same element. In the case of uranium, there are two naturally occurring isotopes. Uranium-238 (U-238) makes up 99.3 percent of natural uranium. Its nuclei contain 92 protons and 146 neutrons. The remaining 0.7 percent is made up of uranium-235 (U-235), whose nuclei contain 92 protons and 143 neutrons. The Bohr–Wheeler theory incorporated the finding that low-energy neutrons could cause fission in U-235, causing the atom to split and releasing energy in the process.

When the news reached the West Coast, Oppenheimer, now at Berkeley, was captivated. He gave a series of lectures and seminars on the brand new theory and quickly »

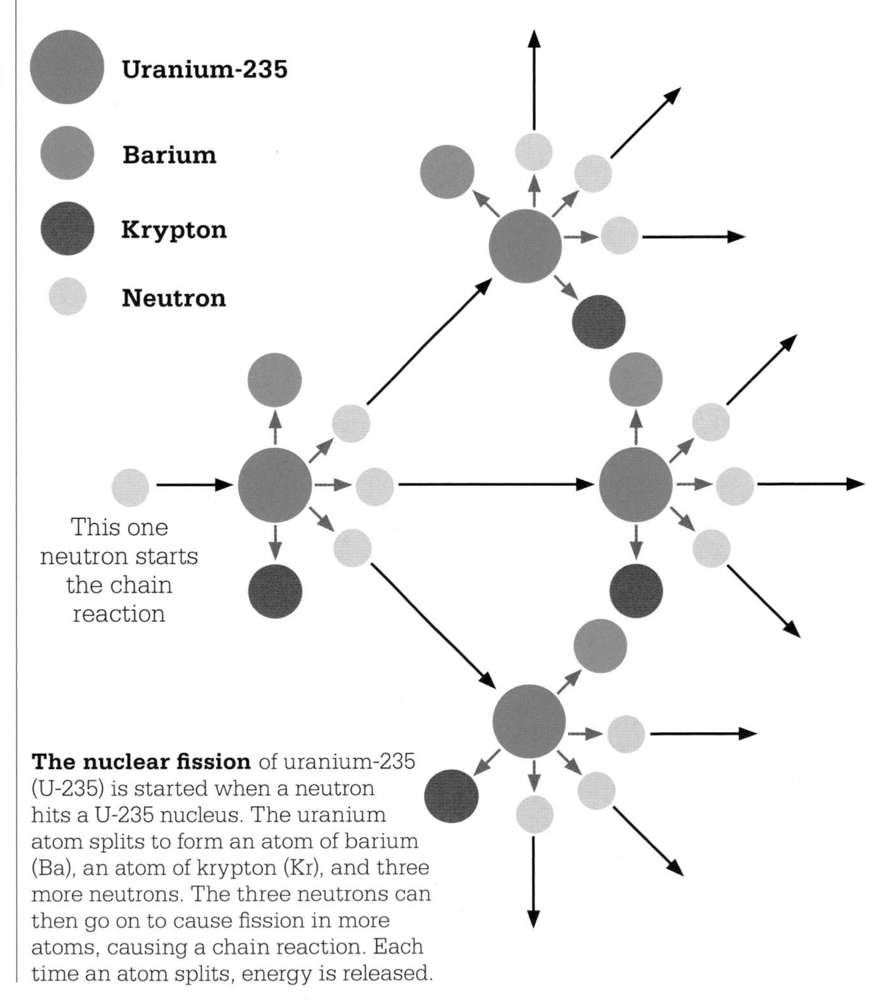

The nuclear fission of uranium-235 (U-235) is started when a neutron hits a U-235 nucleus. The uranium atom splits to form an atom of barium (Ba), an atom of krypton (Kr), and three more neutrons. The three neutrons can then go on to cause fission in more atoms, causing a chain reaction. Each time an atom splits, energy is released.

saw the potential for making a weapon of awesome power—to his mind "a good, honest, practical way" to use the new science. But while laboratories in East Coast universities raced to replicate the results of early fission experiments, Oppenheimer concentrated on his research into stars contracting and collapsing under their own gravity to form black holes.

Birth of the idea

The idea of a nuclear weapon was already in the air. As early as 1913, H. G. Wells wrote of "tapping the internal energy of atoms" to make "atomic bombs." In his novel *The World Set Free*, the innovation was set to happen in the year 1933. In 1933 itself, Ernest Rutherford touched on the large amount of energy released during nuclear fission in a speech printed in *The Times* of London. However, Rutherford dismissed the idea of harnessing this energy as "moonshine," since the process was so inefficient it required much more energy than it released.

It took a Hungarian living in Britain named Leó Szilárd to see how it could be done, and also to realize the horrific consequences for a world heading toward war. Pondering Rutherford's lecture, Szilárd saw that the "secondary neutrons" emerging from the first fission event could themselves create further fission events, resulting in an escalating chain reaction of nuclear fission. Szilárd later recalled, "There was little doubt in my mind that the world was headed for grief."

Experiments in Germany and the US showed that the chain reaction was indeed possible, prompting Szilárd and another Hungarian emigré, Edward Teller, to approach Albert Einstein with a letter. Einstein passed the letter on to US President Roosevelt on October 11, 1939 and just ten days later the Advisory Committee on Uranium was set up to investigate the possibility of developing the bomb in the United States first.

Birth of Big Science

The Manhattan Project that arose from this resolution was science on the grandest scale imaginable. A multiarmed organization that spread over several large sites in the US and Canada and countless

We have made a thing, a most terrible weapon, that has altered abruptly and profoundly the nature of the world. And by so doing we have raised again the question of whether science is good for man.

J. Robert Oppenheimer

smaller facilities, it employed 130,000 people and by its close had swallowed in excess of US$2 billion (more than US$26 bn, or £16 bn, in 2014 money)—all in top secrecy.

Early in 1941, the decision was taken to pursue five separate methods of producing fissionable material for a bomb: electromagnetic separation, gaseous diffusion, and thermal diffusion to separate isotopes of uranium-235 from uranium-238; and two lines of research into

J. Robert Oppenheimer

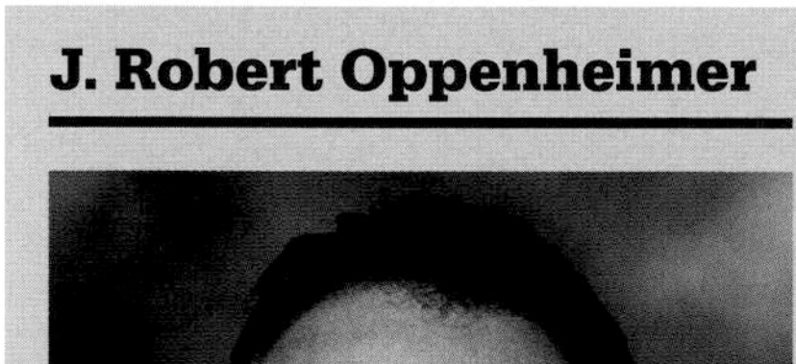

Educated at the Ethical Culture school of New York City, Julius Robert Oppenheimer was a thin, highly-strung boy with a quick grasp of concepts. After graduating from Harvard University, he spent two years at Cambridge University under Ernest Rutherford, followed by a move to Göttingen in Germany, where he was taken under the wing of Max Born.

Oppenheimer was a complex character whose great talent was to be at the center of things, and he made influential friends wherever he went. However, he had a notoriously sharp tongue and a desire to be regarded as a superior intellect. Although he is best known for his work on the Manhattan Project, his most lasting contribution to science was his prewar research at the University of California, Berkeley, on neutron stars and black holes.

Key works

1927 *On the Quantum Theory of Molecules*
1939 *On Continued Gravitational Contraction*

On August 9, 1945, the plutonium bomb "Fat Man" was dropped over Nagasaki in southern Japan. About 40,000 people were killed instantly, and many more died in the following weeks.

nuclear reactor technology. On December 2, 1942, the very first controlled chain reaction involving nuclear fission was carried out on a squash court at the University of Chicago. Enrico Fermi's Chicago Pile-1 was the prototype for the reactors that would enrich uranium and create the newly discovered plutonium—an unstable element that is even heavier than uranium, can also cause a rapid chain reaction, and can be used to create an even deadlier bomb.

The Magic Mountain

Selected to head up the Manhattan Project's research into secret weapons, Oppenheimer approved a disused boarding school at Los Alamos Ranch in New Mexico as the site for research facilities for the project's final stages—the construction of an atomic bomb. "Site Y" would see the highest concentration of Nobel laureates ever gathered in one place.

Since much of the important science had already been done, many of the Los Alamos scientists dismissed their work in the New Mexico desert as merely a "problem of engineering." However, it was Oppenheimer's coordination of 3,000 scientists that made the construction of the bomb possible.

Change of heart

The successful Trinity test on July 16, 1945 and subsequent detonation of a bomb called "Little Boy" above Hiroshima in Japan on August 6, 1945 left Oppenheimer jubilant. However, the event was to cast a long shadow over the Los Alamos director. Germany had already surrendered by the time the bomb was dropped, and many Los Alamos scientists felt a public demonstration of the bomb was all that was necessary—after seeing its awesome power, Japan would be sure to surrender. However, while Hiroshima was believed by some to be a necessary evil, the detonation of a plutonium device—called "Fat Man"—over Nagasaki on August 9 was hard to justify. A year later, Oppenheimer publicly stated his opinion that the atom bombs had been dropped on a defeated enemy. In October 1945, Oppenheimer met President Harry S. Truman and told him, "I feel I have blood on my hands." Truman was furious. Congressional hearings stripped the scientist of his security clearance in 1954, ending his ability to influence public policy.

By then, Oppenheimer had overseen the advent of the military–industrial complex and ushered in a new era of Big Science. In presiding over the creation of a new scientific terror, he became a symbol for the moral consequences of their actions that scientists must now consider. ■

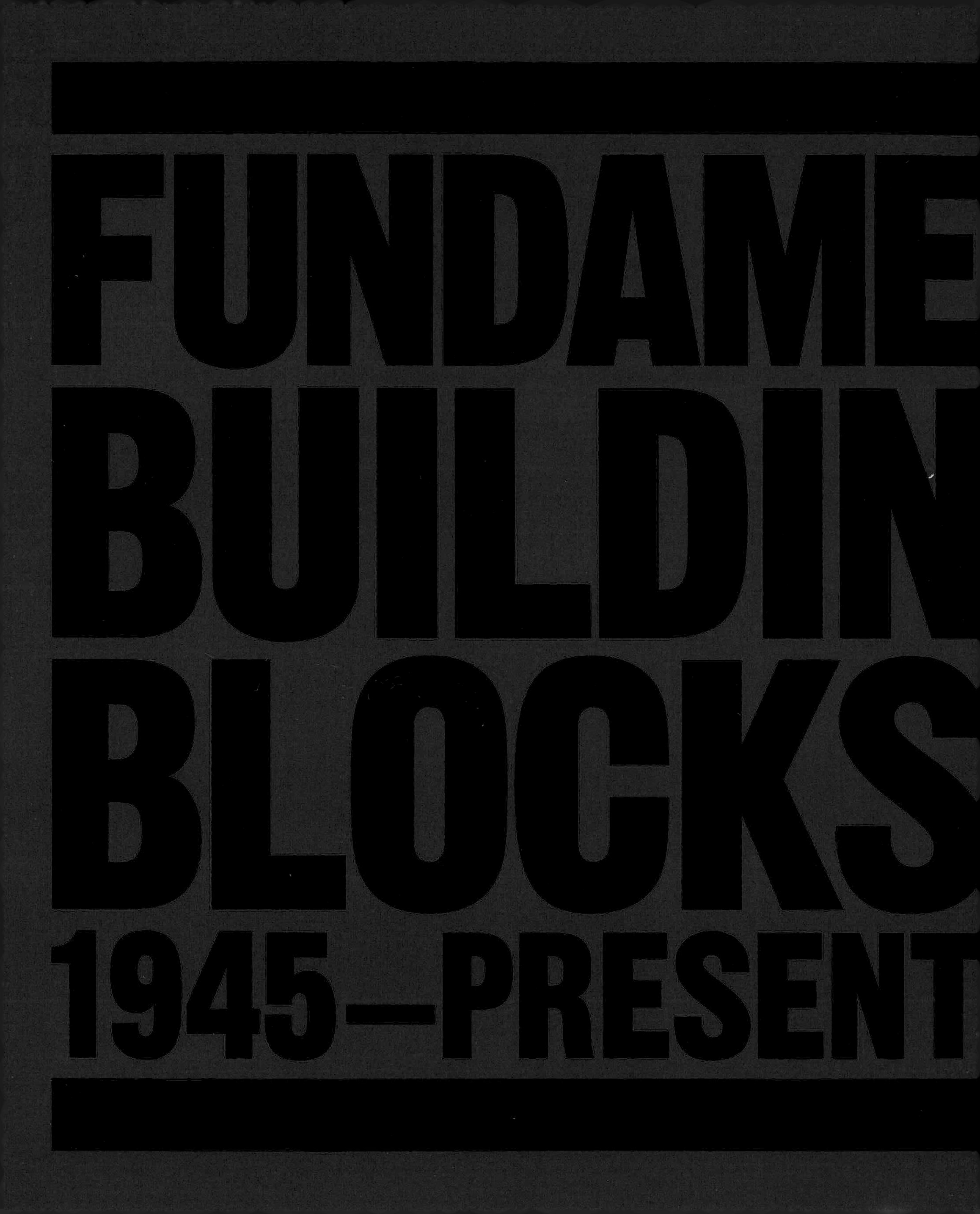
FUNDAME
BUILDIN
BLOCKS
1945–PRESENT

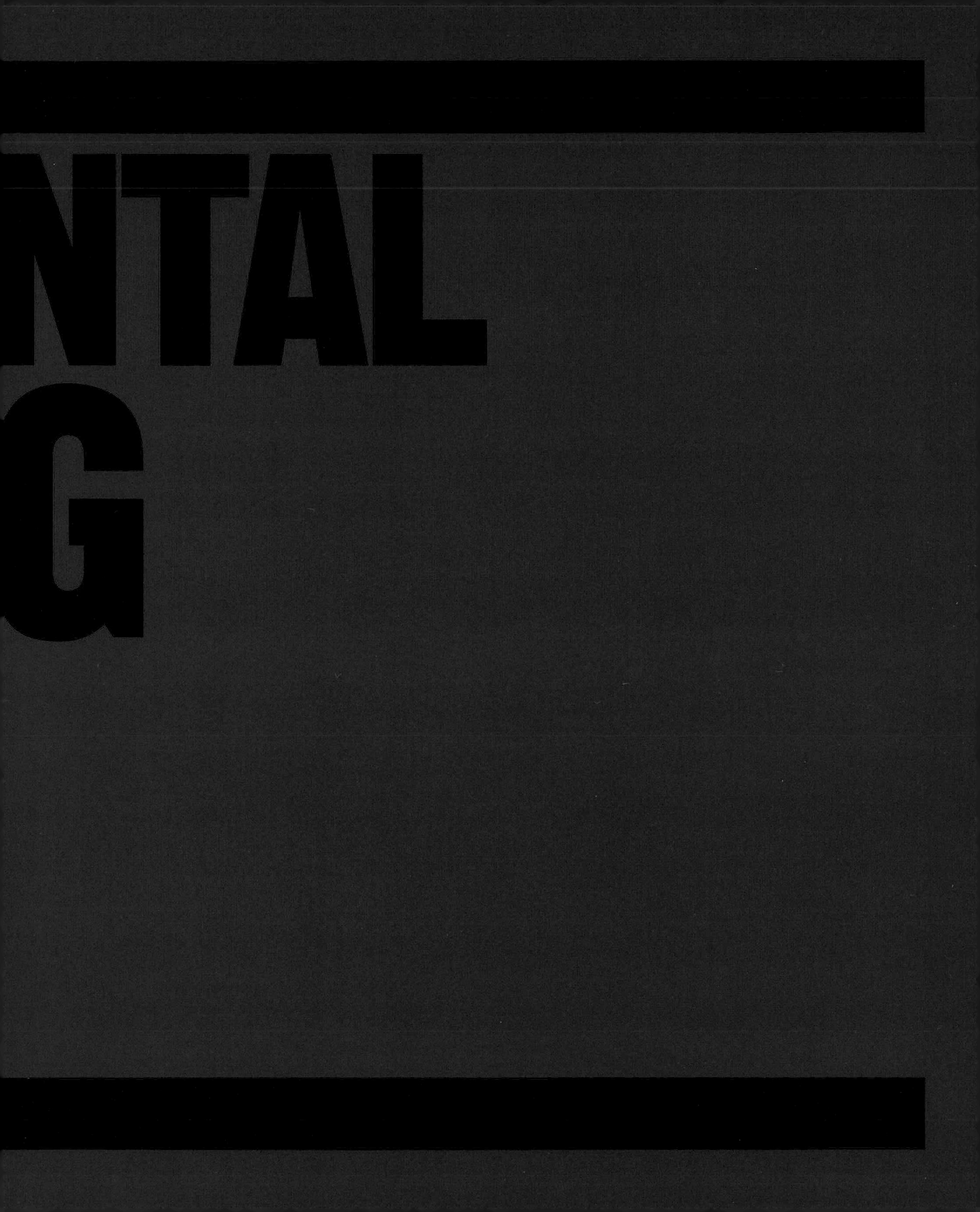
NTAL
G

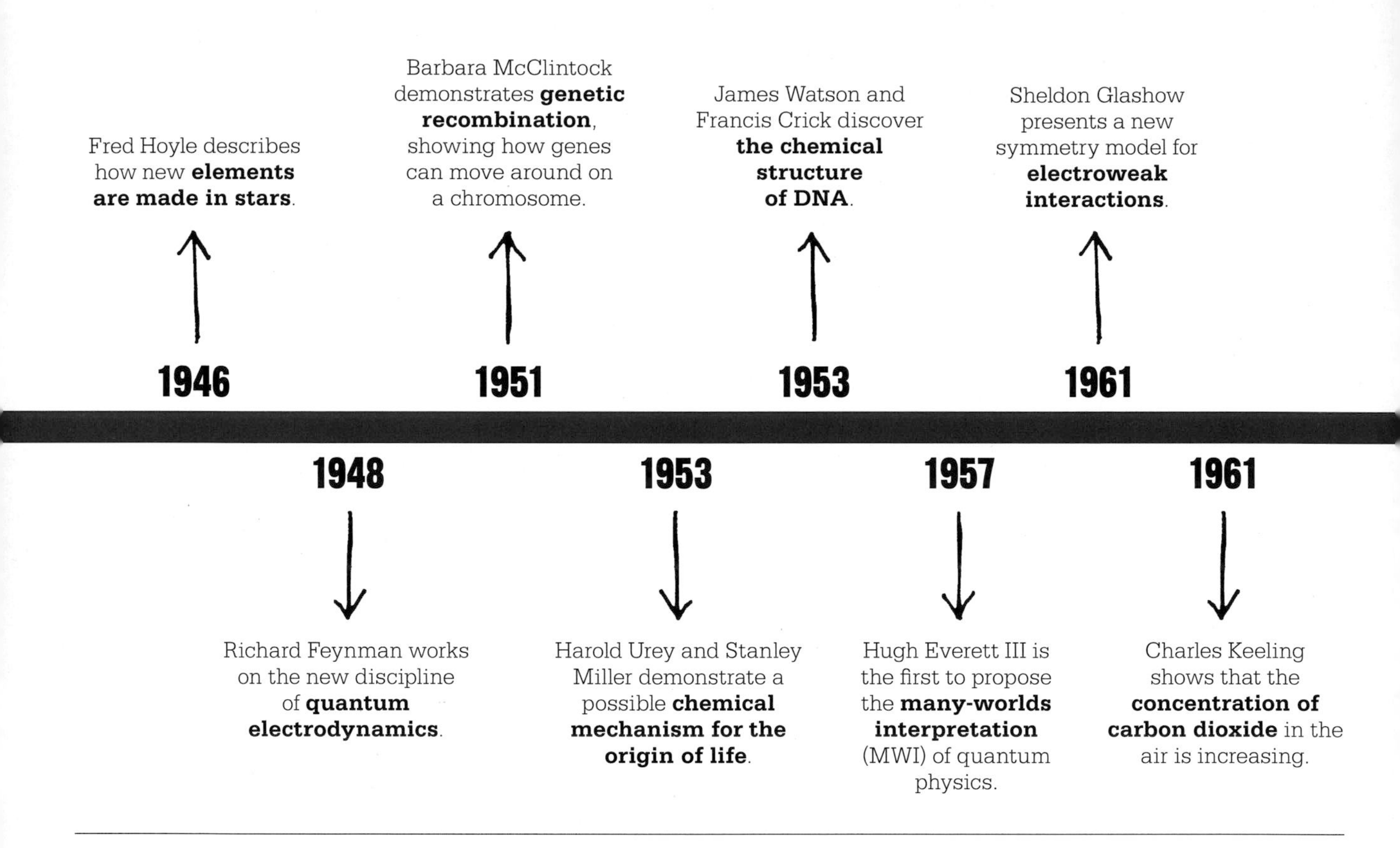

The second half of the 20th century saw rapidly improving technology being employed in almost every field of science, from telescopes to chemical analysis. New technology has widened the possibilities for calculation and experiment. The first computers were built in the 1940s, and a new science, Artificial Intelligence, has emerged. CERN's Large Hadron Collider—a particle accelerator—is the biggest piece of scientific equipment ever made. Powerful microscopes have allowed the first direct glimpses of atoms, while new telescopes have revealed planets beyond our solar system. By the 21st century, science has become largely a team activity, involving ever more expensive apparatus and interdisciplinary cooperation.

The code of life

At the University of Chicago in 1953, American chemists Harold Urey and Stanley Miller set up an ingenious experiment to find out whether life could have started on Earth when lightning sparked chemical reactions in the atmosphere. In the same year, two molecular biologists—American James Watson and Briton Francis Crick—in a race against rival teams in the US and Soviet Union, figured out the molecular structure of deoxyribonucleic acid, or DNA, providing the key to the genetic code of life, which would lead less than half a century later to the complete mapping of the human genome.

Armed with new knowledge about the genetic mechanism, American biologist Lynn Margulis proposed the apparently absurd theory that some organisms can be absorbed by others, while both continue to flourish, and that this process had produced the complex cells of all multicellular life forms. After years of scepticism, she was vindicated by discoveries in genetics made 20 years after her proposal. American microbiologist Michael Syvanen showed how genes can jump from one species to another, while in the 1990s, the old Lamarckian idea that acquired characteristics can be passed on gained new traction with the discovery of epigenetics. Knowledge of the mechanisms by which evolution can take place was becoming far richer.

By the end of the century, American Craig Venter, fresh from running his own human genome

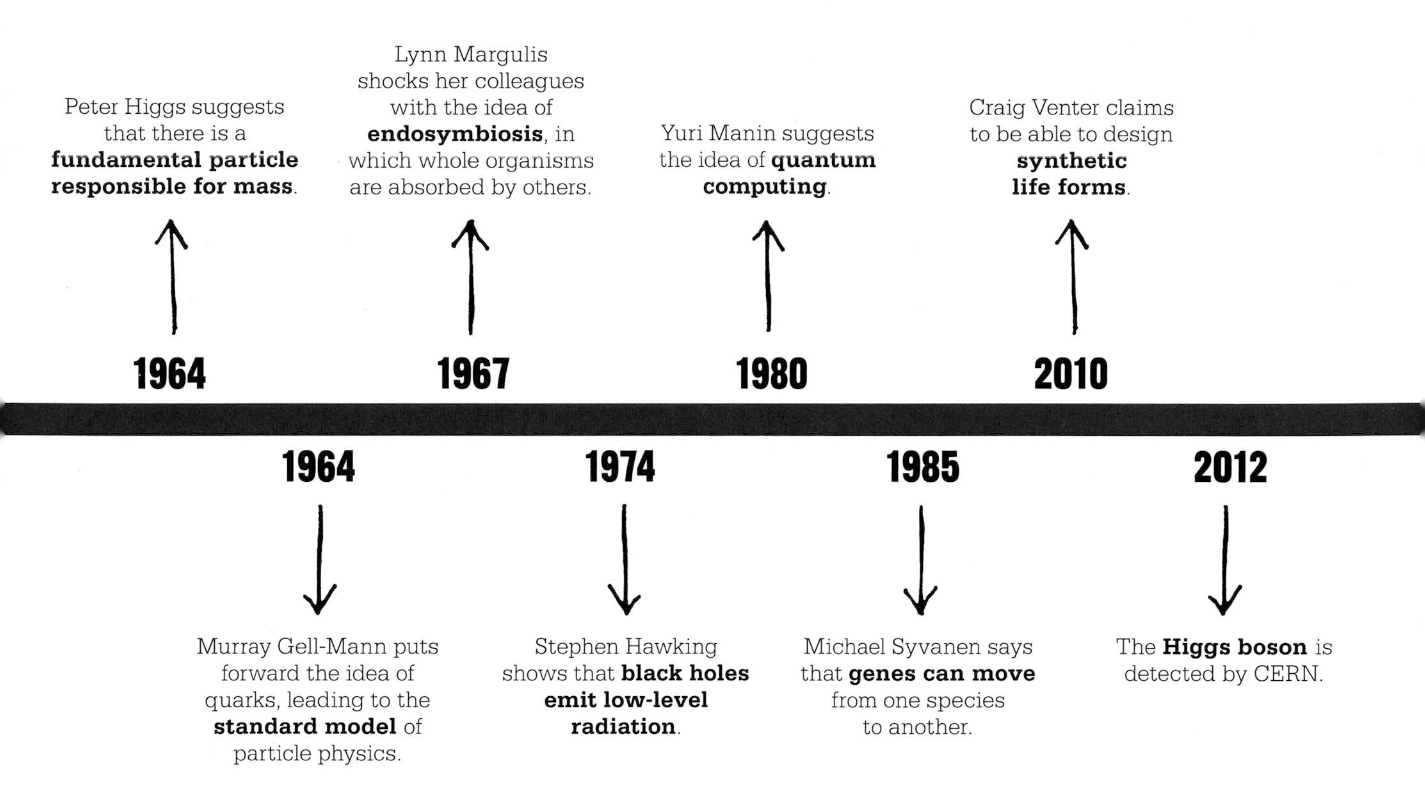

project, had created artificial life by planning its DNA on his computer. In Scotland, after many setbacks, Ian Wilmut and colleagues had succeeded in cloning a sheep.

New particles

In physics, the strangeness of quantum mechanics was further explored by American Richard Feynman and others, who explained quantum interactions in terms of exchange of "virtual" particles. Paul Dirac had correctly predicted the existence of antimatter in the 1930s, and in subsequent decades, more new subatomic particles emerged from the collisions of ever more powerful particle colliders. From this menagerie of exotic particles, the standard model of particle physics emerged, arranging the fundamental particles of nature according to their properties. Not all physicists were convinced, but the power of the standard model received a huge boost in 2012 when the Higgs boson it had predicted was detected by CERN's Large Hadron Collider.

Meanwhile, the search for a "theory of everything"—a theory that would unite all four fundamental forces of nature (gravity, electromagnetism, and the strong and weak nuclear forces)—took many new directions. American Sheldon Glashow united electromagnetism with the weak nuclear force into one "electroweak" theory, while string theory attempted to unite every theory of physics into one by proposing the existence of six hidden dimensions in addition to the three of space and one of time. American physicist Hugh Everett III suggested that there may be a mathematical basis for the existence of more than one universe. Everett's theory of a constantly splitting multiverse was at first ignored, but has gained supporters over the last few years.

Future directions

Deep puzzles remain to be solved, including an elusive theory that would unite quantum mechanics with general relativity. But tantalizing possibilities are also opening up, including a potential revolution in computing courtesy of the quantum mechanical qubit. It is probable that new problems we cannot even imagine will emerge. If the history of science is a guide, we should expect the unexpected. ■

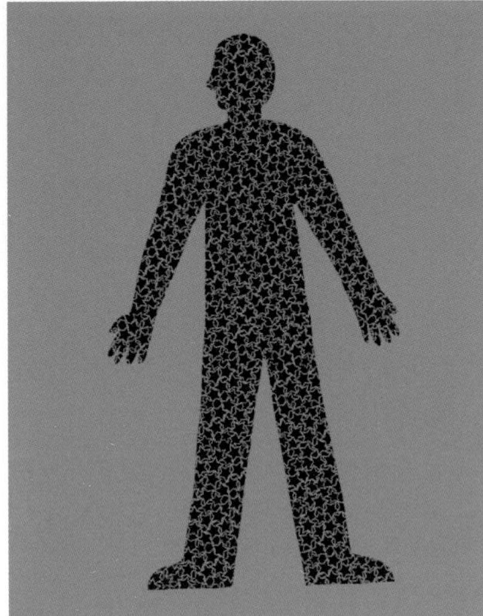

WE ARE MADE OF STARDUST

FRED HOYLE (1915–2001)

IN CONTEXT

BRANCH
Astrophysics

BEFORE
1854 German physicist Hermann von Helmholtz suggests that the Sun generates heat through slow gravitational contraction.

1863 English astronomer William Huggins' spectrum analysis of stars shows they share elements found on Earth.

1905–10 Astronomers in the US and Sweden analyze stars' luminosity and group them into dwarfs and giants.

1920 Arthur Eddington argues that stars turn hydrogen into helium through nuclear fusion.

1934 Fritz Zwicky coins the term "supernova" for a massive star's explosive end.

AFTER
2013 Deep-sea fossils reveal what may be biological traces of iron from a supernova.

The idea that stars generate energy through the process of nuclear fusion was first proposed by British astronomer Arthur Eddington in 1920. Stars, he argued, were factories for fusing nuclei of hydrogen into helium. A helium nucleus contains slightly less mass than the four hydrogen nuclei required to create it. This mass is converted into energy in accordance with the equation $E = mc^2$. Eddington developed a model of star structure in terms of the balance between the inward pull of gravity and the outward pressure of escaping radiation, but he did not figure out the physics of the nuclear reactions involved.

Making heavier elements

In 1939, German-born US physicist Hans Bethe published a detailed analysis of the different pathways that hydrogen fusion might take. He identified two routes—a slow, low-temperature chain that dominates in stars like our Sun, and a rapid, high-temperature cycle that dominates in more massive stars. Between 1946 and 1957, British astronomer Fred Hoyle and others developed Bethe's ideas to show how further fusion reactions involving helium could generate carbon and heavier elements up to and including the mass of iron. This explained the origin of many of the universe's heavier elements. We now know that elements heavier than iron form in supernova explosions—the death throes of massive stars. The elements needed for life are made in stars. ■

Space isn't remote at all. It's only an hour's drive away if your car could go straight upwards.
Fred Hoyle

See also: Marie Curie 190–95 ▪ Albert Einstein 214–21 ▪ Ernest Rutherford 206–13 ▪ Georges Lemaître 242–45 ▪ Fritz Zwicky 250–51

JUMPING GENES

BARBARA McCLINTOCK (1902–1992)

IN CONTEXT

BRANCH
Biology

BEFORE
1866 Gregor Mendel describes inheritance as a phenomenon determined by "particles"—later called genes.

1902 Theodor Boveri and Walter Sutton independently conclude that chromosomes are involved in inheritance.

1915 Thomas Hunt Morgan's fruit fly experiments confirm earlier theories and show that genes can be linked together on the same chromosome.

AFTER
1953 James Watson and Francis Crick's double-helix model of the DNA that makes up chromosomes shows how genetic material is replicated.

2000 The first human genome is published, cataloguing the location of 20,000–25,000 genes on humans' 23 pairs of chromosomes.

In the early 20th century, the laws of inheritance that had been described by Gregor Mendel in 1866 were refined as new discoveries were made about the particles of inheritance, identified as genes, and the microscopic threads that carry them, called chromosomes. In the 1930s, American geneticist Barbara McClintock first realized that chromosomes were not the stable structures previously imagined, and that the position of genes in chromosomes could alter.

Exchanging genes

McClintock was studying inheritance in corn plants. A corncob has hundreds of kernels, each colored yellow, brown, or streaked, according to the cob's genes. A kernel is a seed—a single offspring—so studying many cobs gives a range of data on the inheritance of kernel color. McClintock combined breeding experiments with microscope work on chromosomes. In 1930, she found that, during sexual reproduction, chromosomes paired up when sex cells were formed, creating an X shape. She realized that these X-shaped structures marked locations where chromosome pairs were exchanging segments. Genes that were once linked together on the same chromosome were shuffled around, which resulted in new traits, including variable colors.

Variable colors in corn prompted McClintock to trace the genetic recombinations responsible for this variety, which she reported in 1951.

This shuffling of genes—called genetic recombination—produces a far greater genetic variety in the offspring. As a result, the chances of survival in different environments are enhanced. ■

See also: Gregor Mendel 166–71 ▪ Thomas Hunt Morgan 224–25 ▪ James Watson and Francis Crick 276–83 ▪ Michael Syvanen 318–19

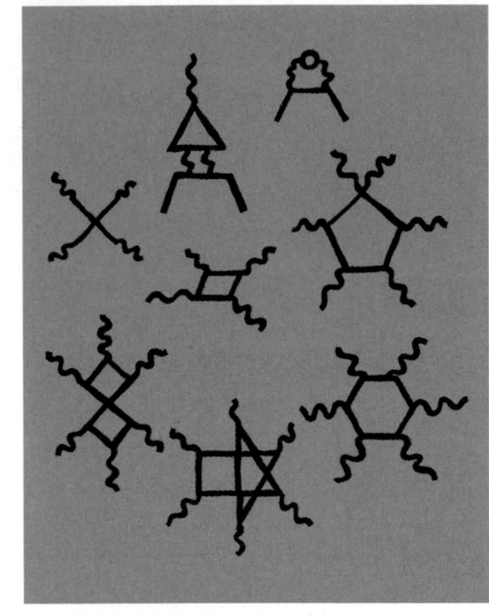

THE STRANGE THEORY OF LIGHT AND MATTER

RICHARD FEYNMAN (1918–1988)

IN CONTEXT

BRANCH
Physics

BEFORE
1925 Louis de Broglie suggests that any particle with mass can behave like a wave.

1927 Werner Heisenberg shows there is an inherent uncertainty in certain pairs of values at the quantum level, such as the position and momentum of a particle.

1927 Paul Dirac applies quantum mechanics to fields rather than single particles.

AFTER
Late 1950s Julian Schwinger and Sheldon Glashow develop the electroweak theory, which unites the weak nuclear force with electromagnetism.

1965 Moo-Young Han, Yoichiro Nambu, and Oscar Greenberg explain the interaction of particles under the strong force in terms of a property now known as "color charge."

One of the questions to arise from the quantum mechanics of the 1920s was how particles of matter interacted by means of forces. Electromagnetism also needed a theory that worked on the quantum scale. The theory that emerged, quantum electrodynamics (QED), explained the interaction of particles through the exchange of electromagnetism. It has proved very successful, although one of its pioneers, Richard Feynman, called it a "strange" theory because the picture of the universe that it describes is hard to visualize.

Messenger particles

Paul Dirac made the first step toward a theory of QED based on the idea that electrically charged particles interacted through the exchange of quanta, or "photons," of electromagnetic energy—the same electromagnetic quanta that comprise light. Photons can be created out of nothing for very brief periods of time in accordance with Heisenberg's uncertainty principle, and this allows fluctuations in the amount of energy available in "empty" space. Such photons are sometimes called "virtual" particles, and physicists have subsequently confirmed their involvement in electromagnetism. More generally, the messenger particles in quantum field theories are known as "gauge bosons."

However, there were problems with QED. Most significantly, its equations often generated nonsensical infinite values.

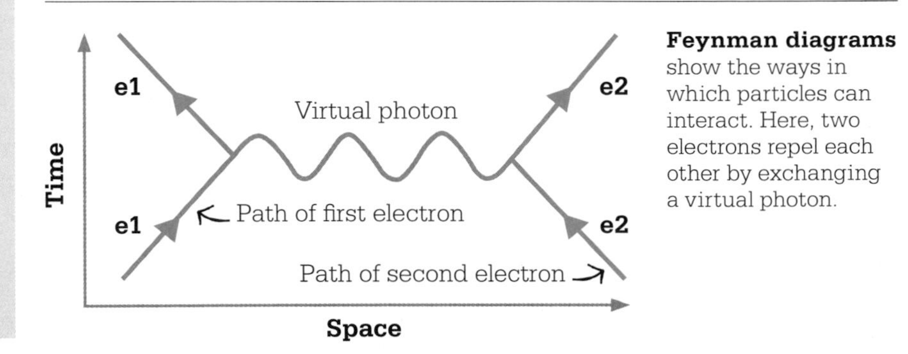

Feynman diagrams show the ways in which particles can interact. Here, two electrons repel each other by exchanging a virtual photon.

See also: Erwin Schrödinger 226–33 ▪ Werner Heisenberg 234–35 ▪ Paul Dirac 246–47 ▪ Sheldon Glashow 292–93

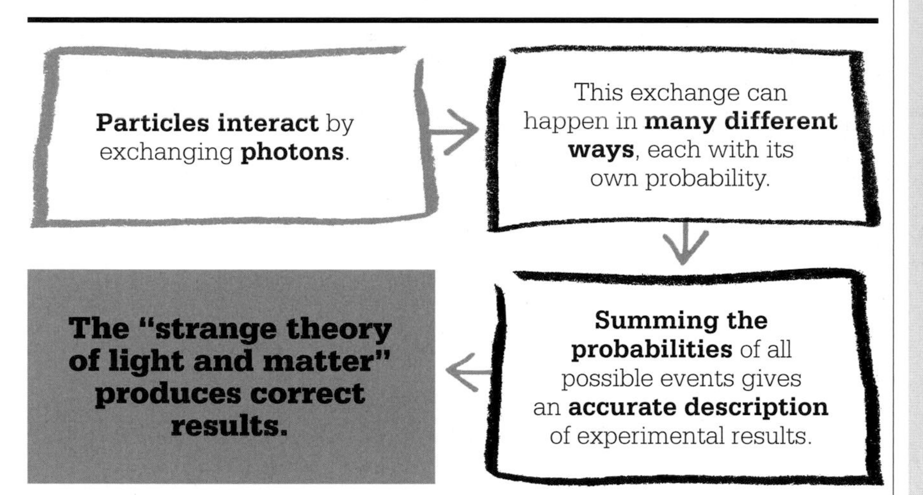

Summing probabilities

In 1947, German physicist Hans Bethe suggested a way of fixing the equations so that they mirrored real laboratory results. In the late 1940s, Japanese physicist Sin-Itiro Tomonaga, Americans Julian Schwinger and Richard Feynman, and others took Bethe's ideas and developed them to produce a mathematically sound version of QED. It produced meaningful results by considering all the possible ways that interactions could take place according to quantum mechanics.

Feynman made this complex subject approachable through his invention of "Feynman diagrams"—simple pictorial representations of possible electromagnetic interactions between particles, which provide an intuitive description of the processes at work. The key breakthrough was to find a mathematical way of modeling an interaction as a sum of the probabilities of each individual pathway, which include pathways in which particles move backward in time. When summed, many of the probabilities cancel each other out: for example, the probability of a particle traveling in a particular direction may be the same as the probability of it traveling in the opposite direction, so adding these probabilities gives a sum of zero. Summing every possibility, including the "strange" ones involving backward time travel, produces familiar results such as light appearing to travel in straight lines. However, under certain conditions, the summed probabilities do produce strange results, and experiments have shown that light does not always necessarily travel in straight lines. As such, QED provides an accurate description of reality even if it feels alien to the world we perceive.

QED proved so successful that it has become a model for similar theories of other fundamental forces—the strong nuclear force has been successfully described by quantum chromodynamics (QCD), while the electromagnetic and weak nuclear forces have been unified in a combined electroweak gauge theory. Only gravitation so far refuses to conform to this kind of model. ▪

Richard Feynman

Born in New York in 1918, Richard Feynman showed a talent for mathematics at an early age, and earned a degree at Massachusetts Institute of Technology (MIT) before attaining a perfect score in mathematics and physics for his graduate entrance exam to Princeton. After receiving his PhD in 1942, Feynman worked under Hans Bethe in the Manhattan Project to develop the atomic bomb. Following the end of World War II, he continued his work with Bethe at Cornell University, where he did his most important work on QED.

Feynman showed a flair for communicating his ideas. He promoted the potential of nanotechnology, and late in his life wrote bestselling accounts of QED and other aspects of modern physics.

Key works

1950 *Mathematical Formulation of the Quantum Theory of Electromagnetic Interaction*
1985 *QED: The Strange Theory of Light and Matter*
1985 *Surely You're Joking, Mr. Feynman?*

LIFE IS NOT A MIRACLE

HAROLD UREY (1893–1981)
STANLEY MILLER (1930–2007)

IN CONTEXT

BRANCH
Chemistry

BEFORE
1871 Charles Darwin suggests that life might have begun in "some warm little pond."

1922 Russian biochemist Alexander Oparin proposes that complex compounds might have formed in a primitive atmosphere.

1952 In the US, Kenneth A. Wilde passes 600-volt sparks through a mixture of carbon dioxide and water vapor, and obtains carbon monoxide.

AFTER
1961 Spanish biochemist Joan Oró adds further likely chemicals to the Urey–Miller mix and obtains molecules vital for DNA, among others.

2008 Miller's former student Jeffrey Bada and others use newer, more sensitve techniques to obtain many more organic molecules.

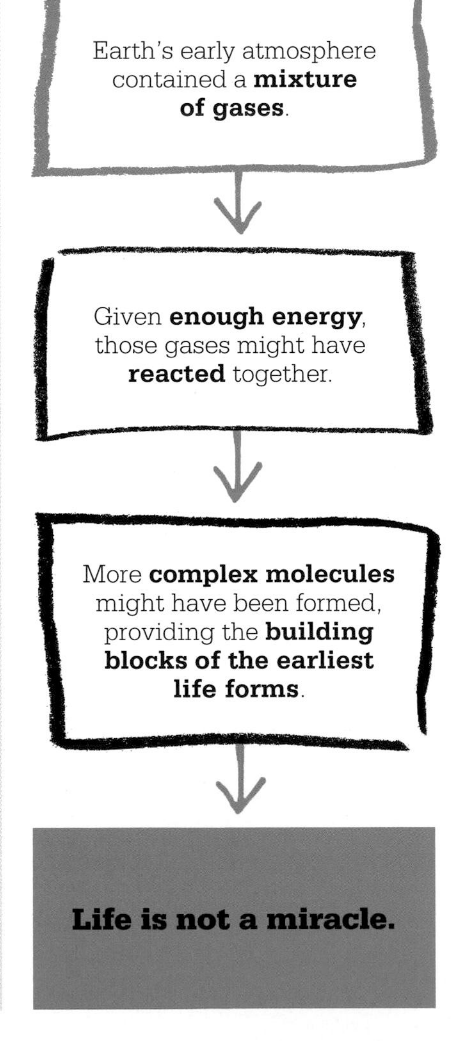

Scientists have long pondered the origin of life. In 1871, Charles Darwin wrote in a letter to his friend Joseph Hooker, "But if…we could conceive in some warm little pond, with all sorts of ammonia and phosphoric salts, lights, heat, electricity etc present, that a protein compound was chemically formed ready to undergo still more complex changes…" In 1953, American chemist Harold Urey and his student Stanley Miller found a way to replicate Earth's early atmosphere in the laboratory, and generated from inorganic matter organic (carbon-based) compounds that are essential to life.

Before the Urey–Miller experiment, advances in chemistry and astronomy had analyzed the atmospheres on the other, lifeless planets in the solar system. In the 1920s, Soviet biochemist Alexander Oparin and British geneticist J. B. S. Haldane independently suggested that if conditions on prebiotic (prelife) Earth resembled those planets, then simple chemicals could have reacted together in a primordial soup to form more complex molecules, from which living things might have evolved.

See also: Jöns Jakob Berzelius 119 ▪ Friedrich Wöhler 124–25 ▪ Charles Darwin 142–49 ▪ Fred Hoyle 270

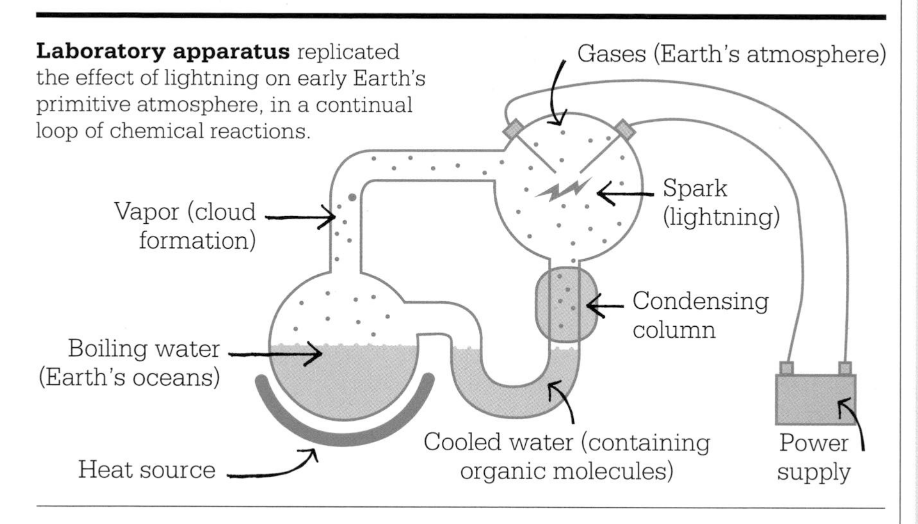

Laboratory apparatus replicated the effect of lightning on early Earth's primitive atmosphere, in a continual loop of chemical reactions.

Recreating Earth's early atmosphere

In 1953, Urey and Miller carried out the first prolonged experiment to test the Oparin–Haldane theory. In a closed series of connected glass flasks, sealed from the atmosphere, they put water and a mixture of gases thought to have been present in Earth's primitive atmosphere—hydrogen, methane, and ammonia. The water was heated so that water vapor formed and wafted its way around all the flasks in the closed loop. In one of the flasks was a pair of electrodes, between which sparks were passed continuously to represent lightning—one of the hypothetical triggers for primordial reactions. The sparks provided enough energy to break up some of the molecules, and generate highly reactive forms that would go on to react with other molecules.

Within a day, the mixture had turned pink, and after two weeks Urey and Miller found that at least 10 percent of the carbon (from the methane) was now in the form of other organic compounds. Two percent of the carbon had formed amino acids, which are the vital building blocks of the proteins in all living things. Urey encouraged Miller to send a paper about the experiment to the journal *Science*, which published it as "Production of amino acids under possible primitive earth conditions." The world could now imagine how Darwin's "warm little pond" may have generated the first life forms.

In an interview, Miller said that "just turning on the spark in a basic prebiotic experiment will yield amino acids." Scientists later found, using better equipment than was available in 1953, that the original experiment had produced at least 25 amino acids—more than are found in nature. Since Earth's early atmosphere almost certainly contained carbon dioxide, nitrogen, hydrogen sulphide, and sulfur dioxide released from volcanoes, a much richer mixture of organic compounds might well have been created then—and was indeed formed in subsequent experiments. Meteorites containing dozens of amino acids, some found on Earth and others not, have also spurred on the search for signs of life on planets beyond the solar system. ■

Harold Urey and Stanley Miller

Harold Clayton Urey was born in Walkerton, IN. His work on the separation of isotopes led to the discovery of deuterium, which won him the Nobel Prize in Chemistry in 1934. He went on to develop enrichment of uranium-235 by gaseous diffusion which was crucial for the Manhattan Project's development of the first atomic bomb. After his prebiotic experiments with Stanley Miller in Chicago he moved to San Diego and studied the Moon rocks brought back by Apollo 11.

Stanley Lloyd Miller was born in Oakland, CA. After studying chemistry at the University of California at Berkeley, he was a teaching assistant at the University of Chicago, and worked with Urey. Later, he became a professor in San Diego.

Key work

1953 *Production of Amino Acids under Possible Primitive Earth Conditions*

> My study [of the universe] leaves little doubt that life has occurred on other planets. I doubt if the human race is the most intelligent form of life.
> **Harold C. Urey**

WE WISH TO SUGGEST A STRUCTURE FOR THE SALT OF DEOXYRIBOSE NUCLEIC ACID (DNA)

JAMES WATSON (1928–)
FRANCIS CRICK (1916–2004)

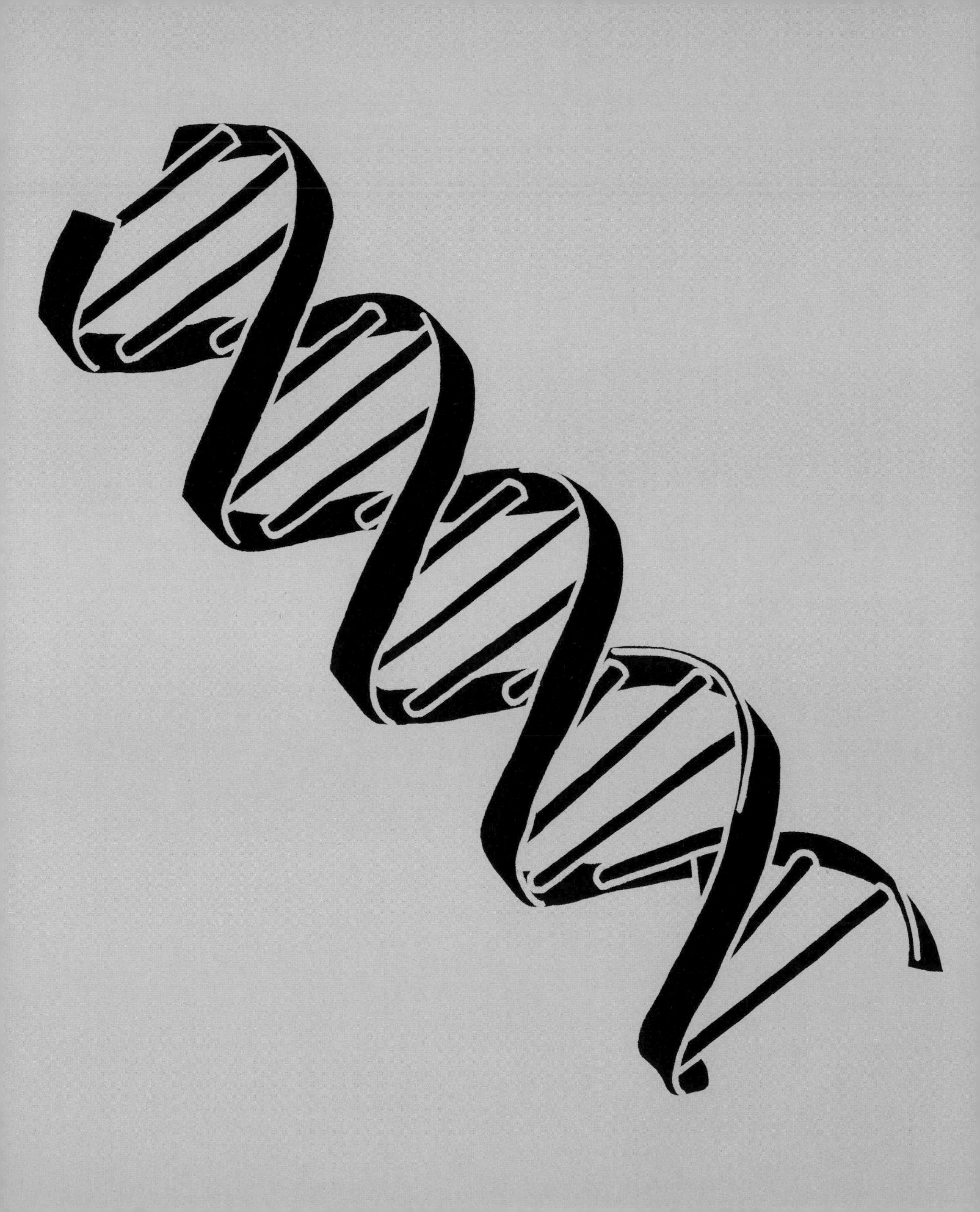

IN CONTEXT

BRANCH
Biology

BEFORE
1869 Friedrich Miescher first identifies DNA, in blood cells.

1920s Phoebus Levene and others analyze the components of DNA as sugars, phosphates, and four types of base.

1944 Experiments show DNA to be a carrier of genetic data.

1951 Linus Pauling proposes the alpha-helix structure for certain biological molecules.

AFTER
1963 Frederick Sanger develops sequencing methods to identify bases along DNA.

1960s DNA's code is cracked: three DNA bases of code for each amino acid in a protein.

2010 Craig Venter and his team implant artificially made DNA into a living bacterium.

In April 1953, the answer to a fundamental mystery about living organisms appeared in a short article published without fanfare in the scientific journal, *Nature*. The article explained both how genetic instructions are held inside organisms and how they are passed on to the next generation. Crucially, it described, for the first time, the double-helix structure of deoxyribose nucleic acid (DNA), the molecule that contains the genetic information.

The article was written by James Watson, a 29-year-old American biologist, and his older British research colleague, biophysicist Francis Crick. Since 1951, they had jointly been working on the challenge of DNA's structure at the Cavendish Laboratory, University of Cambridge, under its director, Sir Lawrence Bragg.

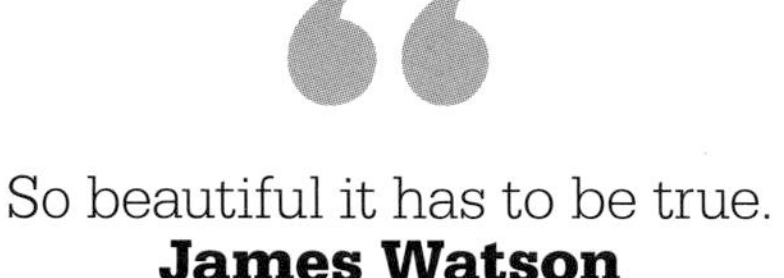

So beautiful it has to be true.
James Watson

DNA was the hot topic of the day, and an understanding of its structure seemed so tantalizingly within reach that by the early 1950s, teams in Europe, the US, and the Soviet Union were vying to be the first to "crack" DNA's three-dimensional shape—the elusive model that allowed DNA simultaneously to carry genetic data in some kind of chemically coded form, and to replicate itself completely and accurately, so that the same genetic data was passed to offspring, or daughter cells, including those of the next generation.

The past in DNA

The DNA molecule was not discovered in 1953, as is often popularly thought, nor were Crick and Watson the first to find out what it was made from. DNA has a much longer history of research. In the 1880s, the German biologist Walther Flemming had reported that "X"-like bodies (later named chromosomes) appeared inside cells as the cells

James Watson and Francis Crick

James Watson (on the right) was born in 1928 in Chicago, IL. At the precocious age of 15 he entered the University of Chicago. After postgraduate study in genetics, Watson moved to Cambridge, England, to team up with Francis Crick. He later returned to the US to work at the Cold Spring Harbor Laboratory in New York. From 1988, he worked on the Human Genome Project, but left after a disagreement over patenting genetic data.

Francis Crick was born in 1916 near Northampton in Britain. He developed antisubmarine mines during World War II. In 1947, he went to Cambridge to study biology and here began work with James Watson. Later, Crick became known for the "central dogma": that genetic data flow in cells in essentially one way. In later life, Crick turned to brain research and developed a theory of consciousness.

Key works

1953 *Molecular Structure of Nucleic Acids: A Structure for Deoxyribose Nucleic Acid*
1968 *The Double Helix*

See also: Charles Darwin 142–49 ▪ Gregor Mendel 166–71 ▪ Thomas Hunt Morgan 224–25 ▪ Barbara McClintock 271 ▪ Linus Pauling 254–59 ▪ Craig Venter 324–25

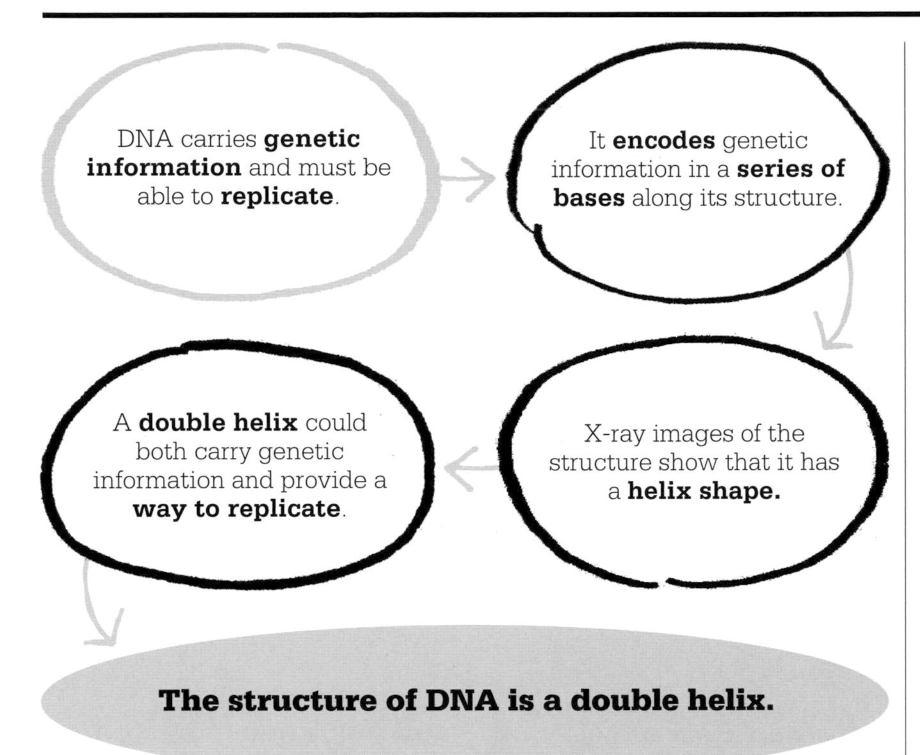

were preparing to divide. In 1900, Gregor Mendel's experiments with heredity in pea plants were rediscovered—Mendel had been the first to suggest that there were units of heredity that came in pairs (which would later be called genes). At about the same time as Mendel was being rediscovered, breeding experiments by American physician Walter Sutton and, independently, by German biologist Theodor Boveri revealed that sets of chromosomes (the rod-shaped structures that carry genes) pass from a dividing cell to each of its daughter cells. The ensuing Sutton–Boveri theory proposed that chromosomes are the carriers of genetic material.

Soon, more scientists were investigating these mysterious X-shaped bodies. In 1915, American biologist Thomas Hunt Morgan showed that chromosomes were indeed the carriers of hereditary information. The next step was to look at the constituent molecules of chromosomes—molecules that might be candidates for genes.

New pairs of genes

In the 1920s, two types of candidate molecules were discovered: proteins called histones, and nucleic acids, which had been described chemically in 1869 as *nuclein* by Swiss biologist Friedrich Miescher. The Russian-American biochemist Phoebus Levene and others gradually identified the main ingredients of DNA in increasing detail as nucleotide units, each made up of a deoxyribose sugar, a phosphate, and one of four subunits called bases. By the end of the 1940s, the basic formula of DNA as a giant polymer—a huge molecule consisting of repeating units, or monomers—was clear. By 1952, experiments with bacteria had shown that DNA itself, and not its rival candidates, the proteins inside chromosomes, was the physical embodiment of genetic information.

Tricky research tools

The competing researchers were using several advanced research tools, including X-ray diffraction crystallography, in which X-rays were passed through a substance's crystals. A crystal's unique geometry in terms of its atomic content made the X-ray beams diffract, or bend, as they passed through. The resulting diffraction patterns of spots, lines, and blurs were captured on photographic film. Working backward from those patterns, it was possible to figure out the structural details within the crystal. This was not an easy task. X-ray crystallography has been »

It is one of the more striking generalizations of biochemistry…that the twenty amino acids and the four bases, are, with minor reservations, the same throughout Nature.

Francis Crick

likened to studying the myriad light patterns cast by a crystal chandelier on the ceiling and walls of a large room, and using them to figure out the shapes and positions of each piece of glass in the chandelier.

Pauling in the lead

The British research team at the Cavendish Laboratory was eager to beat the American researchers, led by Linus Pauling. In 1951, Pauling and his colleagues Robert Corey and Herman Branson had already achieved a breakthrough in molecular biology when they correctly proposed that many biological molecules—including hemoglobin, the oxygen-carrying substance in blood—have a corkscrew-like helix shape. Pauling named this molecular model the alpha-helix.

Pauling's breakthrough had narrowly beaten the Cavendish Laboratory and it looked as though the precise shape of DNA's structure was within his grasp. Then, early in 1953, Pauling proposed that the structure of DNA was in the form of a triple helix.

By this time, James Watson was working at the Cavendish Laboratory. He was only 25 years old, but he had the enthusiasm of youth and two degrees in zoology, and had studied the genes and nucleic acids of bacteriophages—the viruses that infect bacteria. Crick, 37 years old, was a biophysicist with an interest in the brain and neuroscience. He had studied proteins, nucleic acids, and other giant molecules in living things. He had also observed the Cavendish team racing to beat Pauling to the alpha-helix idea, and later analyzed their mistaken suppositions and dead-end exploratory efforts.

Both Watson and Crick also had experience of X-ray crystallography, albeit in different areas, and together they soon began musing on two questions that fascinated them both: how does DNA as a physical molecule encode genetic information, and how is this information translated into the parts of a living system?

Crucial crystal pictures

Watson and Crick knew of Pauling's success with the alpha-helix model of proteins, in which the molecule twisted along a single corkscrew path, repeating its main structure every 3.6 turns. They also knew that the latest research evidence did not seem to support Pauling's triple helix model for DNA. This led them to wonder whether the elusive model was one that was neither a single nor a triple helix. The two conducted hardly any experiments

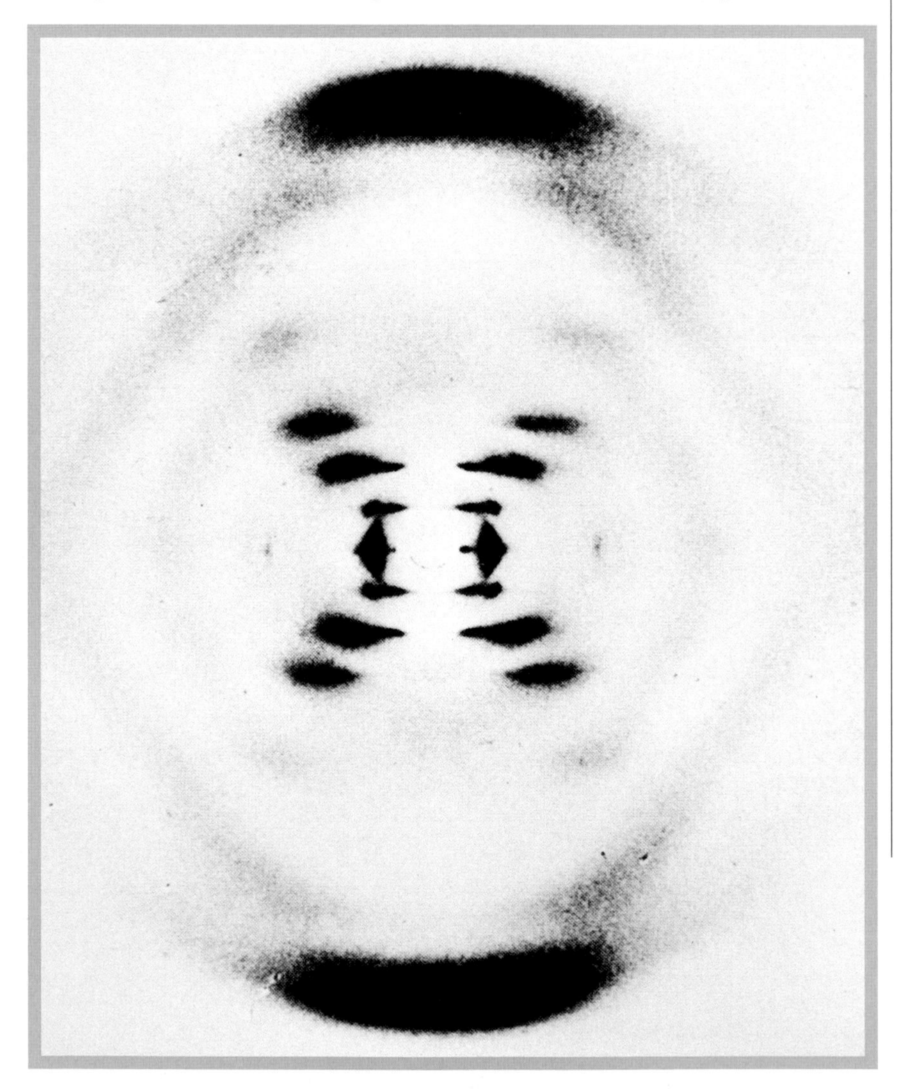

This X-ray diffraction photograph of DNA was obtained by Rosalind Franklin in 1953, and was the biggest clue to cracking DNA. The helical structure of DNA was ascertained from the pattern of spots and bands.

themselves. Instead they collected data from others, including the results of chemical experiments that gave information about the angles of the links, or bonds, between the various ingredient atoms and subgroups of DNA. They also pooled their joint knowledge of X-ray crystallography and approached those researchers who had made the highest-quality images of DNA and other similar molecules. One such image was "photo 51," which became key to their achieving their breakthrough.

Photo 51 was an X-ray diffraction image of DNA that resembled an "X" seen through the slats of a Venetian blind—fuzzy to our eyes, but at that time, among the sharpest and most informative of DNA's X-ray pictures. Some debate surrounds the identity of the photographer who took this historic picture. It came from the laboratory of a British biophysicist named Rosalind Franklin, an expert in X-ray crystallography, and her graduate student Raymond Gosling, at King's College, London. Each has been credited with the image at various times.

Cardboard models

Also working at King's was Maurice Wilkins, a physicist who was interested in molecular biology.

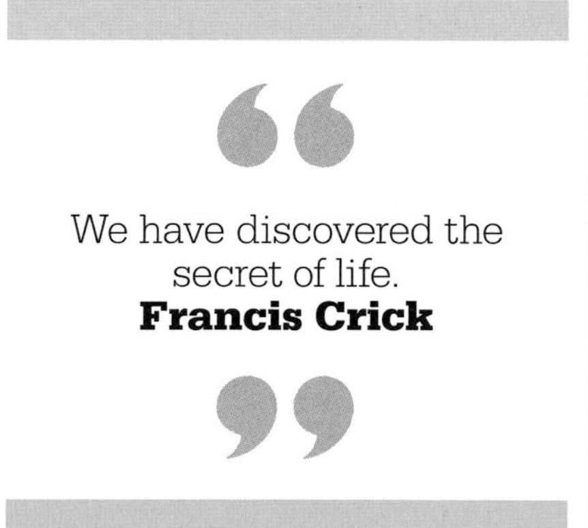

Rosalind Franklin's draft reports on her theoretical models for DNA's structure were key to Watson and Crick's discovery of the double helix, but she received little recognition in her lifetime.

In early 1953, in what was perhaps a break with scientific protocol, Wilkins showed the images taken by Franklin and Gosling, without their permission or knowledge, to James Watson. The American immediately recognized their significance, and took the implications straight back to Crick. Suddenly their work was on the right path.

From this point, the exact sequence of events becomes unclear, and later accounts of the discovery are conflicting. Franklin had described in unpublished draft reports her thoughts about the structure and shape of DNA. These were also incorporated by Watson and Crick as they struggled with their various proposals. The main idea, derived from Pauling's alpha-helix model and supported by Wilkins, centered on some form of repeating helical pattern for the giant molecule. One of Franklin's considerations was whether the structural "backbone," a chain of phosphate and deoxyribose sugar subunits, was in the center with the bases projecting outward, or the other way around. Another colleague who provided help was Austrian-born British biologist Max Perutz, who would win the Nobel prize in Chemistry in 1962 for his work on the structure of hemoglobin and other proteins. Perutz also had access to Franklin's unpublished reports and passed them on to the ever-networking Watson and Crick. They pursued the idea that DNA's backbones were on the outside, with the bases pointing inward and perhaps connecting to each other in pairs. They cut out and shuffled around cardboard shapes that represented these molecular subunits: phosphates and sugars in the backbone, and the four types of base—adenine, thymine, guanine, and cytosine.

In 1952, Watson and Crick had met Erwin Chargaff, an Austrian-born biochemist, who had devised what became known as Chargaff's first rule. This stated that in DNA, the amounts of guanine and cytosine are equal, as are the amounts of adenine and thymine. Experiments had sometimes shown that all four amounts were roughly equal, and sometimes not. The latter findings came to be seen as errors in methodology, and equal amounts of all four bases came to be accepted as the rule of thumb.

Making the pieces fit

By splitting the base quantities into two sets of pairs, Chargaff had shed light on the structure of DNA. Watson and Crick now began to think of adenine as only and always linking to thymine, and guanine to cytosine. »

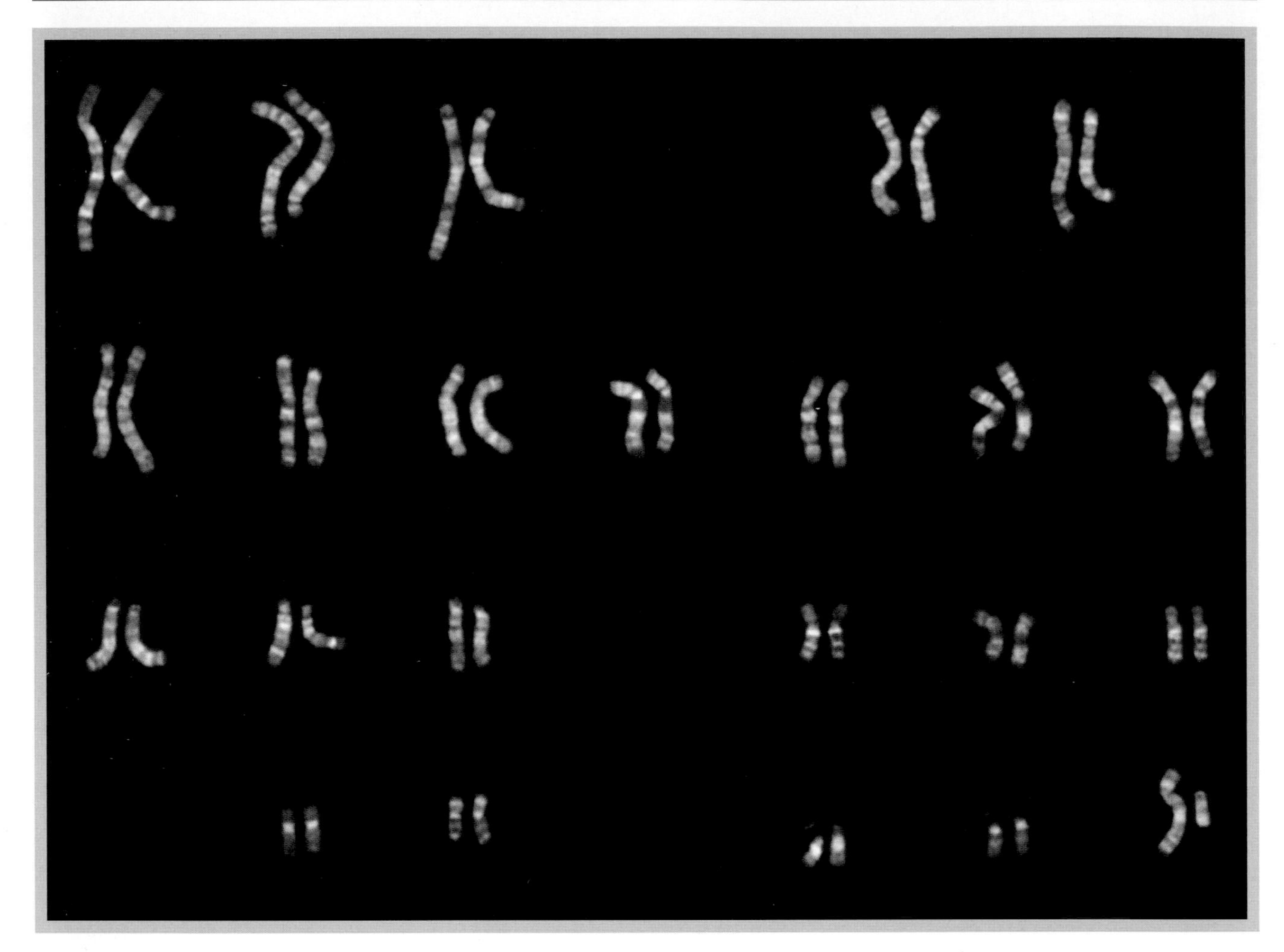

These are human male chromosomes. Before Crick and Watson's discovery, it had been known that chromosomes carry genes that pass from a dividing cell to a daughter cell.

In assembling the cardboard pieces for their 3-D jigsaw, Watson and Crick were juggling a vast amount of data, working from mathematics, X-ray images, their own knowledge of chemical bonds and their angles, and other data—all approximate and subject to ranges of errors. Their final breakthrough came when they realized that making slight adjustments to the configurations of thymine and guanine allowed the pieces to fit together, producing an elegant double helix in which the pairs of bases linked along the middle. Unlike the protein alpha-helix, which had 3.6 subunits in one complete turn, DNA had about 10.4 subunits per turn.

The model that Watson and Crick described consisted of two helical or corkscrew phosphate-sugar backbones curling around each other, like the uprights of a "twisted ladder," connected by pairs of bases serving as rungs. The sequence of bases worked like letters in a sentence, carrying small units of information that combined to make an overall instruction, or gene—which in turn told the cell how to make the particular protein or other molecule that was the physical manifestation of the genetic data and had a particular role in the cell's fabric and function.

Zip and unzip

Each pair of bases is connected by what chemists call hydrogen bonds. These are made and broken relatively easily, so the sections of the double-helix can be "unzipped" by undoing the bonds, which then exposes the code of bases as a template for making a copy.

This zip-unzip allowed two processes to occur. First, a mirror complementary copy of nucleic acid could be made from one unzipped half of the double helix; then, carrying its genetic information as the sequence of bases, it would leave the cell nucleus to become involved in the protein production.

Second, when the whole length of the double helix was unzipped, each part would act as a template to build a new complementary partner—resulting in two lengths of DNA that were identical to the original and to each other. In this way, DNA was copied as cells divided into two for growth and repair throughout an organism's life—and as sperm and eggs, the sex cells, carried their quotient of the genes to make a fertilized egg, so beginning the next generation.

"Secret of life"

On February 28, 1953, elated by their discovery, Watson and Crick went for lunch to The Eagle, one of Cambridge's oldest inns, where colleagues from the Cavendish and other laboratories often met. Crick is said to have startled drinkers by announcing that he and Watson had discovered "the secret of life"—or so Watson later recalled in his book, *The Double Helix*, though Crick denied this really happened.

In 1962, Watson, Crick, and Wilkins were awarded the Nobel Prize in Physiology or Medicine "for their discoveries concerning the molecular structure of nucleic acids and its significance for information transfer in living material." The award, however, was surrounded in controversy. In the preceding years, Rosalind Franklin had received little official credit for producing the key X-ray images and for writing the reports that helped to direct Watson and Crick's research. She died of ovarian cancer in 1958, at only 37, and was therefore ineligible for the Nobel Prize in 1962, since the prizes are not awarded posthumously. Some said the award should have been made earlier, with Franklin as one of the co-recipients, but the rules allow a maximum of three.

Following their momentous work, Watson and Crick became world celebrities. They continued their research in molecular biology and received great numbers of awards and honors. Now that the structure of DNA was known, the next big challenge was to solve the genetic "code." By 1964, scientists figured out how sequences of its bases were translated into the amino acids that make up specific proteins and other molecules that are the building blocks of life.

Today, scientists can identify base sequences for all the genes of an organism, collectively known as its genome. They can manipulate DNA to move genes around, delete them from specific lengths of DNA, and insert them into others. In 2003, the Human Genome Project, the largest international biological research project ever, announced that it had completed the mapping of the human genome—a sequence of more than 20,000 genes. Crick and Watson's discovery had paved the way for genetic engineering and gene therapy. ■

> I never dreamed that in my lifetime my own genome would be sequenced.
> **James Watson**

A DNA molecule is a double helix formed by base pairs attached to a backbone made of sugar-phosphates. The base pairs always match up in combinations of either adenine–thymine or cytosine–guanine.

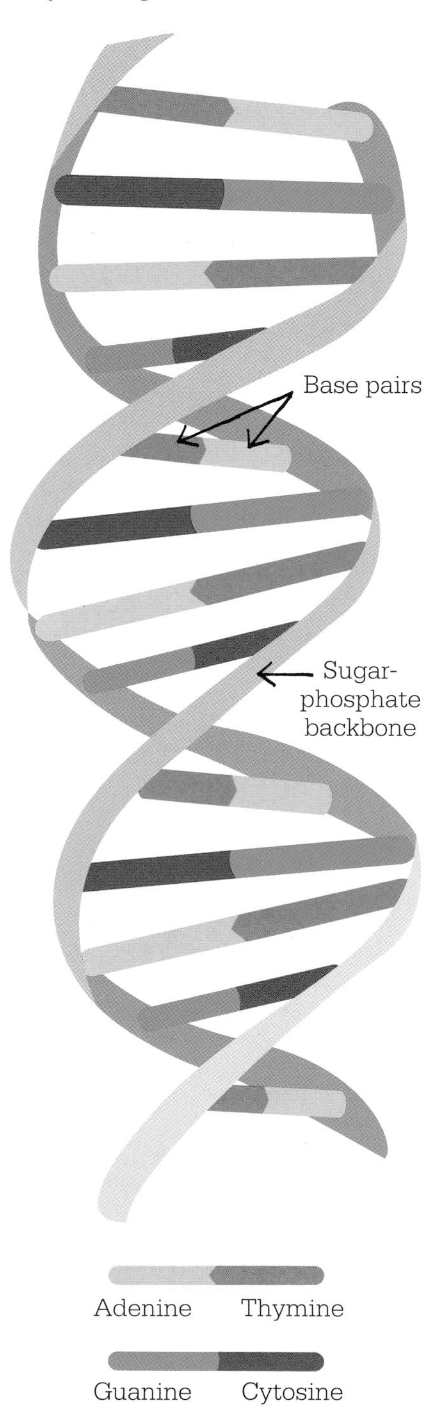

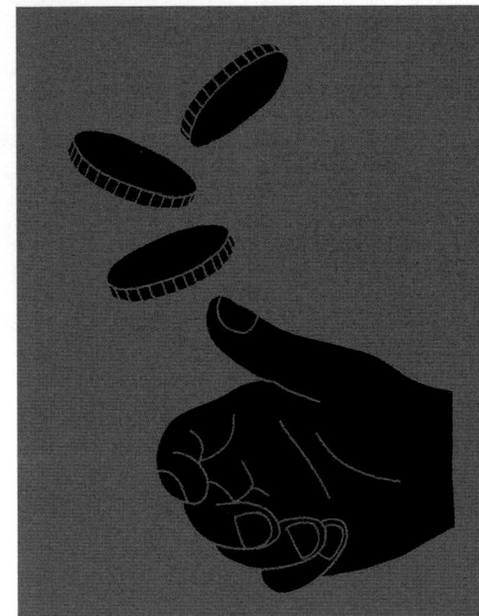

EVERYTHING THAT CAN HAPPEN HAPPENS

HUGH EVERETT III (1930–1982)

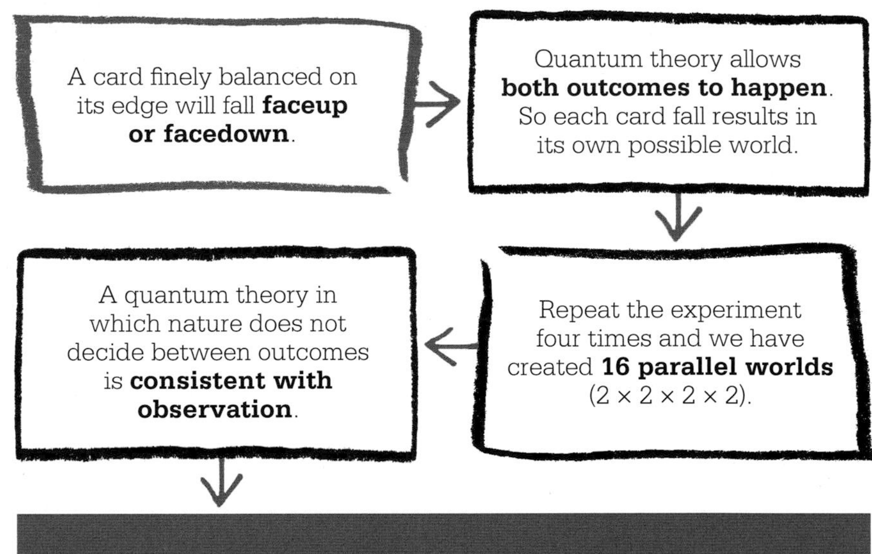

IN CONTEXT

BRANCH
Physics and cosmology

BEFORE
1600 Italian philosopher Giordano Bruno is burned at the stake for his belief in an infinity of inhabited worlds.

1924–27 Niels Bohr and Werner Heisenberg seek to resolve the measurement paradox of wave-particle duality by invoking a wave function collapse.

AFTER
1980s A principle known as decoherence attempts to provide a mechanism by which the many-worlds interpretation may work.

2000s Swedish cosmologist Max Tegmark describes an infinity of universes.

2000s In quantum computer theory, computational power is sourced from superpositions that are not in our universe.

Hugh Everett III is a cult figure to sci-fi enthusiasts because his many-worlds interpretation (MWI) of quantum mechanics changed scientists' ideas about the nature of reality.

Everett's work was inspired by the embarrassing flaw at the heart of quantum mechanics. Although it can explain interactions at the most fundamental level of matter, quantum mechanics also produces bizarre results that seem to be at odds with experiment, a dichotomy at the heart of the measurement paradox (pp.232–33).

In the quantum world, subatomic particles are allowed to exist in any number of possible states of location, velocity, and spin, or "superpositions," as described by Erwin Schrödinger's wave function, but the phenomenon of many possibilities disappears as

See also: Max Planck 202–05 ▪ Werner Heisenberg 234–35 ▪ Erwin Schrödinger 226–33

"Multiverse" is an installation of 41,000 LED lights at the National Gallery of Art in Washington DC. It was inspired by the many-worlds interpretation.

soon as it is observed. The very act of measuring a quantum system seems to "shunt" it into one state or another, forcing it to "choose" its option. In the world we're familiar with, a coin toss results in a definite heads or tails, and not one, the other, and both at once.

Copenhagen fudge

In the 1920s, Niels Bohr and Werner Heisenberg attempted to sidestep the measurement problem with what became known as the Copenhagen interpretation. It holds that the act of making an observation on a quantum system causes the wave function to "collapse" into the single outcome. Although this remains a widely accepted interpretation, many theorists find it unsatisfactory since it reveals nothing about the mechanism of wave function collapse. This bothered Schrödinger, too. For him, any mathematical formulation of the world had to have an objective reality. As Irish physicist John Bell put it, "Either the wave function, as given by the Schrödinger equation, is not everything, or is not right."

Many worlds

Everett's idea was to explain what happens to the quantum superpositions. He presumed the objective reality of the wave function and removed the (unobserved) collapse—why should nature "choose" a particular version of reality every time someone makes a measurement? He then asked another question: what then happens to the various options available to quantum systems?

The MWI says that all possibilities do, in fact, occur. Reality peels itself, or splits, into new worlds, but since we inhabit a world where only one outcome occurs, this is what we see. Other possible outcomes are inaccessible to us, since there can be no interference between worlds and we are fooled into thinking that something is lost every time we measure something.

While Everett's theory is not accepted by all, it removes a theoretical block to interpreting quantum mechanics. MWI does not mention parallel universes, but they are its logical prediction. It has been criticized for being untestable, but this may change. An effect known as "decoherence"—whereby quantum objects "leak" their superposition information—is a mechanism by which MWI might be proved to work. ■

Hugh Everett III

Born in Washington DC, Hugh Everett was a precocious boy. At 12, he wrote to Einstein asking what held the universe together. While he was studying mathematics at Princeton, he drifted into physics. MWI—his answer to the riddle at the heart of quantum mechanics—was the subject of his PhD in 1957, and led to him being pilloried for proposing multiple universes. A trip to Copenhagen in 1959 to discuss the idea with Niels Bohr was a disaster—Bohr rejected everything that Everett said. Discouraged, he left physics for the US defense industry, but today MWI is regarded as a mainstream interpretation of quantum theory—too late for Everett, an alcoholic, who died at just 51. A lifelong atheist, he asked for his ashes to be thrown out with the trash.

Key works

1956 *Wave Mechanics Without Probability*
1956 *The Theory of the Universal Wave Function*

A PERFECT GAME OF TIC-TAC-TOE

DONALD MICHIE (1923–2007)

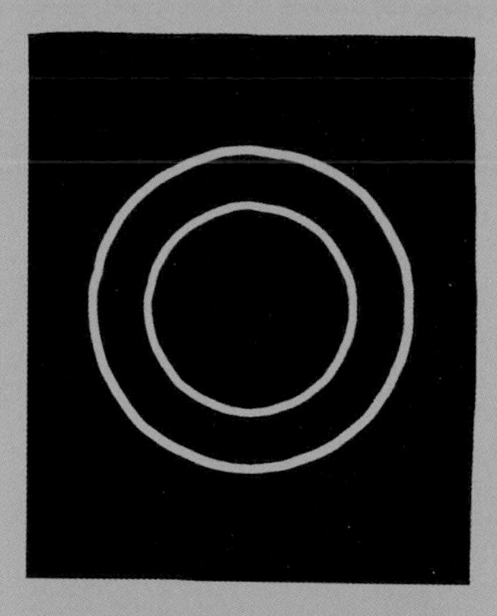

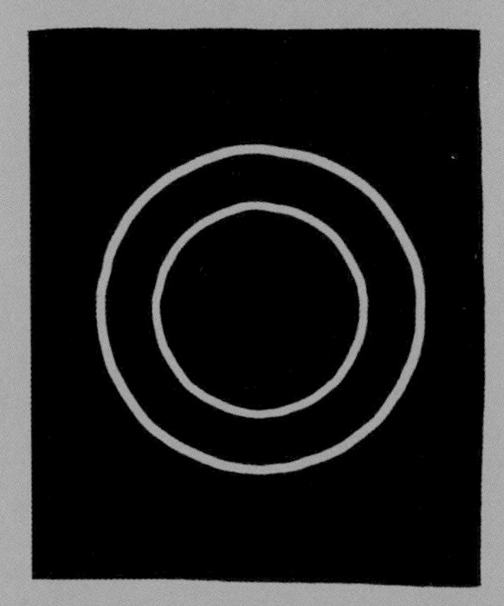

IN CONTEXT

BRANCH
Artificial intelligence

BEFORE
1950 Alan Turing suggests a test to measure machine intelligence (the Turing Test).

1955 American programmer Arthur Samuel improves his program to play tic-tac-toe by writing one that learns to play.

1956 The term "artificial intelligence" is coined by American John McCarthy.

1960 American psychologist Frank Rosenblatt makes a computer with neural networks that learn from experience.

AFTER
1968 MacHack, the first chess program to achieve a good level of skill, is created by American Richard Greenblatt.

1997 World chess champion Garry Kasparov is defeated by IBM's Deep Blue computer.

Computers in 1961 were mostly mainframes the size of a room. Minicomputers would not arrive until 1965 and microchips as we know them today were several years in the future. With computer hardware so huge and specialized, British research scientist Donald Michie decided to use simple physical objects for a small project on machine learning and artificial intelligence— matchboxes and glass beads. He selected a simple task, too—the game of tic-tac-toe, also known as noughts-and-crosses. Or, as Michie called it "tit-tat-to." The result was the Matchbox Educable Noughts And Crosses Engine (MENACE).

Michie's main version of MENACE comprised 304 matchboxes glued together in a chest-of-drawers arrangement. A code number on each box was keyed into a chart. The chart showed drawings of the 3x3 game grid with various arrangements of Os and Xs, corresponding to possible layout permutations as the game progressed. There are actually 19,683 possible layout combinations but some can be rotated to give others, and some are mirror images of, or symmetrical to, each other. This made 304 permutations an adequate working number.

In each matchbox box were beads of nine different kinds, distinguished by color. Each color of bead corresponded to MENACE putting its O on a certain one of the nine squares. For example, a green bead meant O in the lower left square, a red one designated O in the central square, and so on.

Can machines think? The short answer is "Yes: there are machines which can do what we would call thinking, if it were done by a human being."
Donald Michie

Mechanics of the game

MENACE opened the game using the matchbox for no Os or Xs in the grid—the "first move" box. In the tray of each matchbox were two extra pieces of card at one end forming a "V" shape. To play, the tray was removed from the box, jiggled, and tilted so the V was at the lower end. The beads randomly rolled down and one nestled into the apex of the V. Thus chosen, this bead's color determined the position of MENACE's first O in the grid. This bead was then put aside, and the tray replaced in its box but left slightly open.

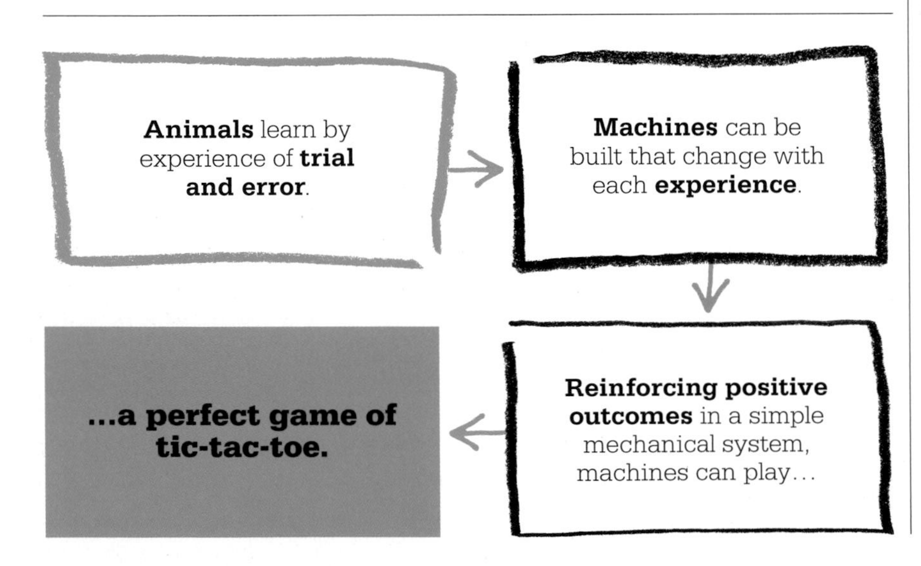

See also: Alan Turing 252–53

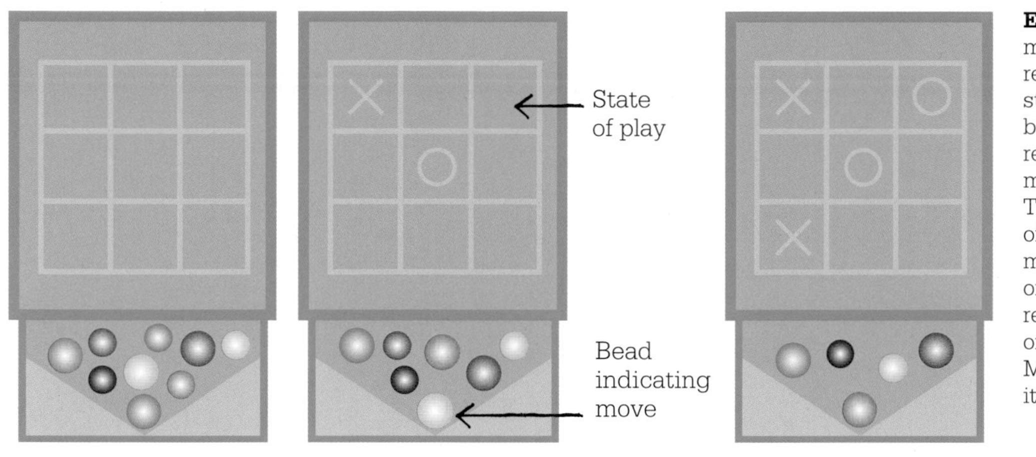

Each of the 304 matchboxes in MENACE represented a possible state of the board. The beads inside the boxes represented each possible move for that state. The bead at the bottom of the "V" determined the move. As games went on, winning beads were reinforced and losing ones removed, allowing MENACE to learn from its experience.

Next, the opponent positioned their first X. For the second turn of MENACE, the matchbox was selected that corresponded to the positions of the X and O on the grid at this time. Again the matchbox was opened, the tray shaken and tilted, and the color of the randomly selected bead determined the position of MENACE's second O. The opponent placed their second X. And so on, recording MENACE's sequence of beads and so moves.

Win, lose, draw

Eventually there came a result. If MENACE won, it received reinforcement or a "reward." The removed beads showed the sequence of winning moves. Each of these beads was put back in its box, identified by the code number and slightly open tray. The tray also received three extra "bonus" beads of the same color. As a consequence, in a future game, if the same permutation of Os and Xs occurred on the grid, this matchbox would come into play again—and it had more of the beads that previously led to a win. The chances of choosing that bead, and so the same move and another possible win, were increased.

If MENACE lost it was "punished" by not receiving back the removed beads, which represented the losing sequence of moves. But this was still positive. In future games, if the same permutation of Xs and Os cropped up, the beads designating the same move as the previous time were either fewer in number or absent, thereby lessening the chance of another loss.

Colossus, the world's first electronic programmable computer, was made in 1943 to crack codes at Bletchley Park in England. Michie trained staff to use the computer.

For a draw, each bead from that game was replaced in its relevant box, along with a small reward, one bonus bead of the same color. This increased the chances of that bead being selected if the same permutation came around again, but not as much as the win with three bonus beads.

Michie's goal was that MENACE would "learn from experience." For given permutations of Os and Xs, when a certain sequence of moves had been successful, it should gradually become more likely, while moves that led to losses would become less likely. It should progress by trial and error, adapt with experience, and with more games, become more successful.

Controlling variables

Michie considered potential problems. What if the selected bead from a tray decreed that MENACE's O should be placed on a square already occupied by an O or X? Michie accounted for this by ensuring that each matchbox contained only beads corresponding to empty squares for its particular permutation. So the » box for the permutation of O top left and X bottom right did not contain beads for putting the next O on those squares. Michie considered that putting beads for all nine possible O positions in every box would "complicate the problem unnecessarily." It meant MENACE would not only learn to win or draw, it would also have to learn the rules as it went along. Such start-up conditions might lead to one or two early disasters that collapsed the whole system. This demonstrated a principle: machine learning works best starting simple and gradually add more sophistication.

Michie also pointed out that when MENACE lost, its last move was the 100 percent fatal one. The move before contributed to the loss, as though backing the machine into a corner, but less so—usually it still left open the possibility of escaping defeat. Working back toward the start of the game, each earlier move contributed less to the final defeat—that is, as moves accumulate, the probability that each becomes the final one increases. Therefore as the total number of moves grows, it becomes more important to get rid of choices that have proved fatal.

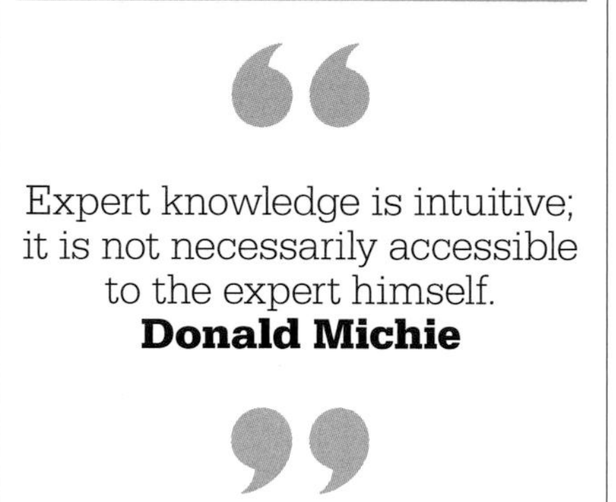

Expert knowledge is intuitive; it is not necessarily accessible to the expert himself.
Donald Michie

Michie simulated this by having different numbers of beads for each move. So for MENACE's second move (third move overall), each box that could be called upon to play—those with permutations of one O and one X already in the grid—had three of each kind of bead. For MENACE's third move, there were two beads of each kind, and for its fourth (seventh move overall), just one. A fatal choice on the fourth move would result in removal of the only bead specifying that position on the grid. Without that bead, the same situation could not recur.

Human vs MENACE

So what were the results? Michie was MENACE's first opponent in a tournament of 220 games. MENACE began shakily but soon settled down to draw more often, then notch up some wins. To counter, Michie began to stray from safe options and employ unusual strategies. MENACE took time to adapt but then began to cope with these too, coming back to achieve more draws, then wins. At one point in a series of 10 games, Michie lost eight.

MENACE provided a simple example of machine learning and how altering variables could affect the outcome. Michie's description of MENACE was, in fact, part of a longer account that went on to compare its performance with trial-and-error animal learning, as Michie explained:

'"Essentially, the animal makes more-or-less random movements and selects, in the sense that it subsequently repeats, those which produced the 'desired' result. This description seems tailor-made for the matchbox model. Indeed, MENACE constitutes a model of trial-and-error learning in so pure

Donald Michie

Born in 1923 in Rangoon, Burma (Myanmar), Michie won a scholarship to Oxford in 1942, but instead assisted in the war effort by joining the code-breaking teams at Bletchley Park, becoming a close colleague of the computing pioneer Alan Turing.

In 1946, he returned to Oxford to study mammalian genetics. However, he had a growing interest in artificial intelligence, and by the 1960s it had become his main pursuit. He moved to the University of Edinburgh in 1967, and became the first Chairman of the Department of Machine Intelligence and Perception. He worked on the FREDDY series of visually-enabled, teachable research robots. In addition, he ran a series of prestigious artificial intelligence projects and founded the Turing Institute in Glasgow.

Michie continued as an active researcher into his eighties. He died in a car accident while traveling to London in 2007.

Key work

1961 *Trial and Error*

a form, that when it shows elements of other categories of learning we may reasonably suspect these of contamination with a trial-and-error component."

Turning point

Before developing MENACE, Donald Michie had pursued a distinguished research career in biology, surgery, genetics, and embryology. After MENACE, he moved into the fast-developing area of artificial intelligence (AI). He developed his machine learning ideas into "industrial-strength tools" applied in hundreds of situations, including assembly lines, factory production, and steel mills. As computers spread, his artificial intelligence work was used to design computer programs and control structures that could learn in ways perhaps not even guessed at by their human originators. Michie demonstrated that careful application of human intelligence empowered machines to make themselves smarter. Recent developments in AI use similar principles to develop networks that mirror the neural networks of animals' brains.

Michie also conceived the notion of memoization, in which the result of each set of inputs in a machine or computer was stored as a reminder or "memo." If the same set of inputs recurred, the device would at once activate the memo and recall the answer, rather than recalculating afresh, thereby saving time and resources. He contributed the memoization technique to computer programming languages such as POP-2 and LISP. ■

New computer technology has led to a rapid development in AI, and in 1997, the chess machine Deep Blue defeated world champion Garry Kasparov. The computer learned strategy by analyzing thousands of past games.

He had this concept that he wanted to try out that he thought might possibly solve computer chess…It was the idea of reaching a steady state.
Kathleen Spracklen

THE UNITY OF FUNDAMENTAL FORCES

SHELDON GLASHOW (1932–)

IN CONTEXT

BRANCH
Physics

BEFORE
1820 Hans Christian Ørsted discovers that magnetism and electricity are aspects of the same phenomenon.

1864 James Clerk Maxwell describes electromagnetic waves in a set of equations.

1933 Enrico Fermi's theory of beta decay describes the weak force.

1954 The Yang–Mills theory lays the mathematical groundwork for unifying the four fundamental forces.

AFTER
1974 A fourth kind of quark, the "charm" quark, is discovered, revealing a new underlying structure to matter.

1983 The force-carrying W and Z bosons are discovered in CERN's Super Proton Synchrotron in Switzerland.

The idea of forces of nature, or fundamental forces, goes back at least to the ancient Greeks. Physicists currently recognize four fundamental forces—gravity, electromagnetism, and the two nuclear forces, weak and strong interactions, which hold together the subatomic particles inside the nucleus of an atom. We now know that the weak force and the electromagnetic force are different manifestations of a single "electroweak" force. Discovering this was an important step on the way to finding a "Theory of Everything" that would explain the relationship between all four forces.

The weak force

The weak force was first invoked to explain beta decay, a type of nuclear radiation in which a neutron turns into a proton inside the nucleus, emitting electrons or positrons in the process. In 1961, a graduate student at Harvard, Sheldon Glashow, was given the ambitious brief to unify the theories of weak and electromagnetic interactions. Glashow fell short of this, but did describe the force-carrying particles that mediate interaction via the weak force.

Messenger particles

In the quantum mechanical description of fields, a force is "felt" by the exchange of a gauge boson, such as the photon, which carries electromagnetic interaction. A boson is emitted by one particle and absorbed by a second. Normally, neither particle is fundamentally changed by this interaction—an electron is still an electron after absorbing or emitting a photon. The weak force breaks this symmetry, changing quarks (the particles that protons and neutrons are made from) from one kind to another.

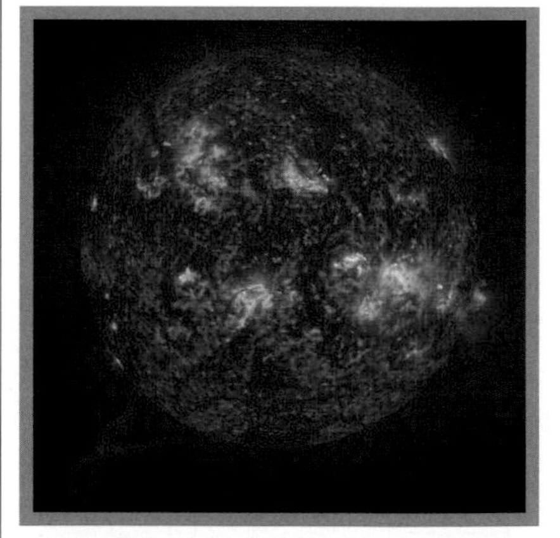

Decay of particles via the weak force drives the Sun's proton–proton fusion reaction, turning hydrogen into helium. Without it, the Sun wouldn't shine.

See also: Marie Curie 190–95 ▪ Ernest Rutherford 206–13 ▪ Peter Higgs 298–99 ▪ Murray Gell-Mann 302–07

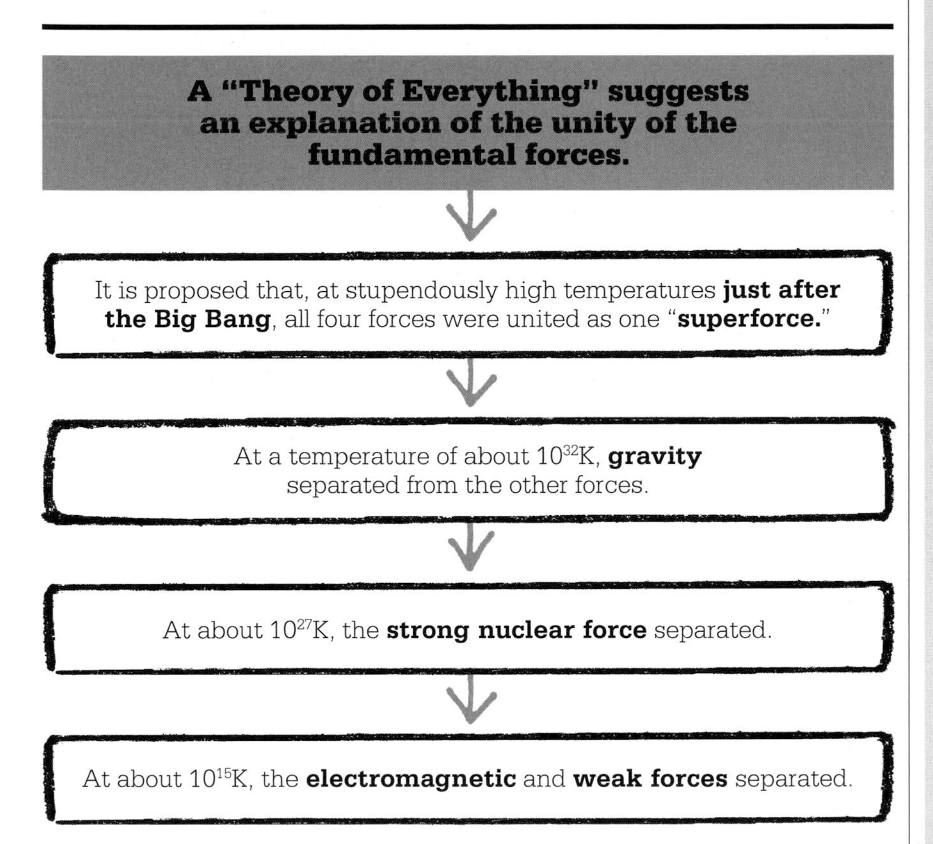

So what kind of boson might be involved? Glashow guessed that the bosons associated with the weak force had to be relatively massive because the force operates over miniscule ranges and heavy particles do not travel far. He proposed two charged bosons, W+ and W–, and a third neutral Z boson. The W and Z force-carriers were detected by CERN's particle accelerator in 1983.

Unification

In the 1960s, two physicists, American Steven Weinberg and Pakistani Abdus Salam, working independently, incorporated the Higgs field (pp.298–99) into Glashow's theory. The resultant Weinberg–Salam model, or unified electroweak theory, brought weak interaction and electromagnetic force together as a single force.

This was an astounding result, since the weak and electromagnetic forces operate in entirely different spheres. The electromagnetic force extends to the very edge of the visible universe (the force is carried by massless photons of light), while the weak force barely reaches across an atomic nucleus and is some 10 million times weaker. Their unification opens up the tantalizing possibility that, under certain high-energy conditions such as those just after the Big Bang, all four fundamental forces may coalesce into one "superforce." The search continues for evidence of such a Theory of Everything. ■

Sheldon Glashow

Sheldon Lee Glashow was born in New York in 1932, the son of Russian Jewish immigrants. He attended high school with his friend Steven Weinberg and upon graduating in 1950, they both studied physics at Cornell University. Glashow earned his PhD from Harvard, where he came up with a description of the W and Z bosons. After Harvard, he went to the University of California at Berkeley in 1961, and later returned to join the faculty at Harvard as a professor of physics in 1967.

In the 1960s, Glashow extended Murray Gell-Mann's quark model, adding a property known as "charm" and predicting a fourth quark, which was discovered in 1974. In recent years, he has been heavily critical of string theory, disputing its place in physics due to its lack of testable predictions, and describing it as a "tumor."

Key works

1961 *Partial Symmetries of Weak Interactions*
1988 *Interactions: A Journey Through the Mind of a Particle Physicist*
1991 *The Charm of Physics*

WE ARE THE CAUSE OF GLOBAL WARMING

CHARLES KEELING (1928–2005)

IN CONTEXT

BRANCH
Meteorology

BEFORE
1824 Joseph Fourier suggests that Earth's atmosphere makes the planet warmer.

1859 Irish physicist John Tyndall proves that carbon dioxide (CO_2), water vapor, and ozone trap heat in Earth's atmosphere.

1903 Swedish chemist Svante Arrhenius suggests that the CO_2 released by burning fossil fuel might be causing atmospheric warming.

1938 British engineer Guy Callendar reports that Earth's average temperature increased by 1°F (0.5°C) between 1890 and 1935.

AFTER
1988 The Intergovernmental Panel on Climate Change (IPCC) is set up to assess scientific research and guide global policy.

Carbon dioxide is a **greenhouse gas** that traps heat in Earth's atmosphere.

↓

Its **concentration** in the air is **rising** in line with fossil fuel consumption.

↓

Earth's **temperature** is **rising**.

↓

We are the cause of global warming.

The realization that carbon dioxide (CO_2) levels in the atmosphere are not only rising but might also cause disastrous warming first came to widespread scientific and public attention in the 1950s. Past scientists had assumed that the concentration of CO_2 in the atmosphere varied from time to time, but was always around 0.03 percent, or 300 parts per million (ppm). In 1958, American geochemist Charles Keeling began to measure the concentration of CO_2 using a sensitive instrument he had developed. It was his findings that alerted the world to the relentless rise of CO_2 and, by the late 1970s, to the human role in accelerating the so-called greenhouse effect.

Regular measurements

Keeling measured CO_2 in several places: Big Sur in California, the Olympic peninsula in Washington State, and the high mountain forests of Arizona. He also recorded measurements at the South Pole and from aircraft. In 1957, Keeling founded a meteorological station at 10,000 ft (3,000 m) above sea level on the top of Mauna Loa in Hawaii.

See also: Jan Ingenhousz 85 ▪ Joseph Fourier 122–23 ▪ Robert FitzRoy 150–55

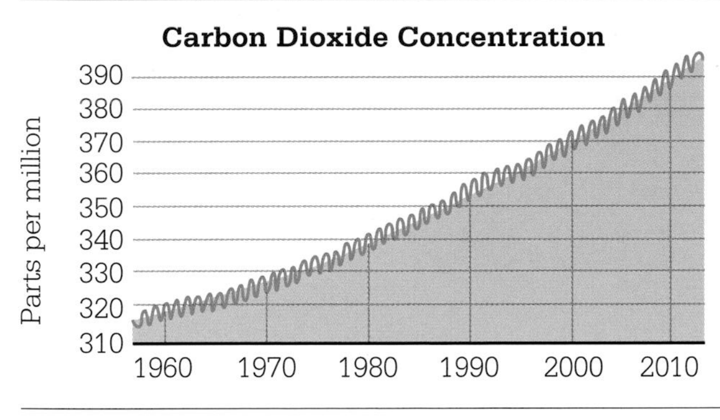

Keeling's graph plots the rising levels of CO_2 in the atmosphere year after year. The small annual fluctuation (shown by the blue line) is due to seasonal changes in CO_2 uptake by plants.

Keeling measured the carbon dioxide level at the station regularly, and discovered three things.

First, there was a daily variation locally. The concentration was at a minimum midafternoon, when green plants were at their most active in soaking up CO_2. Second, there was annual variation globally. The northern hemisphere had more land for plants to grow, and the level of CO_2 rose slowly during the northern winter when plants were not growing. It reached a peak in May before plants started to grow and began soaking up CO_2 again. The level dropped to a minimum in October, when northern plants died back for winter. Third, crucially, the concentration was increasing inexorably. Cores of polar ice contained bubbles of air, which showed that during most of the time since 9000 BCE, the CO_2 concentration varied from 275 to 285 ppm by volume. In 1958, Keeling measured 315 ppm; by May 2013, the average concentration exceeded 400 ppm for the first time. The increase from 1958 to 2013 was 85 ppm, meaning that the concentration had increased by 27 percent in 55 years. This was the first concrete evidence that the concentration of CO_2 in Earth's atmosphere is increasing.

CO_2 is a greenhouse gas, helping to trap heat from the Sun, so increasing CO_2 concentration is likely to lead to global warming. Keeling found the following: "At the South Pole the concentration has increased at the rate of about 1.3 ppm per year…the observed rate of increase is nearly that to be expected from the combustion of fossil fuel (1.4 ppm)." In other words, humans are at least part of the cause. ■

The demand for energy is certain to increase…as an ever larger population strives to improve its standard of living.
Charles Keeling

”

Charles Keeling

Born in Scranton, Pennsylvania, Charles Keeling was an accomplished pianist as well as a scientist. In 1954, as a postdoctoral fellow in geochemistry at the California Institute of Technology (Caltech), he developed a new instrument to measure carbon dioxide in atmospheric samples. He found that the concentration varied hour by hour at Caltech, probably because of all the traffic, so he went camping in the wilderness at Big Sur and found small but significant variations there, too. This inspired him to begin what was to be a lifetime's work. In 1956, he joined the Scripps Institution of Oceanography in La Jolla, California, where he worked for 43 years.

In 2002, Keeling received the National Medal of Science, America's highest award for lifetime achievement in science. Since his death, his son Ralph has taken over his work monitoring the atmosphere.

Key work

1997 *Climate Change and Carbon Dioxide: An Introduction*

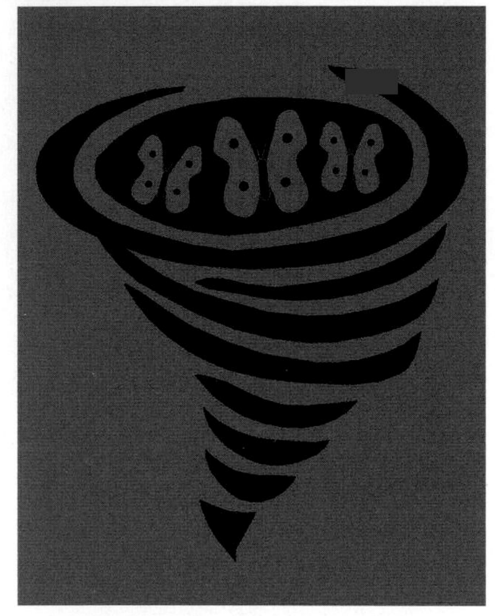

THE BUTTERFLY EFFECT

EDWARD LORENZ (1917–2008)

IN CONTEXT

BRANCH
Meteorology

BEFORE
1687 Newton's three laws of motion hold that the universe is predictable.

1880s Henri Poincaré shows that the motion of three or more bodies interacting gravitationally is generally chaotic and unpredictable.

AFTER
1970s Chaos theory is used to model traffic flow, digital encryption, function, and in designs for cars and aircraft.

1979 Benoît Mandelbrot discovers the Mandelbrot set, which shows how complex patterns can be created using very simple rules.

1990s Chaos theory is thought of as a subset of complexity science, which seeks to explain complex natural phenomena.

Much of the history of science has been devoted to developing simple models that predict the behavior of systems. Certain phenomena in nature, such as planetary motion, lend themselves readily to this schema. With a description of the initial conditions—the mass of a planet, its position, velocity, and so on—future configurations can be

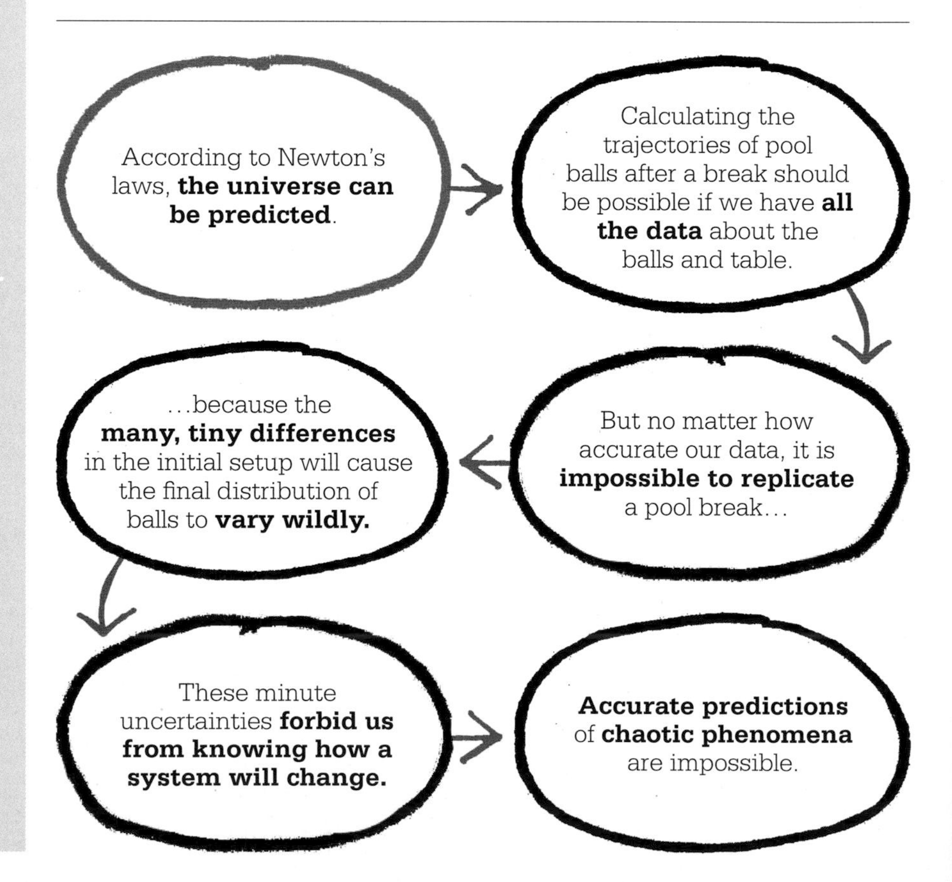

See also: Isaac Newton 62–69 ▪ Benoît Mandelbrot 316

calculated. However, the behavior of many processes, such as waves crashing on a beach, smoke rising from a candle, or weather patterns, is chaotic and unpredictable. Chaos theory seeks to explain such unpredictable phenomena.

Three-body problem

The first strides toward chaos theory were taken in the 1880s, when French mathematician Henri Poincaré worked on the "three-body problem." Poincaré showed that for a planet with a satellite orbiting a star—an Earth-Moon-Sun system—there is no solution for a stable orbit. Not only was the gravitational interaction between bodies far too complex to calculate, Poincaré found that tiny differences in initial conditions resulted in large and unpredictable changes. However, his work was largely forgotten.

A surprise discovery

Few further developments occurred in the field until the 1960s, when scientists began to use new, powerful computers to predict the weather. Surely, they reasoned, given enough data on the state of the atmosphere at a given time and enough computational power to crunch the data, it should be possible to know how weather systems evolve. Working on the assumption that ever-larger computers would increase the range of predictions, Edward Lorenz, an American meteorologist at the Massachusetts Institute of Technology (MIT), tested simulations involving just three simple equations. He ran the simulation several times, each time inputting the same initial state and expecting to see the same results. Lorenz was astounded when the computer returned hugely different outcomes each time. Checking his figures again, he found that the program had rounded up the numbers from six decimal places to three. This tiny alteration to the initial state had a major impact on the end result. This sensitive dependence on initial conditions was named the "butterfly effect"—the idea that a small change in a system, as trivial as a teaspoonful of air molecules moved by a butterfly flapping its wings in Brazil, can be amplified over time to create unpredictable outcomes, such as a tornado in Texas.

Edward Lorenz defined the limits of predictability, explaining that the impossibility of knowing what will happen is actually written into the rules that govern a chaotic system. Not only weather, but many real-world systems are chaotic—traffic systems, stock market fluctuations, the flow of fluids and gases, the growth of galaxies—and they have all been modeled using chaos theory. ▪

Here, turbulence forms at the tip of a vortex left in the wake of an aircraft's wing. Study of the critical point beyond which a system creates turbulence was key to the development of chaos theory.

Edward Lorenz

Born in West Hartford, Connecticut, in 1917, Edward Norton Lorenz received his MSc in mathematics from Harvard in 1940. During World War II he served as a meteorologist, forecasting the weather for the US Army Air Corps. After the war, he studied meteorology at Massachusetts Institute of Technology (MIT).

Lorenz's discovery of sensitive dependency on initial conditions (SDIC) was accidental—and one of the great "eureka" moments in science. Running simple computer simulations of weather systems he found that his model was churning out wildly different outcomes, despite being supplied with almost identical starting conditions. His seminal 1963 paper showed that perfect weather prediction was a pipe dream. Lorenz remained physically and academically active all his life, contributing academic papers, and hiking and skiing until shortly before his death in 2008.

Key work

1963 *Deterministic Nonperiodic Flow*

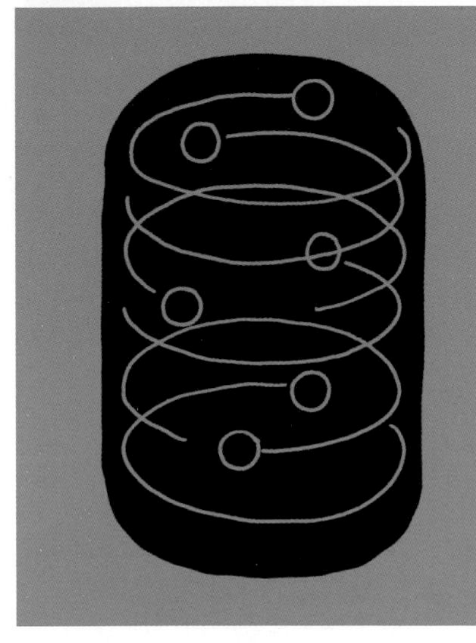

A VACUUM IS NOT EXACTLY NOTHING

PETER HIGGS (1929–)

IN CONTEXT

BRANCH
Physics

BEFORE
1964 Peter Higgs, François Englert, and Robert Brout describe a field that gives mass to all elementary and force-carrying particles.

1964 Three separate teams of physicists predict the existence of a new massive particle (the Higgs boson).

AFTER
1966 Physicists Steven Weinberg and Abdus Salam use the Higgs field to formulate the electroweak theory.

2010 CERN's Large Hadron Collider reaches full power. The search begins for the Higgs boson.

2012 Scientists at CERN announce the discovery of a new particle matching the description of the Higgs boson.

Imagine a room of physicists at a **cocktail party**. This is like the Higgs field, which fills everything, **even a vacuum**.

A **tax collector** enters the party, and travels unimpeded to the bar at the far end of the room.

In walks **Peter Higgs**. The physicists would like to talk to him, so they gather around, impeding his progress.

The taxman has **little interaction** with the "field" of physicists and is analogous to a **particle of low mass**.

Peter Higgs **interacts strongly** with the "field" and moves slowly through the room. He is like a **high-mass particle**.

A vacuum is not exactly nothing.

The great scientific event of 2012 was the announcement from scientists at the Large Hadron Collider (LHC) at CERN in Switzerland that a new particle had been found, and that it might be the elusive Higgs boson. The Higgs boson gives mass to all things in the universe, and is the missing piece that completes the standard model of physics. Its existence had been hypothesized by six physicists, among them Peter Higgs, in 1964. Finding the Higgs boson was of

See also: Albert Einstein 214–21 ▪ Erwin Schrödinger 226–33 ▪ Georges Lemaître 242–45 ▪ Paul Dirac 246–47 ▪ Sheldon Glashow 292–93

fundamental importance because it answered the question "why are some force-carrying particles massive while others are massless?"

Fields and bosons

Classical (pre-quantum) physics imagines electrical or magnetic fields as continuous, smoothly changing entities spread through space. Quantum mechanics rejects the notion of a continuum, so fields become distributions of discrete "field particles" where the strength of the field is the density of the field particles. Particles passing through a field are influenced by it via exchange of "virtual" force-carrying particles called gauge bosons.

The Higgs field fills space—even a vacuum—and elementary particles gain mass by interacting with it. How this effect occurs can be explained by analogy. Imagine a field covered in thick snow that skiers and people in snowshoes must cross. Each person will take more or less time, depending on how strongly they "interact" with the snow. Those that glide across on skis are like low-mass particles, while those that sink into the snow experience a greater mass as they travel. Massless particles, such as photons and gluons—the force-carriers of the electromagnetic and strong nuclear forces respectively—are unaffected by the Higgs field and sail straight through, like geese flying over the field.

The Higgs boson destroys itself within trillionths of a second of being born. It is created when other particles interact with the Higgs field.

The hunt for the Higgs

In the 1960s, six physicists, including Peter Higgs, François Englert, and Robert Brout, developed the theory of "spontaneous symmetry breaking," which explained how the particles that mediate the weak force, the W and Z bosons, are massive, while protons and gluons have no mass. This symmetry breaking was crucial in the formulation of the electroweak theory (pp.292–93). Higgs showed how the Higgs boson (or rather the decay products of the boson) should be detectable.

The search for the Higgs boson spawned the world's largest science project, the Large Hadron Collider—a giant proton collider with a 17-mile (27-km) circumference, buried 300 ft (100 m) underground. When running full tilt, the LHC generates energies similar to those that existed just after the Big Bang—enough to create one Higgs boson every billion collisions. The difficulty is spotting its traces among a vast shower of debris—and the Higgs is so massive that, on appearing, it decays instantly. However, after nearly 50 years of waiting, the Higgs has finally been confirmed. ■

Peter Higgs

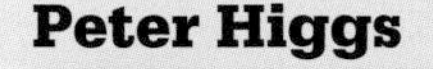

Born in Newcastle-upon-Tyne, England, in 1929, Peter Higgs earned undergraduate and doctoral degrees from King's College, London before joining the University of Edinburgh as a Senior Research Fellow. After a stint in London, he returned to Edinburgh in 1960. Walking in the Cairngorm Mountains, Higgs had his "one big idea"—a mechanism that would enable a force field to generate both high-mass and low-mass gauge bosons. Others were working along similar lines, but we talk of the "Higgs field" today, rather than the Brout–Englert–Higgs field, because his 1964 article described how the particle could be spotted. Higgs claims to have an "underlying incompetence" since he did not study particle physics at the PhD level. This handicap did not stop him from sharing the 2013 Nobel Prize in Physics with François Englert for their work in 1964.

Key works

1964 *Broken Symmetry and the Mass of Gauge Vector Mesons*
1964 *Broken Symmetries and the Mass of Gauge Bosons*

SYMBIOSIS IS EVERYWHERE

LYNN MARGULIS (1938–2011)

IN CONTEXT

BRANCH
Biology

BEFORE
1858 German doctor Rudolf Virchow proposes that cells arise only from other cells, and are not formed spontaneously.

1873 German microbiologist Anton de Bary coins the term "symbiosis" for different kinds of organisms living together.

1905 According to Konstantin Mereschkowsky, chloroplasts and nuclei originated by a process of symbiosis, but his theory lacks evidence.

1937 French biologist Edouard Chatton divides life forms by cell structure, into eukaryote (complex) and prokaryote (simple). His theory is rediscovered in 1962.

AFTER
1970–75 US microbiologist Carl Woese discovers that chloroplast DNA is similar to that of bacteria.

Charles Darwin's theory of evolution coincided with a cellular theory of life that emerged in the 1850s, asserting that all organisms were made of cells, and new cells could only come from existing ones by a process of division. Some of their internal components, such as food-making chloroplasts, apparently reproduced by division too.

This last discovery led Russian botanist Konstantin Mereschkowsky to the idea that chloroplasts may once have been independent life forms. Evolutionary and cellular biologists asked: how did complex cells arise? The answer lay in endosymbiosis—a theory that was first proposed by Mereschkowsky in 1905, but was only accepted after an American biologist called Lynn Sagan (later Margulis) furnished the evidence in 1967.

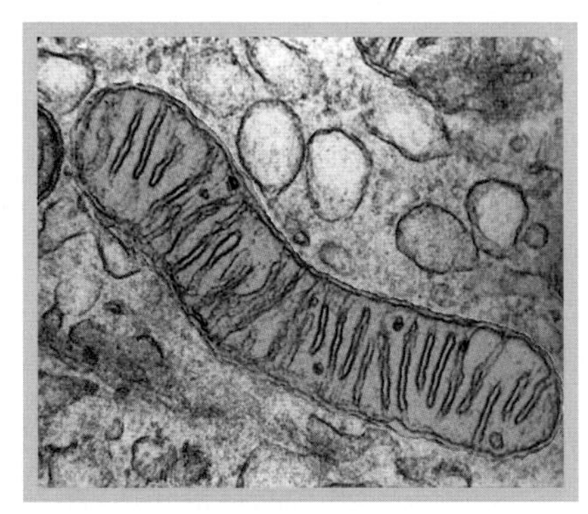

Mitochondria are organelles that make the energy-carrying chemical adenosine triphosphate (ATP) inside a eukaryotic cell. This mitochondrion has been artificially colored blue.

Complex cells with internal structures called organelles—the nucleus (which controls the cell), mitochondria (which release energy), and chloroplasts (which conduct photosynthesis)—are found in animals, plants, and many microbes. These cells, now called eukaryotic, evolved from simpler bacterial cells, which lack organelles and are now called prokaryotic. Mereschkowsky imagined primordial communities of the simpler cells—some making food by photosynthesis, others preying on their neighbors and engulfing them whole. Sometimes the engulfed cells were left undigested and, he suggested, became chloroplasts—but without proof, this theory of endosymbiosis (living together and within) faded.

New evidence

The invention of the electron microscope in the 1930s, combined with advances in biochemistry,

See also: Charles Darwin 142–49 ▪ James Watson and Francis Crick 276–83 ▪ James Lovelock 315

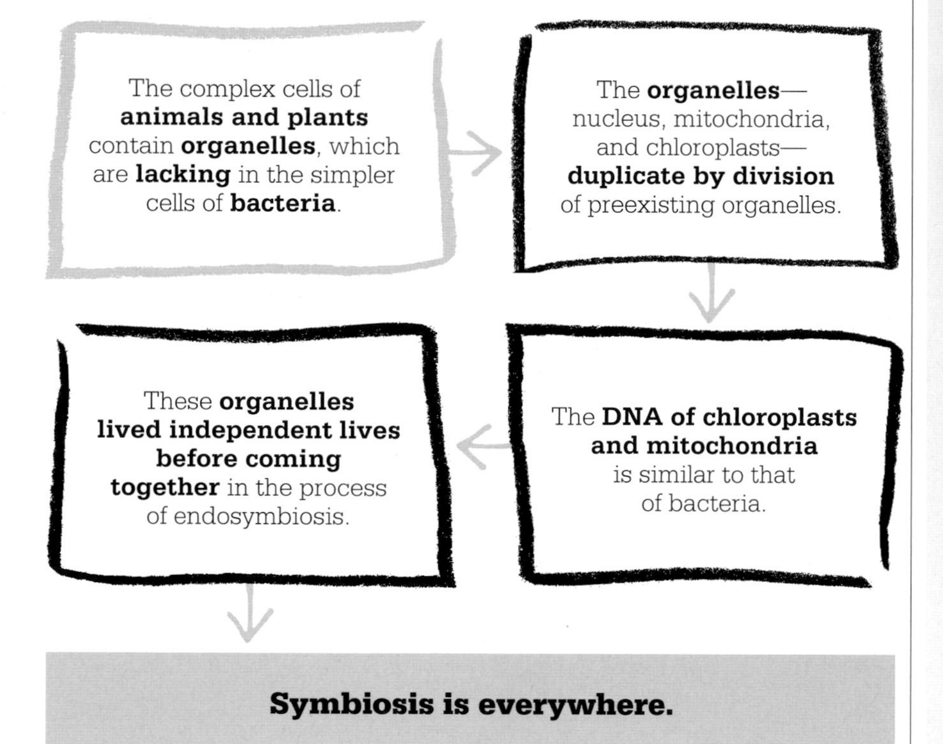

helped biologists to unlock the inner working of cells. By the 1950s, scientists knew that DNA provided genetic instructions for carrying out life processes and was relayed from generation to generation. In eukaryotic cells, DNA is packaged in the nucleus, but it is also found in chloroplasts and mitochondria.

In 1967, Margulis used this discovery as evidence to revive and substantiate the endosymbiosis theory. She included the suggestion that there had been an oxygen "holocaust" in the early history of life on Earth. About two billion years ago, as photosynthesizers flourished, they saturated the world with oxygen, which poisoned many of the microbes around at the time. Predatory microbes survived by engulfing others that could "soak up" the oxygen in their energy-releasing processes. These became mitochondria: the "power packs" of cells today. At first, this appeared farfetched to most biologists, but the evidence for Margulis's theory gradually became persuasive, and it has now been widely accepted. For example, the DNA of mitochondria and chloroplasts are made from circular molecules—just like the DNA of living bacteria.

Evolution by cooperation was not something new: Darwin himself had conceived the idea to explain the mutually beneficial interplay between nectar-giving plants and pollinating insects. But few had thought it could happen so intimately—and fundamentally—as when cells merged together at the very dawn of life. ■

Lynn Margulis

Lynn Alexander (later Sagan, then Margulis) entered Chicago University at just 14, before earning a PhD at the University of California, Berkeley. Her interests in the cellular diversity of organisms led her to revive and champion the evolutionary theory of endosymbiosis, which biologist Richard Dawkins has described as "one of the great achievements of 20th-century evolutionary biology."

For Margulis, cooperative interactions were as important as competition in driving evolution—and she viewed living things as self-organizing systems. She later supported James Lovelock's Gaia hypothesis that Earth, too, could be viewed as a self-regulating organism. In recognition of her work, she was made a member of the US National Academy of Science and received the National Medal of Science.

Key works

1967 *On the Origin of Mitosing Cells*
1970 *Origin of Eukaryotic Cells*
1982 *Five Kingdoms: An Illustrated Guide to the Phyla of Life on Earth*

QUARKS COME IN THREES

MURRAY GELL-MANN (1929–)

IN CONTEXT

BRANCH
Physics

BEFORE
1932 A new particle, the neutron, is discovered by James Chadwick. There are now three known subatomic particles with mass: the proton, neutron, and electron.

1932 The first antiparticle, the positron, is discovered.

1940s–50s Increasingly powerful particle accelerators—which smash particles together at high speeds—produce large numbers of new subatomic particles.

AFTER
1964 The discovery of the omega (Ω–) particle confirms the quark model.

2012 The Higgs boson is discovered at CERN, adding weight to the standard model.

Understanding of the structure of the atom has changed dramatically since the end of the 19th century. In 1897, J. J. Thomson made the bold suggestion that cathode rays are streams of particles far smaller than the atom; he had discovered the electron. In 1905, building on the light quanta theory of Max Planck, Albert Einstein suggested that light should be thought of as a stream of tiny massless particles, which we now call photons. In 1911, Thomson's protégé Ernest Rutherford deduced that an atom's nucleus is small and dense, with electrons in orbit around it. The image of an atom as an indivisible whole had been destroyed.

In 1920, Rutherford named the nucleus of the lightest element, hydrogen, the proton. Twelve years later, the neutron was discovered, and a more complex picture of nuclei made of protons and neutrons emerged. Then, in the 1930s, a glimpse of further realms of particles came from studies of cosmic rays—high-energy particles that are thought to originate in supernovae. The studies revealed new particles associated with high

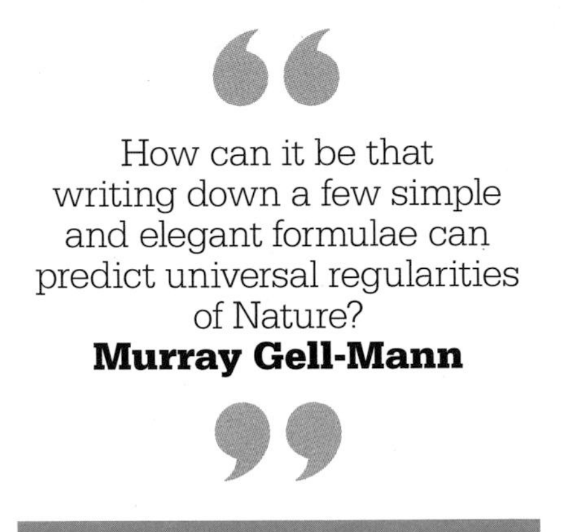

energies, and hence with greater masses according to Einstein's principle of mass–energy equivalence ($E = mc^2$).

Seeking to explain the nature of interactions inside the atomic nucleus, scientists in the 1950s and 1960s produced an enormous body of work providing the conceptual framework for all matter in the universe. Many figures contributed to this process, but American physicist Murray Gell-Mann played a pivotal role in the construction of a taxonomy of fundamental particles and force-carriers called the standard model.

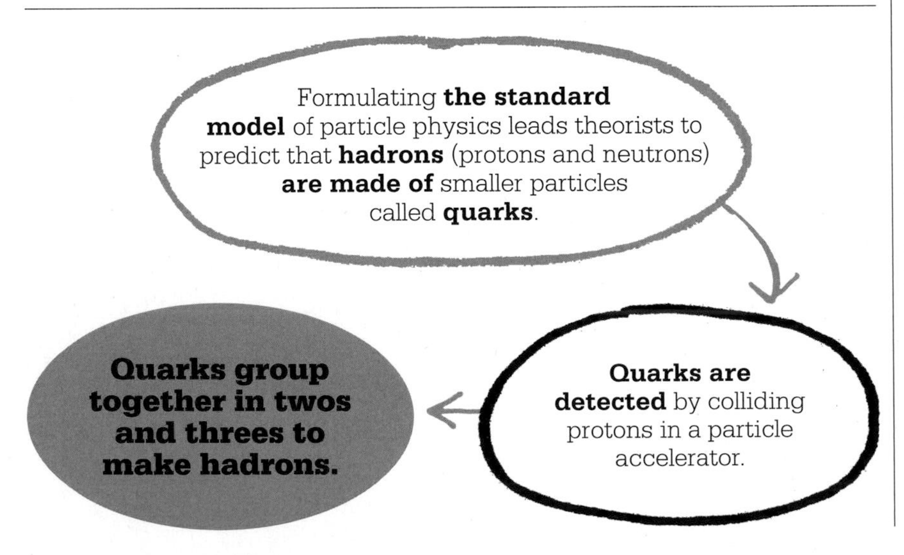

The particle zoo

Gell-Mann jokes that the goals of the theoretical elementary particle physicist are "modest"—they merely aim to explain the "fundamental laws that govern all matter in the universe." Theorists, he says, "work with pencil, paper, and wastebasket, and the most important of those is the last." By contrast, the experimentalist's principal tool is the particle accelerator, or collider.

In 1932, the first atomic nuclei—of the element lithium—were blown apart by physicists Ernest Walton

See also: Max Planck 202–05 ▪ Ernest Rutherford 206–13 ▪ Albert Einstein 214–21 ▪ Paul Dirac 246–47 ▪ Richard Feynman 272–73 ▪ Sheldon Glashow 292–93 ▪ Peter Higgs 298–99

and John Cockcroft using a particle accelerator in Cambridge, England. Since then, ever more powerful particle accelerators have been constructed. These machines boost tiny subatomic particles to nearly the speed of light before slamming them into targets or each other. Research is now driven by theoretical predictions—the largest particle accelerator, the Large Hadron Collider (LHC) in Switzerland, was built primarily to find the theoretical Higgs boson (pp.298–99). The LHC is a 17-mile (27-km) ring of superconducting magnets that took 10 years to build. Collisions between subatomic particles splinter them into their core units. The energy released is sometimes enough to produce new generations of particles that cannot exist under everyday conditions. Showers of short-lived, exotic particles spray off these pileups, before swiftly annihilating or decaying. With ever-increasing energies at their disposal, researchers attempt to probe the mysteries of matter by getting even closer to the conditions at the birth of matter itself—the Big Bang. The process has been likened to smashing two watches together and then sifting through the wrecked pieces in an attempt to find out how the timepiece works.

By 1953, with colliders achieving ever-increasing energies, exotic particles not found in ordinary matter seemed to tumble out of thin air. More than 100 strongly interacting particles were detected, all thought at the time »

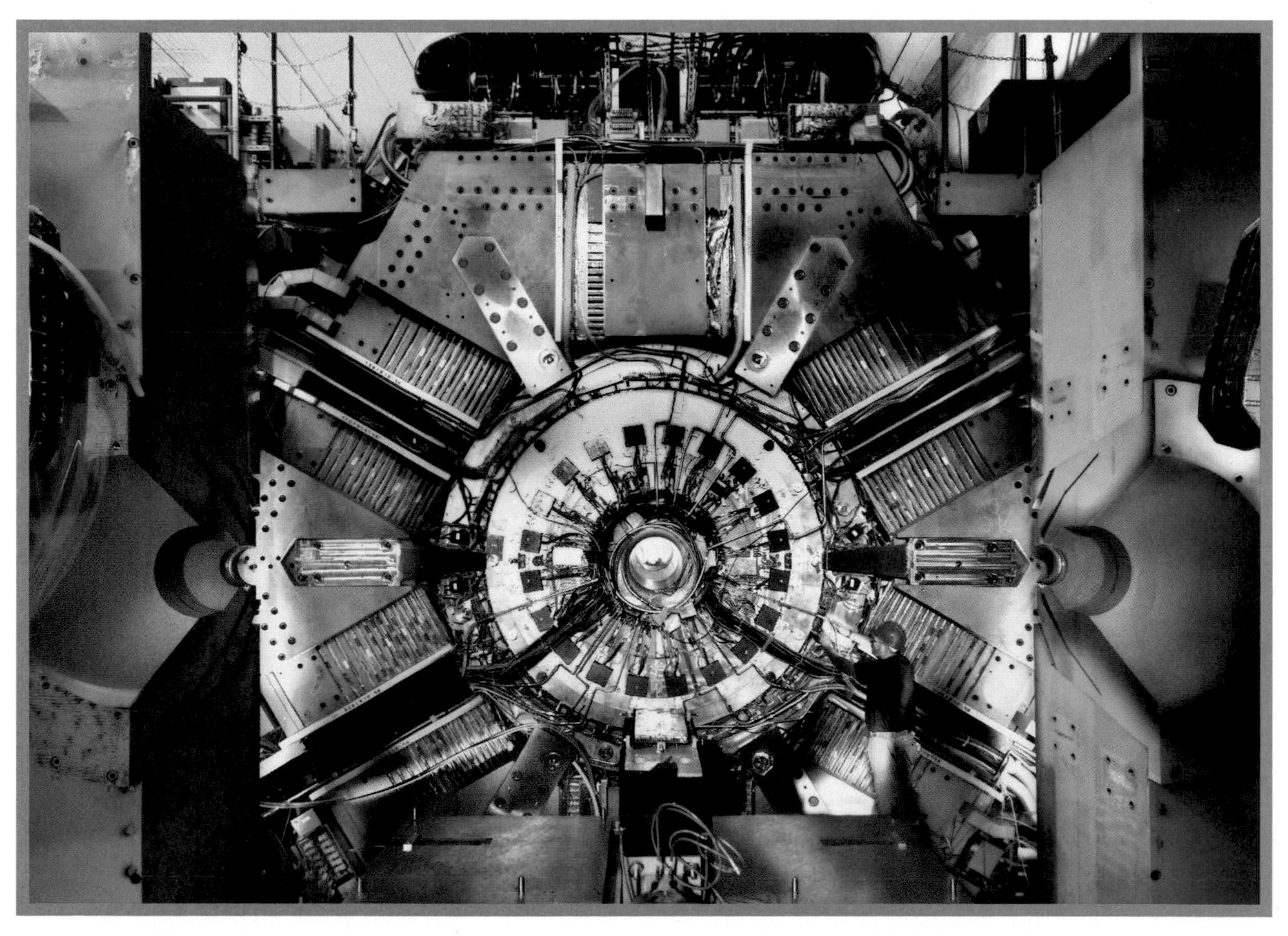

The Stanford Linear Accelerator in California, built in 1962, is 2 miles (3 km) long—the longest linear accelerator in the world. It was here, in 1968, that it was first demonstrated that protons are composed of quarks.

to be fundamental. This merry circus of new species was dubbed the "particle zoo."

The Eightfold Way

By the 1960s, scientists had grouped particles according to how they were affected by the four fundamental forces: gravity, electromagnetic force, and the weak and strong nuclear forces. All particles with mass are influenced by gravity. The electromagnetic force acts on any particle with an electric charge. The weak and strong forces operate over the miniscule ranges found within the atomic nucleus. Heavyweight particles called "hadrons," which include the proton and neutron, are "strongly interacting" and influenced by all four fundamental forces, while the lightweight "leptons," such as the electron and neutrino, are unaffected by the strong force.

Gell-Mann made sense of the particle zoo with an octet ordering system he called the "Eightfold Way," a pun on the Buddhist Noble Eightfold Path. Just as Mendeleev had done when arranging the chemical elements into a periodic table, Gell-Mann imagined a chart into which he placed the elementary particles, leaving spaces for as yet undiscovered pieces. In an effort to make the most economical design, he proposed that hadrons contained a new and as-yet-unseen fundamental subunit. Since the heavier particles were no longer fundamental, this change reduced the number of fundamental particles down to a manageable number—hadrons were now simply combinations of multiple elementary components. Gell-Mann, with his penchant for wacky names dubbed this particle a "quark" (pronounced "kwork"), after a favorite line from James Joyce's novel *Finnegans Wake*.

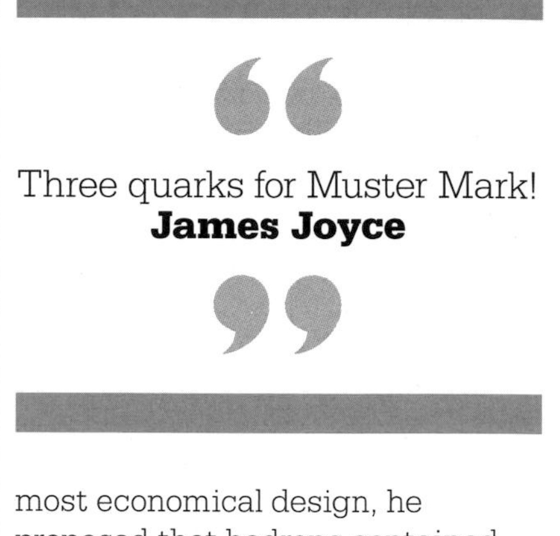

Three quarks for Muster Mark!
James Joyce

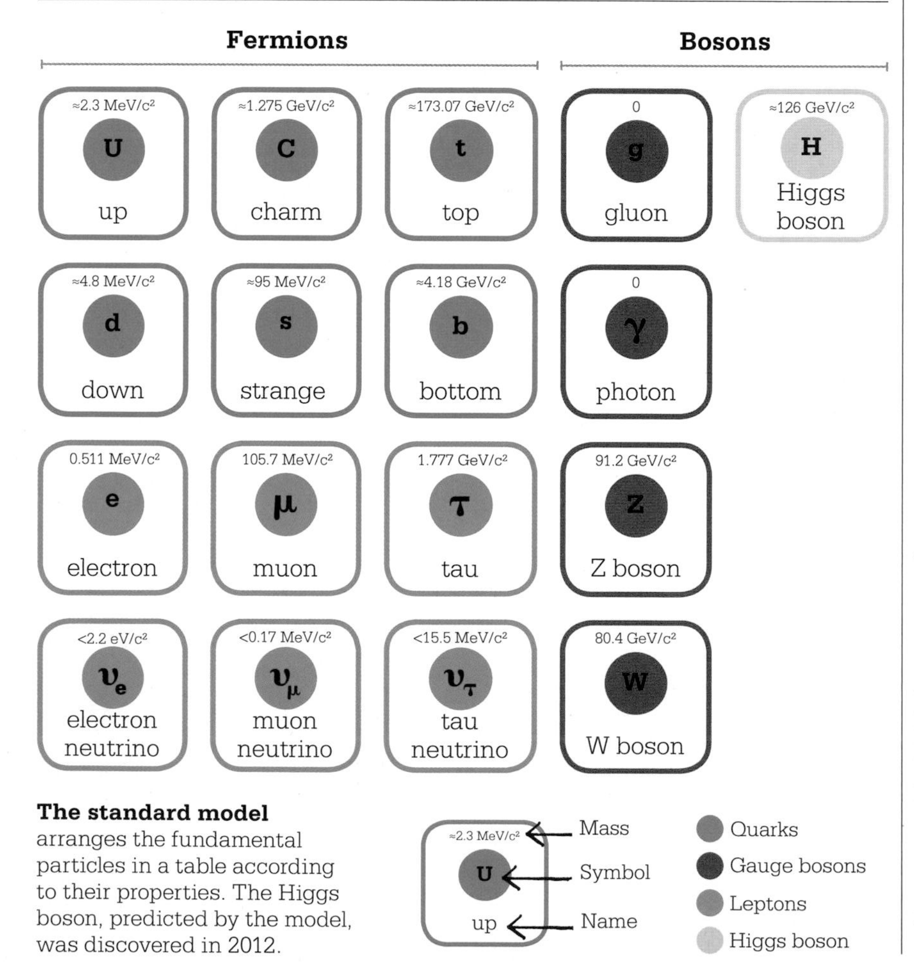

The standard model arranges the fundamental particles in a table according to their properties. The Higgs boson, predicted by the model, was discovered in 2012.

Real or not real?

Gell-Mann was not the only person to suggest this idea. In 1964, a student at Caltech, Georg Zweig, had suggested that hadrons were made of four basic parts, which he called "aces." The CERN journal *Physics Letters* refused Zweig's paper, but that same year published a paper by the more senior Gell-Mann outlining the same idea.

Gell-Mann's paper may have been published because he did not suggest that there was any underlying reality to the pattern—he was simply proposing an organizing design. However, this design appeared unsatisfactory, since it required quarks to have fractional charges, such as $-^1/_3$ and $+^2/_3$. These were nonsensical to accepted theory, which only allowed for whole-number charges. Gell-Mann realized that if these subunits remained hidden, trapped inside hadrons, this didn't matter. The predicted omega particle (Ω–), made up of three quarks, was detected at Brookhaven National Laboratory, New York, soon after Gell-Mann's publication. This confirmed the new model, which Gell-Mann has insisted should be credited both to him and to Zweig.

Initially, Gell-Mann was doubtful that quarks could ever be isolated. However, he now emphasizes that although he initially saw his quarks as mathematical entities, he never ruled out the possibility that quarks might be real. Experiments at the Stanford Linear Accelerator Center (SLAC) between 1967 and 1973 scattered electrons off hard granular particles within the proton, revealing the reality of quarks in the process.

The standard model

The standard model developed from Gell-Mann's quark model. In this model, particles are divided into fermions and bosons. Fermions are the building blocks of matter, while bosons are force-carrying particles.

The fermions are further split into two families of elementary particles—quarks and leptons. Quarks group together in twos and threes to make up the composite particles called hadrons. Subatomic particles with three quarks are known as baryons, and include protons and neutrons. Those made of a quark and antiquark pair are called mesons, and include pions and kaons. In total there are six quark "flavors"—up, down, strange, charm, top, and bottom. The defining characteristic of quarks is that they carry something called "color charge," which allows them to interact via the strong force. The leptons do not carry color charge and are not affected by the strong force. There are six leptons—the electron, muon, tau, and the electron, muon, and tau neutrinos. Neutrinos have no electrical charge and only interact via the weak force, making them extremely hard to detect. Each particle also has a corresponding "antiparticle" of antimatter.

The standard model explains forces at the subatomic level as the result of an exchange of force-carrying particles known as "gauge bosons." Each force has its own gauge boson: the weak force is mediated by the W+, W–, and Z bosons; the strong electromagnetic force by photons; and the strong force by gluons.

The standard model is a robust theory and has been verified by experiment, notably with the discovery of a Higgs boson—the particle that gives other particles mass—at CERN in 2012. However, many consider the model inelegant and there are problems with it, such as its failure to incorporate dark matter or explain gravity in terms of boson interaction. Other questions that remain unanswered are why there is a preponderance of matter (rather than antimatter) in the universe, and why there appear to be three generations of matter. ■

Our work is a delightful game.
Murray Gell-Mann

Murray Gell-Mann

Born in Manhattan, Murray Gell-Mann was a child prodigy. He taught himself calculus at 7 years old and entered Yale at 15. He earned a doctoral degree from the Massachusetts Institute of Technology (MIT), graduating in 1951, and then decamped to the California Institute of Technology (Caltech), where he worked with Richard Feynman to develop a quantum number called "strangeness." Japanese physicist Kazuhiko Nishijima had made the same discovery, but called it "eta-charge."

With wide-ranging interests and speaking some 13 languages fluently, Gell-Mann enjoys displaying his polymath's breadth of knowledge with plays on words and arcane references. He is perhaps the originator of the trend for giving new particles funny names. His discovery of the quark won him the 1969 Nobel Prize.

Key works

1962 *Prediction of the Ω– Particle*
1964 *The Eightfold Way: A Theory of Strong Interaction Symmetry*

A THEORY OF EVERYTHING?

GABRIELE VENEZIANO (1942–)

IN CONTEXT

BRANCH
Physics

BEFORE
1940s Richard Feynman and other physicists develop quantum electrodynamics (QED), which describes quantum-level interactions due to the electromagnetic force.

1960s The standard model of particle physics reveals the full range of subatomic particles known so far and the interactions that affect them.

AFTER
1970s String theory falls out of favor temporarily as quantum chromodynamics appears to offer a better explanation of the strong nuclear force.

1980s Lee Smolin and Italian Carlo Rovelli develop the theory of loop quantum gravity, which removes the need to theorize hidden extra dimensions.

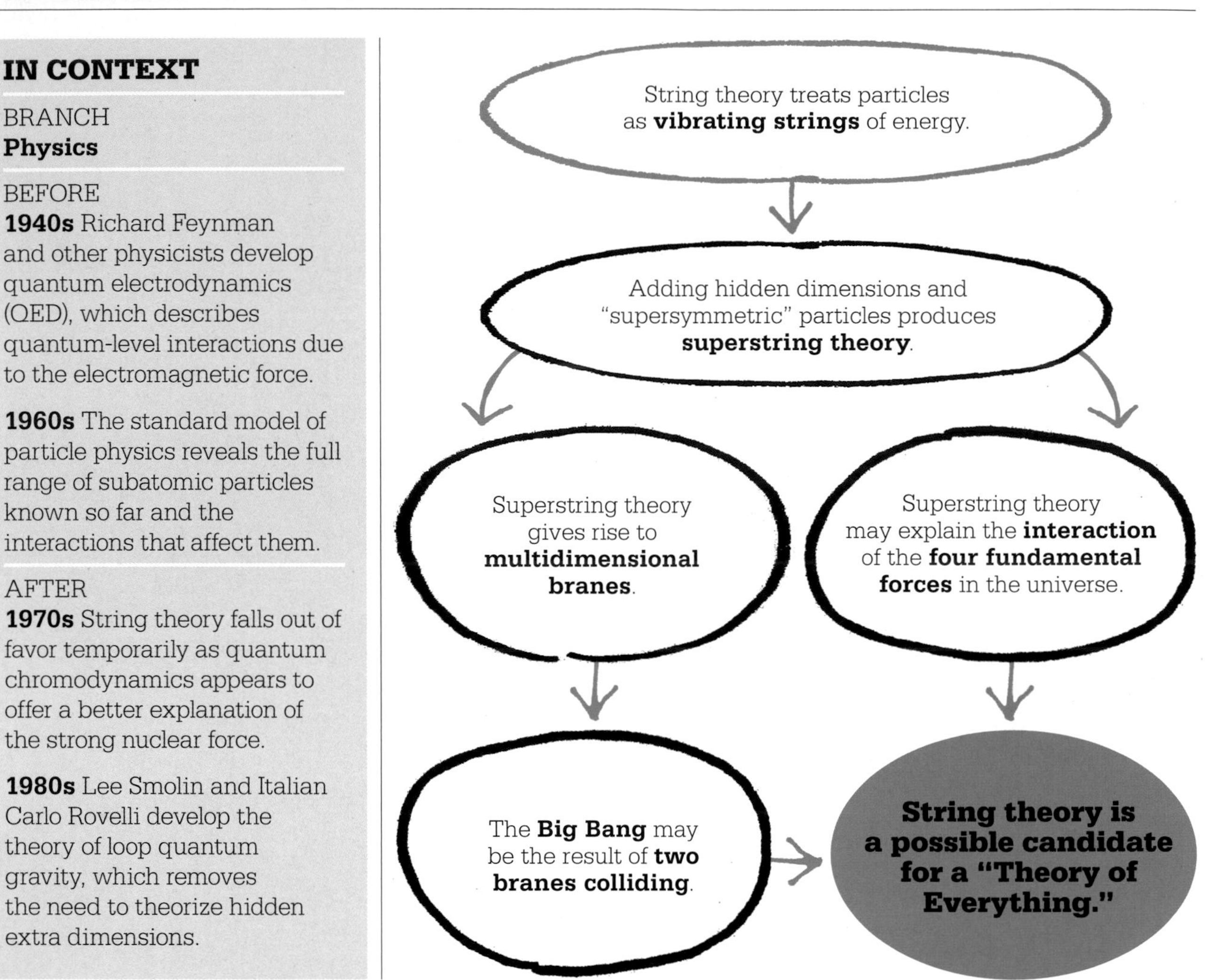

Put simply, string theory is the remarkable—and still controversial—idea that all matter in the universe is made up not of pointlike particles, but of tiny "strings" of energy. The theory lays out a structure that we cannot detect, but that explains all the phenomena that we see. Waves of vibration within these strings give rise to the quantized behaviors (discrete properties such as electric charge and spin) that are found in nature, and mirror the harmonics that can be produced, for example, by plucking a violin string.

The development of string theory has had a long and bumpy road, and it is still not accepted by many physicists. But work on the theory continues—not least because it is currently the only theory trying to unite the "quantum gauge" theories of the electromagnetic, weak, and strong nuclear forces with Einstein's theory of gravity.

Explaining the strong force

String theory began life as a model to explain the strong force that binds together the particles in the nuclei of atoms, and the behavior of hadrons, the composite particles that are subject to the influence of the strong force.

In 1960, as part of an ongoing study of the properties of hadrons, American physicist Geoffrey Chew proposed a radical new approach—abandoning the preconception that hadrons were particles in the traditional sense, and modeling their interactions in terms of a mathematical object called an S-matrix. When Italian physicist Gabriele Veneziano investigated the results of Chew's model, he found patterns

See also: Albert Einstein 214–21 ▪ Erwin Schrödinger 226–33 ▪ Georges Lemaître 242–45 ▪ Paul Dirac 246–47 ▪ Richard Feynman 272–73 ▪ Hugh Everett III 284–85 ▪ Sheldon Glashow 292–93 ▪ Murray Gell-Mann 302–07

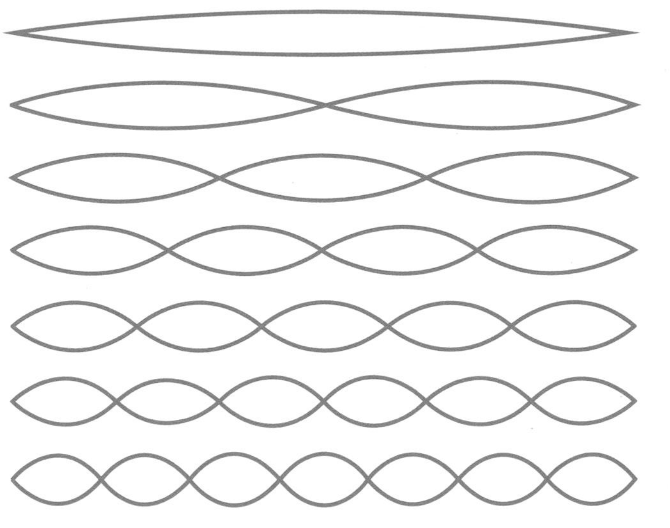

According to string theory, the quantized properties we observe arise when a string takes on different vibrational states, similar to the harmonic notes played on a violin.

suggesting that particles would appear at points along straight one-dimensional lines—the first hint of what we now call strings. In the 1970s, physicists continued to map these strings and their behavior, but their work began to bring up annoyingly complex and counterintuitive results. For example, particles have a property called spin (analogous to angular momentum), which can only take certain values. The initial drafts of string theory could produce bosons (particles with zero or whole-number spins, typically the "messenger" particles in models of quantum forces), but not fermions (particles with half-integer spins, including all matter particles). The theory also predicted the existence of particles that move faster than the speed of light, thus traveling backward in time.

One final complication was that the theory could not properly work without assuming the existence of no fewer than 26 separate dimensions (instead of the usual four—three dimensions of space, plus time). The concept of extra dimensions had been around for a long time: German mathematician Theodor Kaluza had attempted to unify electromagnetism and gravity through the use of an extra (fifth) dimension. This was not a problem mathematically, but did pose the question as to why we do not experience all dimensions. In 1926, Swedish physicist Oscar Klein explained how such extra dimensions might remain invisible on everyday macroscopic scales by suggesting they might "roll up" into quantum-scale loops.

String theory suffered a fall from grace in the mid-1970s. The theory of quantum chromodynamics (QCD), which introduced the concept of "color charge" for quarks to explain their interaction via the strong nuclear force, offered a much better description. But »

> String theory is an attempt at a deeper description of nature by thinking of an elementary particle not as a little point but as a little loop of vibrating string.
> **Edward Witten**

Gabriele Veneziano

Born in Florence, Italy, in 1942, Gabriele Veneziano studied in his home city before obtaining his PhD from Israel's Weizmann Institute of Science, where he returned in 1972 as professor of physics following some time at the European particle physics laboratory CERN. While at the Massachusetts Institute of Technology (MIT) in 1968, he hit upon string theory as a model for describing the strong nuclear force, and began to pioneer research into the topic. From 1976 onward, Veneziano worked mainly at CERN's Theory Division in Geneva, rising to become its director between 1994 and 1997. Since 1991, he has focused on investigating how string theory and QCD can help to describe the hot, dense conditions just after the Big Bang.

Key work

1968 *Construction of a Cross-Symmetric, Regge-behaved Amplitude for Linearly Rising Trajectories*

even before this, some scientists had been murmuring that the theory was conceptually flawed. The more work they did, the more it seemed as though strings were not describing the strong force at all.

The rise of superstrings

Groups of physicists continued to work on string theory, but they needed to find solutions to some of its problems before the wider scientific community would take it seriously again. A breakthrough came in the early 1980s with the idea of supersymmetry. This is the suggestion that each of the known particles found in the standard model of particle physics (pp.302–05) has an undiscovered "superpartner"—a fermion to match every boson, and a boson to match every fermion. If this were the case, then many of the outstanding problems with strings would promptly vanish, and the number of dimensions required to describe them would be reduced to ten. The fact that these additional particles remain undetected might be due to the fact that they are only capable of independent existence at energies far above those produced in even the most powerful modern particle accelerators.

This revised "supersymmetric string theory" soon became known more simply as "superstring theory." However, major issues remained—particularly the fact that five rival interpretations of superstrings emerged. Evidence also began to mount that superstrings should give rise not only to 2-dimensional strings and 1-dimensional points, but also to multidimensional structures, collectively known as "branes." Branes can be thought of as analogous to 2-dimensional membranes moving in our 3-dimensional world: similarly, a 3-dimensional brane could move in a 4-dimensional space.

String theory envisions a multiverse in which our universe is one slice of bread in a big cosmic loaf. The other slices would be displaced from ours in some extra dimension of space.
Brian Greene

M-theory

In 1995, US physicist Edward Witten presented a new model known as M-theory, which offered a solution to the problem of competing superstring theories. He added a single additional dimension, bringing the total up to 11, and this allowed all five superstring approaches to be described as aspects of a single theory. The 11 dimensions of space-time required by M-theory mirrored the 11 dimensions required by then-popular models of "supergravity" (supersymmetric gravity). According to Witten's theory, the seven additional dimensions of space required would be "compactified"—curled up into tiny structures analogous to spheres that would effectively act and appear as points on all but the most microscopic of scales.

The major problem of M-theory, however, is that the detail of the theory itself is currently unknown. Rather, it is a prediction of the existence of a theory with certain characteristics that would neatly fulfill a number of observed or predicted criteria.

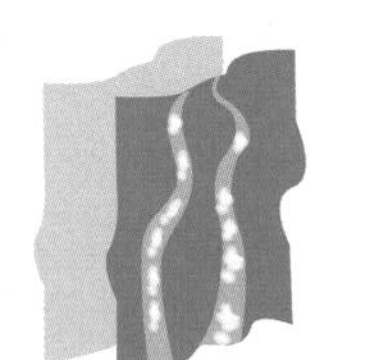

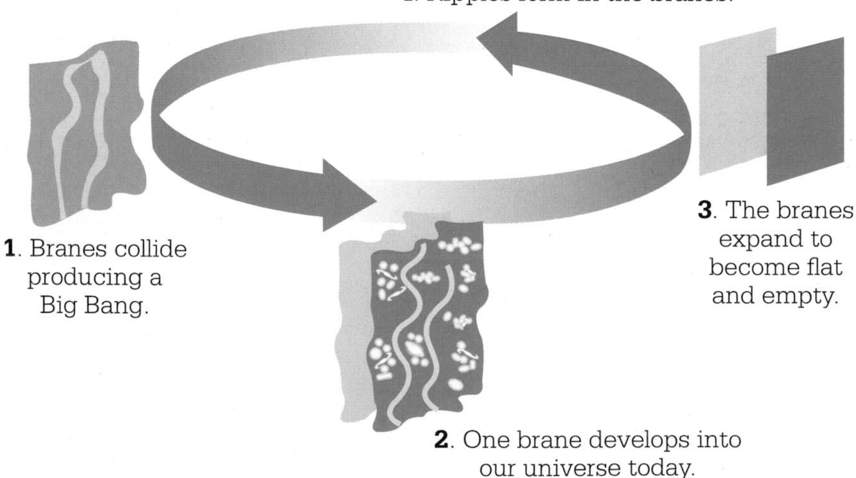

Superstring theory predicts the existence of multidimensional branes. Our universe might be one such brane. It is suggested that a Big Bang event occurs when two branes collide, producing a "cyclic universe" model.

Despite its current limitations, M-theory has proved a huge inspiration to various fields of physics and cosmology. Black hole singularities can be interpreted as string phenomena, as can the early stages of the Big Bang. One intriguing upshot of M-theory is the "cyclic universe" model proposed by cosmologists such as Neil Turok and Paul Steinhardt. In this theory, our universe is just one of many separate branes separated from each other by minute distances in 11-dimensional space-time, and drifting minutely in relation to one another on trillion-year time scales. Collisions between branes, it has been argued, could result in huge releases of energy and trigger new Big Bangs.

Theories of everything

M-theory has been proposed as a possible "Theory of Everything"—a means of uniting the quantum field theories that successfully describe electromagnetism and the weak and strong nuclear forces with the description of gravity provided by Einstein's general theory of relativity. Hitherto, a quantum description of gravitation has remained elusive. Gravity appears to be radically different in nature from the other three forces. These three forces all act between individual particles but only on relatively small scales, while gravity is insignificant except when huge numbers of particles conglomerate, but acts across enormous distances. One possible explanation of gravity's unusual behavior is that its influence at the quantum level may "leak out" into the higher dimensions, so that only a small fraction is perceived within the familiar dimensions of our universe.

If string theory is a mistake, it's not a trivial mistake. It's a deep mistake and therefore kind of worthy.
Lee Smolin

String theory is not the only candidate for a Theory of Everything. Loop quantum gravity (LQG) was developed by Lee Smolin and Carlo Rovelli from the late 1980s. In this theory, the quantized properties of particles arise not from their stringlike nature, but rather from the small-scale structure of space-time itself, which is quantized into tiny loops. LQG and its various developments offer several intriguing advantages over string theory, removing the need for additional dimensions, and it has been applied successfully to several major cosmological problems. However, the case for either string particles or looped space-time as the "Theory of Everything" remains inconclusive. ■

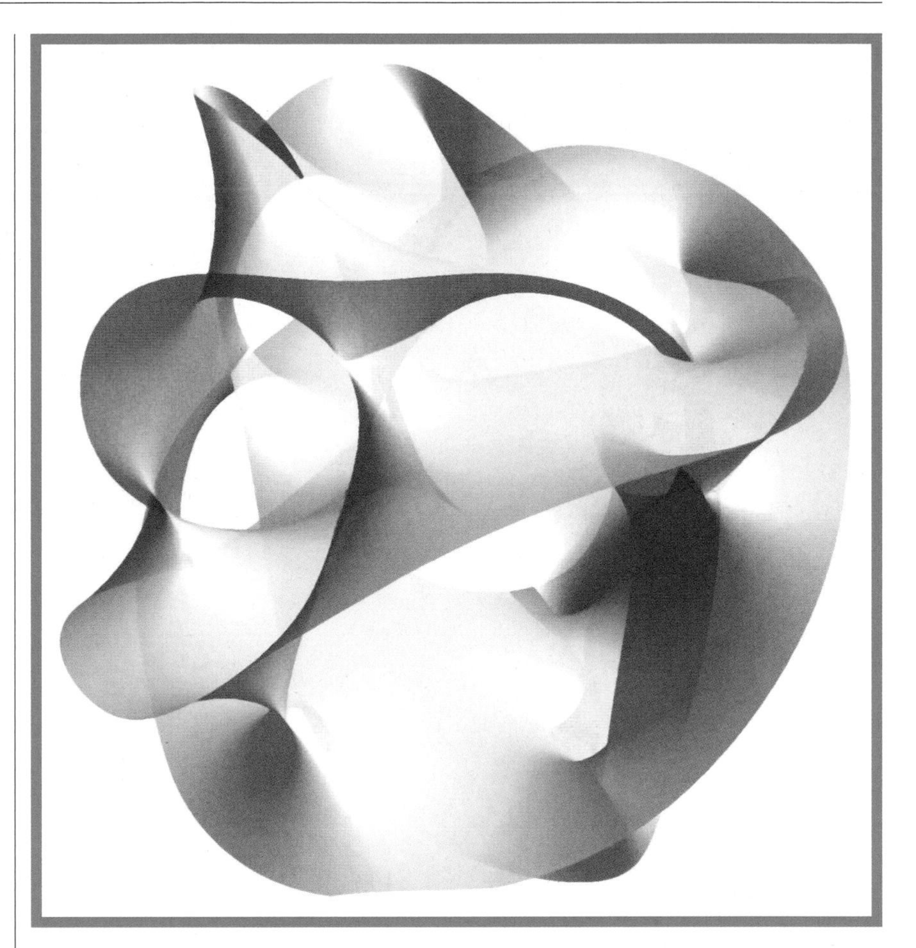

This is a 2-dimensional slice of a 6-dimensional mathematical structure called a Calabi-Yau manifold. It is suggested that string theory's six hidden dimensions may take this form.

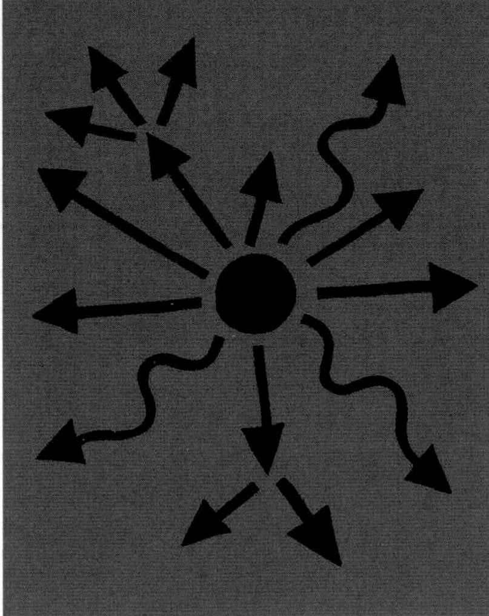

BLACK HOLES EVAPORATE

STEPHEN HAWKING (1942–)

IN CONTEXT

BRANCH
Cosmology

BEFORE
1783 John Michell theorizes objects whose gravity is so great that they trap light.

1930 Subrahmanyan Chandrasekhar proposes that a collapsing stellar core above a certain mass would give rise to a black hole.

1971 The first likely black hole is identified—Cygnus X-1.

AFTER
2002 Observations of stars orbiting close to the center of our galaxy suggest the presence of a giant black hole.

2012 American string theorist Joseph Polchinski suggests that quantum entanglement produces a super-hot "firewall" at a black hole's event horizon.

2014 Hawking announces that he no longer thinks black holes can exist.

In the 1960s, British physicist Stephen Hawking was one among several brilliant researchers who became interested in the behavior of black holes. He wrote his doctoral thesis on the cosmological aspects of a singularity (the point in space-time at which all of a black hole's mass is concentrated), and drew parallels between the singularities of stellar-mass black holes and the initial state of the universe during the Big Bang.

My goal is simple. It is a complete understanding of the universe, why it is as it is and why it exists at all.
Stephen Hawking

Around 1973, Hawking became interested in quantum mechanics and the behavior of gravity on a subatomic scale. He made an important discovery—that despite their name, black holes do not just swallow up matter and energy but emit radiation. So-called Hawking radiation is emitted at the black hole's event horizon—the outer boundary at which the black hole's gravity becomes so strong that not even light can escape. Hawking showed that in the case of a rotating black hole, the intense gravity would give rise to the production of virtual, subatomic particle-antiparticle pairs. On the event horizon, it would be possible for one element of each pair to be pulled into the black hole, effectively boosting the survivor into a sustained existence as a real particle. The result of this to a distant observer is that the event horizon emits low-temperature thermal radiation. Over time, the energy carried away by this radiation causes the black hole to lose mass and evaporate away. ■

See also: John Michell 88–89 ▪ Albert Einstein 214–21 ▪ Subrahmanyan Chandrasekhar 248

EARTH AND ALL ITS LIFE FORMS MAKE UP A SINGLE LIVING ORGANISM CALLED GAIA

JAMES LOVELOCK (1919–)

IN CONTEXT

BRANCH
Biology

BEFORE
1805 Alexander von Humboldt declares that nature can be represented as one whole.

1859 Charles Darwin argues that life forms are shaped by their environment.

1866 German naturalist Ernst Haeckel coins the term ecology.

1935 British botanist Arthur Tansley describes Earth's life forms, landscape, and climate as a giant ecosystem.

AFTER
1970s Lynn Margulis describes the symbiotic relationship of microbes and Earth's atmosphere; she later defines Gaia as a series of interacting ecosystems.

1997 The Kyoto Protocol sets targets for the reduction of greenhouse gases.

During the early 1960s, a team was assembled by NASA in Pasadena, California, to think about how to look for life on Mars. British environmental scientist James Lovelock was asked how he would tackle the problem, which prompted him to think about life on Earth.

Lovelock soon discovered a range of necessary features for life. All life on Earth depends on water. The average surface temperature must stay within 50–60°F (10–16°C) for enough liquid water to be present, and it has remained within this range for 3.5 million years. Cells require a constant level of salinity and generally cannot survive levels above 5 percent, and ocean salinity has remained at about 3.4 percent. Since oxygen first appeared in the atmosphere, about two billion years ago, its concentration has remained close to 20 percent. If it were to drop below 16 percent, there would not be enough to breathe—if it rose to 25 percent, forest fires would never go out.

> Evolution is a tightly coupled dance, with life and the material environment as partners. From the dance emerges the entity Gaia.
> **James Lovelock**

The Gaia hypothesis

Lovelock suggested that the entire planet makes up a single, self-regulating, living entity, which he called Gaia. The very presence of life itself regulates the temperature of the surface, the concentration of oxygen, and the chemical composition of the oceans, optimizing conditions for life. However, he warned that human impact on the environment may disrupt this delicate balance. ■

See also: Alexander von Humboldt 130–35 ▪ Charles Darwin 142–49 ▪ Charles Keeling 294–95 ▪ Lynn Margulis 300–01

A CLOUD IS MADE OF BILLOWS UPON BILLOWS

BENOÎT MANDELBROT (1924–2010)

IN CONTEXT

BRANCH
Mathematics

BEFORE
1917–20 In France, Pierre Fatou and Gaston Julia build mathematical sets using complex numbers—that is, combinations of real and imaginary numbers (multiples of the square root of –1). The resulting sets are either "regular" (Fatou sets) or "chaotic" (Julia sets) and are the precursors of fractals.

1926 British mathematician and meteorologist Lewis Fry Richardson publishes *Does the Wind Possess a Velocity*, pioneering mathematical models for chaotic systems.

AFTER
Present-day Fractals form part of the field of complexity science. They are used in marine biology, earthquake modeling, population studies, and oil and fluid mechanics.

Belgian mathematician Benoît Mandelbrot used computers to model the patterns in nature in the 1970s. In doing so, he launched a new field of mathematics—fractal geometry—which has since found uses in many fields.

Fractional dimensions

Whereas conventional geometry uses whole-number dimensions, fractal geometry employs fractional dimensions, which can be thought of as a "roughness measure." To understand what this means, think of measuring Britain's coastline with a stick. The longer the stick, the shorter the measurement, as it will smooth out any roughness along its length. The British coast has a fractional dimension of 1.28, which is an index of how much the measurement increases as the length of the stick decreases.

A characteristic of fractals is self-similarity—meaning that there is an equal amount of detail at all scales of magnification. The fractal nature of clouds, for example, makes it impossible to tell how close they are to us without external clues—clouds look the same from all distances. Our bodies contain many examples of fractals, such as the way the lungs branch out to fill space efficiently. Like chaotic functions, fractals show sensitivity to small changes in initial conditions, and they are used to analyze chaotic systems such as the weather. ■

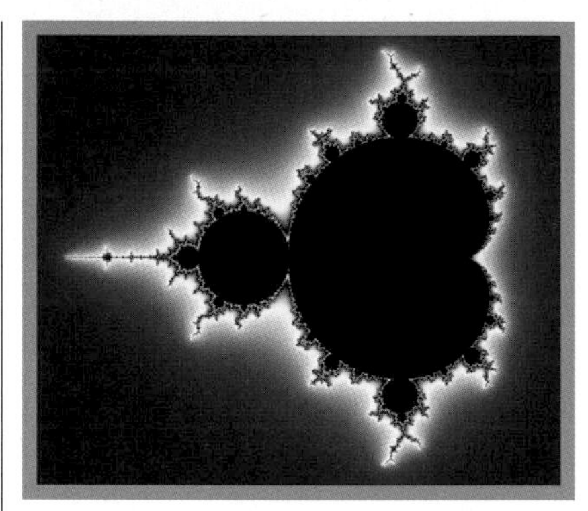

The Mandlebrot set is a fractal generated using a set of complex numbers, and conceals limitless representations of itself at every scale. When visualized graphically, it produces the distinctive shape shown here.

See also: Robert FitzRoy 150–55 ▪ Edward Lorenz 296–97

A QUANTUM MODEL OF COMPUTING

YURI MANIN (1937–)

IN CONTEXT

BRANCH
Computer science

BEFORE
1935 Albert Einstein, Boris Podolsky, and Nathan Rosen develop the "EPR paradox," providing the first description of quantum entanglement.

AFTER
1994 American mathematician Peter Shor develops an algorithm that can achieve the factorization of numbers using quantum computers.

1998 Using Hugh Everett's many-worlds interpretation of quantum mechanics, theorists imagine a superposition state in which a quantum computer is both on and off.

2011 A research team from the University of Science and Technology in Hefei, China, correctly finds the prime factors of 143 using a quantum array of four qubits.

Quantum information processing is one of the newest fields in quantum mechanics. It operates in a fundamentally different way from conventional computing. The Russian-German mathematician Yuri Manin was among the very first pioneers developing the theory.

The bit is the fundamental carrier of information in a computer, and can exist in two states: 0 and 1. The fundamental unit of information in quantum computing is called a qubit. It is made of "trapped" subatomic particles, and also has two possible states. An electron, for example, can be spin-up or spin-down, and photons of light can be polarized horizontally or vertically. However, the quantum mechanical wave function allows qubits to exist in a superposition of both states, increasing the amount of information that they can carry. Quantum theory also permits qubits to become "entangled," which exponentially increases the data carried with each additional qubit. This parallel processing could theoretically produce extraordinary computing power.

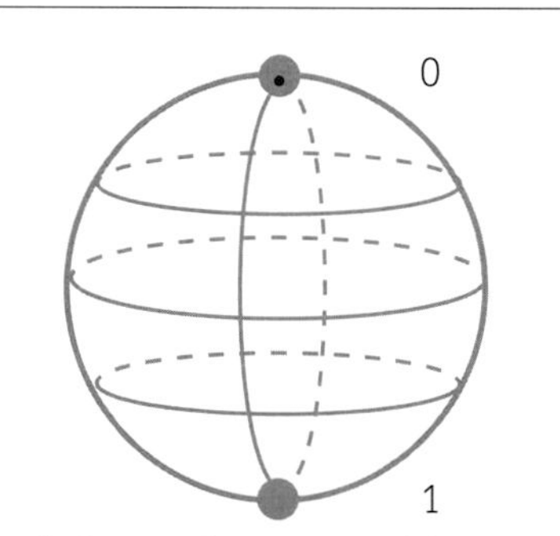

The information on a qubit can be represented as any point on the surface of a sphere—a 0, a 1, or a superposition of the two.

Demonstrating the theory

First aired in the 1980s, quantum computers seemed just theoretical. However, calculations have recently been achieved on arrays with only a few qubits. To provide a useful machine, quantum computers must achieve hundreds or thousands of entangled qubits, and there are problems scaling up to this size. Work on these problems continues. ■

See also: Albert Einstein 214–21 ▪ Erwin Schrödinger 226–33 ▪ Alan Turing 252–53 ▪ Hugh Everett III 284–85

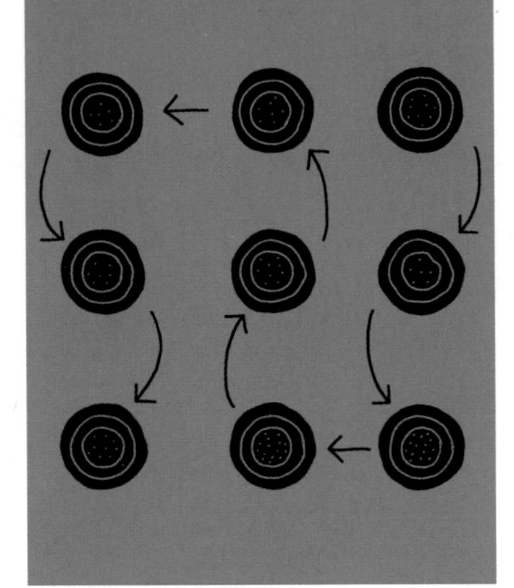

GENES CAN MOVE FROM SPECIES TO SPECIES

MICHAEL SYVANEN (1943–)

IN CONTEXT

BRANCH
Biology

BEFORE
1928 Frederick Griffith shows that one strain of bacteria can be transformed into another, by the transfer of what is later found to be DNA.

1946 Joshua Lederberg and Edward Tatum discover the natural exchange of genetic material in bacteria.

1959 Tomoichiro Akiba and Kunitaro Ochia report that antibiotic-resistant plasmids (rings of DNA) can move between bacteria.

AFTER
1993 American geneticist Margaret Kidwell identifies instances where genes have crossed species boundaries in complex organisms.

2008 American biologist John K. Pace and others present evidence of horizontal gene transfer in vertebrates.

Heat-killed **bacteria** can **transfer their characteristics** to living bacteria.

This happens because **genes can move** between bacterial cells.

Similar genes have been identified in **distantly related species** of organisms, including vertebrates.

Genes can move from species to species.

The continuity of life—the growth, reproduction, and evolution of organisms—is widely seen as a vertical process, driven by genes passed down from parents to offspring. But in 1985, American microbiologist Michael Syvanen proposed that, rather than being simply passed down, genes could also be passed horizontally between species, independently of reproduction, and that horizontal gene transfer (HGT) plays a key role in evolution.

Back in 1928, British physician Frederick Griffith was studying the bacteria implicated in pneumonia. He found that a harmless strain could be made dangerous simply by mixing its living cells with the dead remnants of a heat-killed virulent one. He attributed his results to a transforming "chemical principle" that had leaked from the dead cells into the living ones. A quarter of a century before DNA's structure was unlocked by James Watson and Francis Crick, Griffith

See also: Charles Darwin 142–49 ▪ Thomas Hunt Morgan 224–25 ▪ James Watson and Francis Crick 276–83 ▪ William French Anderson 322–23

The flow of genes between different species represents a form of genetic variation whose implications have not been fully appreciated.
Michael Syvanen

had found the first evidence that DNA could pass horizontally between cells of the same generation, as well as vertically between generations.

In 1946, American biologists Joshua Lederberg and Edward Tatum demonstrated that bacteria exchange genetic material as part of their natural behavior. In 1959, a team of Japanese microbiologists led by Tomoichiro Akiba and Kunitaro Ochia showed that this kind of DNA transfer explains how resistance to antibiotics can spread through bacteria so quickly.

Transforming microbes

Bacteria have small, mobile rings of DNA called plasmids that pass from cell to cell when they come into direct contact—taking their genes with them. Some bacteria contain genes that make them resist the action of certain types of antibiotics. The genes are copied whenever the DNA replicates, and can spread through a population of bacteria as the DNA is transferred. This sort of horizontal gene transfer can also happen via viruses, as Lederberg's student Norton Zinder discovered. Viruses are even smaller than bacteria and can invade living cells—including bacteria. They may interfere with the host genes, and when they move from host to host, they may take host genes with them.

Genes for development

From the mid-1980s, Syvanen set HGT in a wider context. He noted similarities in how the development of embryos is genetically controlled at a cellular level—even between distantly related species—and attributed this to genes moving between different organisms in evolutionary history. He argued that the genetic control of animal development had evolved to be similar in different groups because this maximized the chances that gene-swapping would work.

As genome sequences are completed for more species, and as the fossil record is reexamined, evidence suggests that HGT may occur in not only microbes but also more complex organisms, in both plants and animals. Darwin's tree of life may look more like a net, with multiple ancestors rather than a last universal common ancestor. With potential implications for taxonomy, disease and pest control, and genetic engineering, HGT's full significance is still unfolding. ■

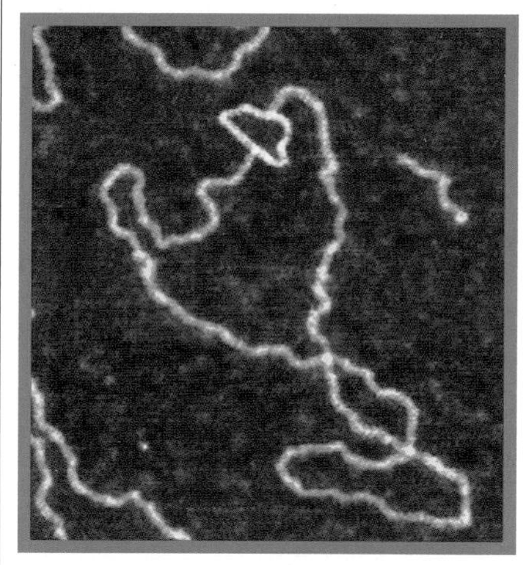

DNA plasmids, colored blue in this micrograph, are independent of a cell's chromosomes, yet they can replicate genes and be used to insert new genes into organisms.

Michael Syvanen

Michael Syvanen trained in chemistry and biochemistry at the University of Washington and the University of California at Berkeley before going on to specialize in the field of microbiology. He was appointed professor of microbiology and molecular genetics at Harvard Medical School in 1975, where he conducted research in the development of antibiotic resistance in bacteria, and insecticide resistance in flies. His findings led him to publish his theory of horizontal gene transfer (HGT) and its role in adaptation and evolution.

Since 1987, Syvanen has been professor of medical microbiology and immunology at the School of Medicine in the University of California at Davis.

Key works

1985 *Cross-species Gene Transfer: Implications for a New Theory of Evolution*
1994 *Horizontal Gene Transfer: Evidence and Possible Consequences*

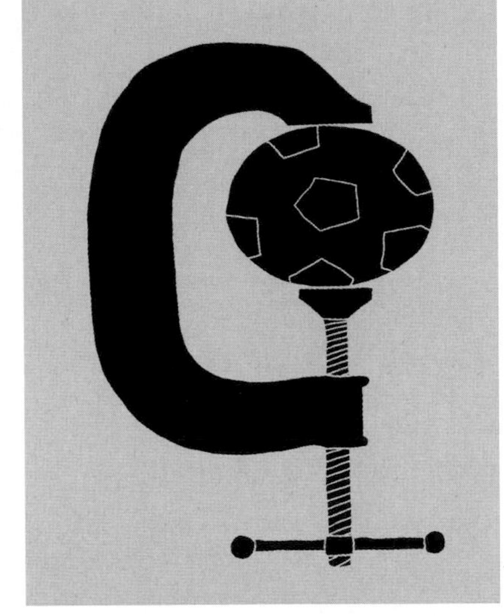

THE SOCCER BALL CAN WITHSTAND A LOT OF PRESSURE

HARRY KROTO (1939–)

IN CONTEXT

BRANCH
Chemistry

BEFORE
1966 British chemist David Jones predicts the creation of hollow carbon molecules.

1970 Scientists in Japan and Britain independently predict the existence of the carbon-60 (C_{60}) molecule.

AFTER
1988 C_{60} is found in soot from candles.

1993 German physicist Wolfgang Krätschmer and American physicist Don Huffman develop a method for synthesizing "fullerenes."

1999 Austrian physicists Markus Arndt and Anton Zeilinger demonstrate that C_{60} has wavelike properties.

2010 The spectrum of C_{60} is seen in cosmic dust 6,500 light years from Earth.

We've made a **molecule** that is so **tough** and **resilient** that...

...it has **multiple applications** in many fields of technology and medicine.

It is shaped like a soccer ball.

The soccer ball can withstand a lot of pressure.

For more than two centuries, scientists thought that elemental carbon (C) existed in only three forms, or allotropes: diamond, graphite, and amorphous carbon—the main constituent of soot and charcoal. That changed in 1985 with the work of British chemist Harry Kroto and his American colleagues Robert Curl and Richard Smalley. The chemists vaporized graphite with a laser beam to produce various carbon clusters, forming molecules with an even number of carbon atoms. The most abundant clusters had the formulae C_{60} and C_{70}. These were molecules that had never been seen before.

C_{60} (or carbon-60) soon turned out to have remarkable properties. The chemists realized that it had a structure like a soccer ball—a complete spherical cage of carbon

See also: August Kekulé 160–65 ▪ Linus Pauling 254–59

atoms, each bonded to three others in such a way that all the faces of the polyhedron are either pentagons or hexagons. C_{70} is more like a football; it has an extra ring of carbon atoms around its equator.

Both C_{70} and C_{60} reminded Kroto of the futuristic geodesic domes designed by American architect Buckminster Fuller, so he named the compounds buckminsterfullerene, but they are also called buckyballs, or fullerenes.

Properties of buckeyballs

The team found that the C_{60} compound was stable and could be heated to high temperatures without decomposing. It turned into a gas at about 1,202° F (650° C). It was odorless, and was insoluble in water, but slightly soluble in organic solvents. The buckyball is also one of the largest objects ever found to exhibit the properties both of a particle and of a wave. In 1999, Austrian researchers sent molecules of C_{60} through narrow slits and observed the interference pattern of wavelike behavior.

Solid C_{60} is as soft as graphite, but when highly compressed, it changes into a superhard form of diamond. The soccer ball, it seems, can withstand a lot of pressure.

Pure C_{60} is a semiconductor of electricity, meaning that its conductivity is between that of an insulator and a conductor. But when atoms of alkali metals such as sodium or potassium are added to it, it becomes a conductor, and even a superconductor at low temperatures, conducting electricity with no resistance at all.

C_{60} also undergoes a wide variety of chemical reactions, resulting in huge numbers of products (chemical substances) whose properties are still being studied.

The new world of nano

Although C_{60} was the first of these molecules to be studied, its discovery has led to an entire new branch of chemistry—the study of fullerenes. Nanotubes have been made—cylindrical fullerenes, only a few nanometers wide, but up to several millimeters long. They are good conductors of heat and electricity, chemically inactive, and enormously strong, which makes them hugely useful for engineering.

There are many others that are being studied for everything from electrical properties to medical treatments for cancer to HIV. The latest spin-off from the fullerenes is graphene, a flat sheet of carbon atoms, like a single layer of graphite. This substance has remarkable properties that are being hotly studied. ■

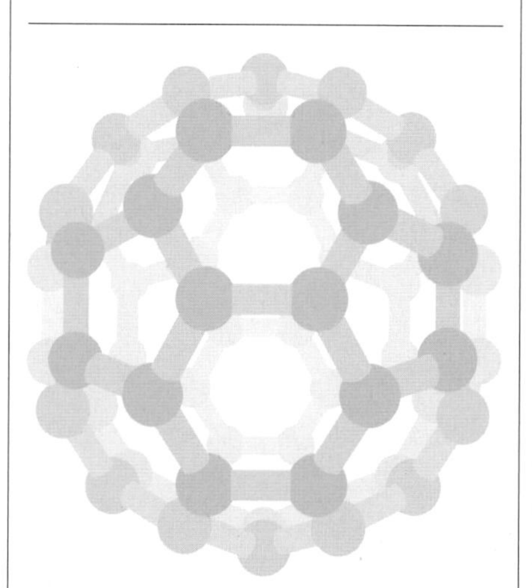

Each carbon atom of a C_{60} molecule bonds to three others. The molecule has 32 faces in total, 12 of which are pentagons and 20 hexagons, forming a distinctive, soccer-ball shape.

Harry Kroto

Harold Walter Krotoschiner was born in Cambridgeshire, England, in 1939. Fascinated by the toy building set Meccano, he chose to study chemistry, and became a professor at Sussex University in 1975. He was interested in looking into space for compounds with multiple carbon-carbon bonds, such as H-C≡C-C≡C-C≡N, and found evidence using spectroscopy (studying the interaction between matter and radiated energy). When he heard of the laser spectroscopy work of Richard Smalley and Robert Curl at Rice University, he joined them in Texas, and together they discovered C_{60}. Since 2004, Kroto has worked on nanotechnology at Florida State University.

In 1995, he set up the Vega Science Trust to make science movies for education and training. They are freely available on the Internet at www.vega.org.uk.

Key works

1981 *The Spectra of Interstellar Molecules*
1985 *60:Buckminsterfullerene* (with Heath, O'Brien, Curl, and Smalley)

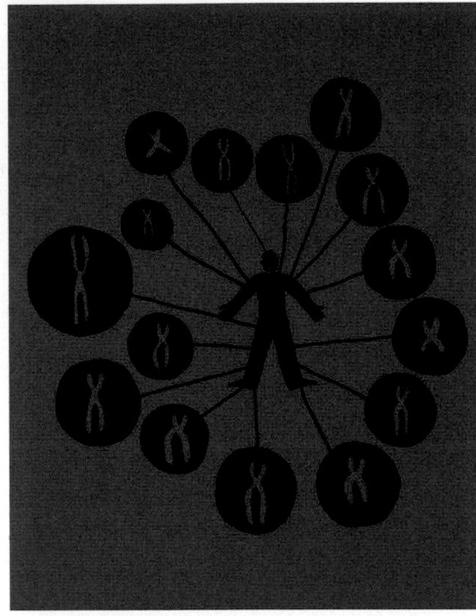

INSERT GENES INTO HUMANS TO CURE DISEASE

WILLIAM FRENCH ANDERSON (1936–)

IN CONTEXT

BRANCH
Biology

BEFORE
1984 US researcher Richard Mulligan uses a virus as a tool for inserting genes into cells taken from mice.

1985 William French Anderson and Michael Blaese show this technique can be used to correct defective cells.

1989 Anderson performs the first safety test in human gene therapy, injecting a harmless marker into a 52-year-old man. He performs the first clinical trial a year later.

AFTER
1993 UK researchers describe the results of successful animal experiments providing gene therapy treatment of cystic fibrosis.

2012 The first multidose trial of cystic fibrosis gene therapy on humans begins.

Many **diseases** are inherited and are **caused by defective genes**.

↓

Functional genes can be isolated from normal cells using enzymes that cut DNA.

↓

Genes can be transferred between cells by using vectors: viruses or rings of DNA called plasmids.

↓

Genes can be inserted into humans to cure disease.

The human genome—the entirety of a human's hereditary information—consists of about 20,000 genes. A gene is a living organism's molecular unit of heredity. However, genes often malfunction. A defective gene is made when a normal gene is not copied properly, and the "error" is passed down from parents to offspring. The symptoms that arise from these so-called genetic diseases depend upon the gene involved. A gene works by controlling the production of a protein—one of many that perform a vast variety of functions in living organisms—but this production fails if there is an error. For example, if a blood-clotting gene malfunctions, the body stops producing the blood protein that makes blood clot—causing the disease hemophilia.

Genetic diseases cannot be cured by conventional drugs, and for a long time, it was only possible to alleviate the symptoms and make a sufferer's life as comfortable as possible. But in the 1970s, scientists began considering the possibility of "gene therapy" to cure disease—using "healthy" genes to replace or override faulty ones.

See also: Gregor Mendel 166–71 ▪ Thomas Hunt Morgan 224–25 ▪ Craig Venter 324–25 ▪ Ian Wilmut 326

1. Cells containing the defective gene are taken from the body.

2. A virus is modified so that it cannot reproduce.

3. The healthy gene is inserted into the virus.

4. The virus is mixed with cells from the body.

5. The cells are genetically altered by the virus.

6. The healthy cells are injected into the body, where they work normally.

Scientists use viruses as a vector to introduce healthy genes into a patient's cells.

Introducing new genes

Genes can be introduced into diseased parts of the body by a vector—a particle that "carries" the gene to its source. Researchers investigated several possibilities for entities that might act as a vector—including viruses, which are more

Gene therapy is ethical because it can be supported by the fundamental moral principle of beneficence: it would relieve human suffering.
William French Anderson

normally associated with causing disease, rather than fighting it. Viruses naturally invade living cells as part of their infection cycle, but could they perhaps carry the therapeutic genes with them?

In the 1980s, a team of American scientists including William French Anderson succeeded in using viruses to insert genes into cultured (laboratory-grown) tissue. They tested it on animals that suffered from a genetic immune deficiency disease. The goal was to get the therapeutic gene into the animals' bone marrow, which would then make healthy red blood cells and cure the deficiency. The test was not very effective, although the procedure worked better when white blood cells were targeted.

In 1990, however, Anderson performed the first clinical trial, treating two girls who both suffered from the same immune deficiency condition, known as bubble-boy disease. Sufferers of this condition are so susceptible to infection that they may have to spend their whole lives in a sterile environment, or "bubble."

Anderson's team took sample cells from the two girls, treated them with the gene-carrying virus, then transfused the cells back into the girls. The treatment was repeated several times over two years—and it worked. However, its effects were only temporary, since new cells made by the body would still inherit the malfunctioning gene. This remains a central problem for gene therapy researchers today.

Future prospects

Remarkable breakthroughs have been made in the treatment of other conditions. In 1989, scientists working in the US identified the gene that causes cystic fibrosis. In this condition, defective cells produce sticky mucus that clogs lungs and the digestive system. Within five years of identifying the defective gene responsible, a technique had been developed to deliver healthy genes using liposomes—a type of oily droplet—as a vector. Results from the first clinical trial are due in 2014.

Considerable challenges still remain to be overcome to extend gene therapy. Cystic fibrosis is caused by a defect in just one gene. However, many conditions with a genetic component—such as Alzheimer's, heart disease, and diabetes—are caused by the interplay of many different genes. Such conditions are far harder to treat, and the search for successful, safe gene therapies is ongoing. ■

DESIGNING NEW LIFE FORMS ON A COMPUTER SCREEN

CRAIG VENTER (1946–)

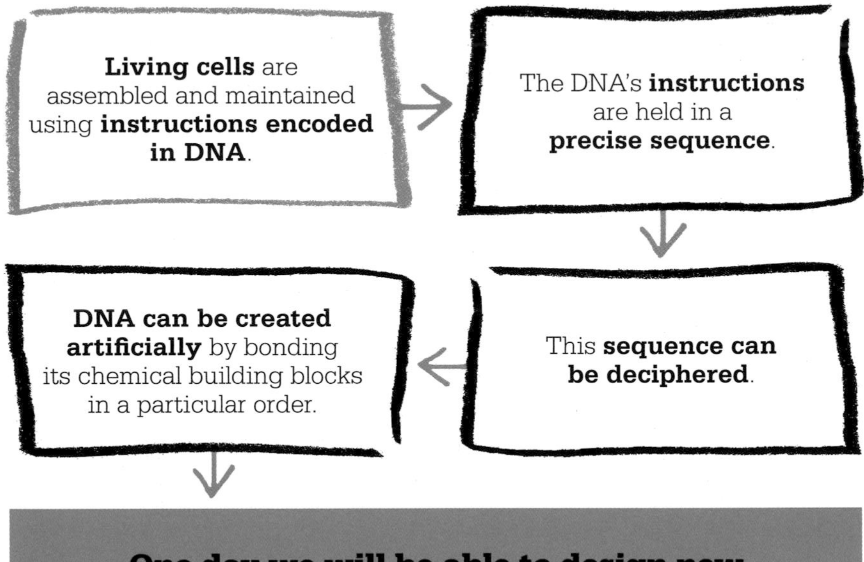

IN CONTEXT

BRANCH
Biology

BEFORE
1866 Gregor Mendel shows that the inherited traits in pea plants follow certain patterns.

1902 American biologist and physician Walter Sutton suggests that chromosomes are the carriers of heredity.

1910–11 Thomas Hunt Morgan proves Sutton's theory in fruit fly experiments.

1953 Francis Crick and James Watson reveal how DNA carries genetic instructions.

1995 A bacterium's genome (complete set of genes) is the first to be sequenced.

2000 The human genome is first sequenced.

2007 Craig Venter synthesizes an artificial chromosome.

AFTER
2010 Venter announces the first synthesis of a life form.

In May 2010, an American team of scientists led by biologist Craig Venter created the first wholly artificial life form. The organism—a single-celled bacterium—was assembled from its raw chemical building blocks. This was a testament to the advance in our understanding of the nature of life itself. The dream of creating life is nothing new. In 1771, Luigi Galvani used electricity to make a dissected frog's leg twitch, inspiring novelist Mary Shelley to write *Frankenstein*. But scientists gradually realized that life depends less on a physical "spark" and more on the chemical processes taking place inside cells.

By the mid-1950s, the real secret of life had been found in a molecule called deoxyribonucleic acid, or

See also: Gregor Mendel 166–71 ▪ Thomas Hunt Morgan 224–25 ▪ Barbara McClintock 271 ▪ James Watson and Francis Crick 276–83 ▪ Michael Syvanen 318–19 ▪ William French Anderson 322–23

DNA, which exists in the nucleus of every cell. The long string of DNA's chemical building blocks was identified as the genetic code that controls the workings of the cell. Creating life would mean creating DNA—and getting the sequence of building blocks, called nucleotides, exactly right. Nucleotides each have one of just four kinds of bases, but combine in countless ways.

Making DNA

The sequence of nucleotides differs in each organism, and is the result of millions of years of evolution. A random sequence would send a nonsense chemical "message" that could not maintain a living thing. In order to create life, scientists had to copy a sequence from a naturally existing organism. By 1990, new technology was available to work this out through a host of complex methods, and the international Human Genome Project was launched to sequence the entire human genetic makeup, or genome.

The first organism—a bacterium—was sequenced in 1995. Three years later, frustrated by the slow pace of the Human Genome Project, Venter left to set up the private company Celera Genomics to sequence the human genome more quickly and to release the data into the public domain. In 2007, his team announced that it had made an artificial chromosome—a complete string of DNA—based on that of a bacterium of the genus *Mycoplasma*. By 2010, his team had inserted an artificial chromosome into another bacterium whose genetic material had been removed, effectively creating a new life form.

Computer-generated life

The genome of even the simplest living thing—such as *Mycoplasma*—consists of sequences of hundreds of thousands of nucleotides. These nucleotides must be artificially bonded together in a specific order, but doing this for a whole genome is a formidable task. The process is automated with the help of computer technology, on machines that can now decode the genetic blueprint of life, identify genetic factors in disease, and even serve to create new life forms. ■

Mycoplasma are bacteria that lack a cell wall. They are the smallest known life forms, and were chosen by Venter to be the first organisms to have their chromosomes artificially sequenced.

Craig Venter

Born in Salt Lake City, Utah, Craig Venter performed poorly at school. Drafted into the Vietnam War, he worked in a field hospital and became drawn to biomedical science. After studying at the University of California, San Diego, he joined the US National Institute of Health in 1984. In the 1990s, he helped develop technology that could locate genes in the human genetic makeup, becoming a pioneer in the growing field of genome research. He left the NIH to set up the not-for-profit Institute of Genomic Research in 1992. He invented a way of sequencing whole genomes, focusing first on the bacterium *Haemophilus influenzae*. Turning to the human genome, he set up the profit-making company Celera and helped build advanced sequencing machines. In 2006, he founded the not-for-profit J. Craig Venter Institute to conduct research into the creation of artificial life forms.

Key works

2001 *The Sequence of the Human Genome*
2007 *A Life Decoded*

A NEW LAW OF NATURE

IAN WILMUT (1944–)

IN CONTEXT

BRANCH
Biology

BEFORE
1953 James Watson and Francis Crick demonstrate that DNA has a double helix structure that carries the genetic code and can replicate.

1958 F. C. Stewart clones carrots from mature (differentiated) tissues.

1984 Danish biologist Steen Willadsen develops a way of fusing embryo cells with egg cells that have had their genetic material removed.

AFTER
2001 The first endangered animal, a gaur (Indian bison) named Noah, is born in the US by reproductive cloning. It dies of dysentery two days later.

2008 Therapeutic cloning of tissue is shown to be effective at curing Parkinson's disease in mice.

Cloning is the production of a new, genetically identical organism from a single parent. It occurs in nature, such as when a strawberry plant sends out runners and the offspring inherit all their genes asexually. However, artificial cloning is tricky, as not all cells have the potential to grow into complete individuals, and mature cells may be reluctant to do so. The first successful cloning of a multicell organism was achieved in 1958 by British biologist F. C. Stewart, who grew a carrot plant from a single mature cell. Cloning animals proved trickier.

The pressures for human cloning are powerful; but we need not assume that it will ever become a common or significant feature of human life.
Ian Wilmut

Cloning animals

In animals, fertilized eggs and the cells of a young embryo are among the few totipotent cells—cells that can grow to form a whole body. By the 1980s, scientists could produce clones by separating young embryo cells, but it was difficult. British biologist Ian Wilmut and his team instead inserted the nuclei of body cells into fertilized eggs that had had their genetic material removed—thereby making them totipotent.

Using udder cells of sheep as the source of nuclei, the team inserted the resultant embryos into sheep to develop normally. In total, 27,729 of these cells grew into embryos, and one, named Dolly, born in 1996, survived into adulthood. Research into cloning for agriculture, conservation, and medicine continues, as does public debate over its ethics. ■

See also: Gregor Mendel 166–71 ▪ Thomas Hunt Morgan 224–25 ▪ James Watson and Francis Crick 276–83

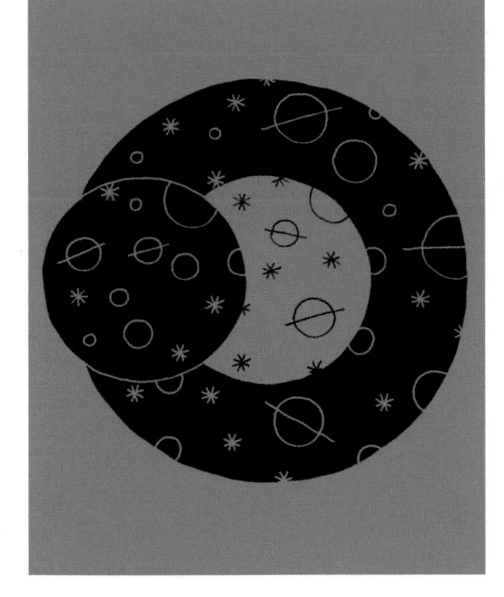

WORLDS BEYOND THE SOLAR SYSTEM

GEOFFREY MARCY (1954–)

IN CONTEXT

BRANCH
Astronomy

BEFORE
1960s Astronomers hope to detect new planets through measurement of "wobbles" in the paths of stars, but such movements remain beyond the range of even the strongest telescopes today.

1992 Polish astronomer Aleksander Wolszczan finds the first confirmed extrasolar planets in orbit around a pulsar (a burned-out stellar core).

AFTER
2009–2013 NASA's Kepler satellite discovers more than 3,000 candidate exoplanets by looking for minute drops in the brightness of stars as planets pass in front of them. Based on Kepler data, astronomers predict there could be as many as 11 billion Earthlike worlds orbiting Sunlike stars in the Milky Way galaxy.

Astronomers have long pondered the possibility of planets orbiting stars other than our Sun, but technology has, until recently, limited our ability to detect them. First to be found were planets that orbited pulsars—rapidly spinning neutron stars whose radio signals vary slightly as their planets pull them this way and that. Then, in 1995, Swiss astronomers Michel Mayor and Didier Queloz discovered 51 Pegasi b—a Jupiter-sized planet orbiting a Sunlike star about 51 light years from Earth. Since then, more than 1,000 other extrasolar planets, or "exoplanets," have been confirmed.

Planet hunter

Astronomer Geoffrey Marcy at the University of California, Berkeley, along with his team, currently holds the record for the most planets found by a human observer, including 70 out of the first 100.

Such distant planets are too faint to be seen directly, but can be revealed indirectly. The effect of a planet's gravity on its host star produces variations in the star's radial velocity—the speed at which it moves toward or away from Earth—which can be measured from changes in its light frequency. Whether any exoplanets support life remains to be seen. ■

The radial velocity method relies on detecting slight Doppler shifts (p.127) in a star's light frequency as it is pulled back and forth in relation to Earth by the gravity of an orbiting planet.

Host star

Blueshift as star moves toward Earth

Redshift as star recedes from Earth

Exoplanet

See also: Nicolaus Copernicus 34–39 ▪ William Herschel 86–87 ▪ Christian Doppler 127 ▪ Edwin Hubble 236–41

DIRECTO

RY

DIRECTORY

From its roots with individuals or small groups working mostly in isolation, often in pursuit of quasi-religious goals, science has been transformed into a practical activity that is central to the working of modern society. Today, many projects are highly collaborative in nature, and it can be hard—and indeed invidious—to pick out particular figures. More areas of research exist than ever before, and the boundaries between disciplines are becoming blurred. Mathematicians provide solutions to the problems of physics and physicists explain the nature of chemical reactions, while chemists delve into the mysteries of life and biologists turn their attention to artificial intelligence. Here, we list just some of the figures who have added to our understanding of the world.

PYTHAGORAS

c.570–495 BCE

Little is known for certain about the life of the Greek mathematician Pythagoras, who did not leave behind any written work. He was born on the Greek island of Samos, but left some time before 518 BCE for Croton in southern Italy, where he founded a secretive philosophical and religious society called the Pythagoreans. The society's inner circle called themselves *mathematikoi*, and held that reality, at its deepest level, is mathematical in nature. Pythagoras believed that the relations between all things could be reduced to numbers, and his group began discovering these relations. Among his many contributions to science and mathematics, Pythagoras studied the harmonics of vibrating strings, and probably provided the first proof of the theorem that now bears his name: that the square of the hypotenuse on a right-angled triangle is equal to the sum of the squares of the other two sides.

See also: Archimedes 24–25

XENOPHANES

c.570–475 BCE

Xenophanes of Colophon was an itinerant Greek philosopher and poet. His wide-ranging interests reflected the knowledge he gained from careful observations made on his extensive travels. He identified the energy of the Sun that heats the oceans to create clouds as the driving force behind physical processes on Earth. Xenophanes thought that clouds were the origin of heavenly bodies: the stars were burning clouds, while the Moon was made of compressed cloud. Upon discovering the fossilized remains of sea creatures far inland, he reasoned that Earth alternated between periods of flood and drought. Xenophanes produced one of the earliest accounts of natural phenomena that did not invoke divine forces to explain them, but his works were largely neglected in the centuries after his death.

See also: Empedocles 21 ▪ Zhang Heng 26–27

ARYABHATA

476–550 CE

Working in Kusumapura, a center of learning in India's Gupta empire, the Hindu mathematician and astronomer Aryabhata wrote a short treatise that was to prove highly influential among later Islamic scholars. Written in verse when he was just 23 years old, the *Arabhatiya* contains sections on arithmetic, algebra, trigonometry, and astronomy. It includes an approximation of pi (π, the ratio of a circle's circumference to its diameter) as 3.1416, which is accurate to four decimal places, and of Earth's circumference as 24,835 miles (39,968km)—very close to the current accepted figure of 24,902 miles (40,075km). Aryabhata also suggested that the apparent movement of the stars was due to the rotation of Earth and that the orbits of the planets were ellipses, but appears to have fallen short of proposing a heliocentric model of the solar system.

See also: Nicolaus Copernicus 34–39 ▪ Johannes Kepler 40–41

BRAHMAGUPTA
598–670

The Indian mathematician and astronomer Brahmagupta introduced the concept of zero into the number system, defining it as the result of subtracting a number from itself. He also detailed the arithmetic rules for dealing with negative numbers. He wrote his major work in 628, while living and working in Bhillamala, the capital city of the Gurjara-Pratihara dynasty. Called *Brahma-sphuta-siddhanta* (*The Correct Treatise of the Brahma*), the work contained no mathematical symbols but included a full description of the quadratic formula, a means of solving quadratic equations. The work was translated into Arabic in Baghdad the following century and was a major influence on later Arab scientists.
See also: Alhazen 28–29

JABIR IBN-HAYYAN
c.722–c.815

The Persian alchemist Jabir Ibn-Hayan, also known by the latinized name Geber, was a practical, experimental scientist, who outlined detailed methods for, among other things, making alloys, testing metals, and fractional distillation. Almost 3,000 different books have been attributed to Jabir, but many were probably written in the century after his death. Few of Jabir's works were known to medieval Europe, but a work attributed to him, called *Summa Perfectionis Magisterii* (*The Sum of Perfection*), appeared in the 13th century. It became the best-known book on alchemy in Europe, but was probably written by the Franciscan monk Paul of Taranto. At the time, it was common practice for an author to adopt the name of an illustrious predecessor.
See also: John Dalton 112–13

IBN-SINA
980–1037

Also known as Avicenna, the Persian physician Abu 'Ali al-Husayn Ibn-Sina was a child prodigy who had memorized the entire Koran by the age of 10. He wrote widely on topics including mathematics, logic, astronomy, physics, alchemy, and music, producing two major works: the *Kitab al-shifa* (*The Book of Healing*), a huge encyclopedia of science; and *Al-Qanun fi al-Tibb* (*The Canon of Medicine*), which was to remain in use as a university textbook into the 17th century. Ibn-Sina outlined not only medical cures but also ways to stay healthy, stressing the importance of exercise, massage, diet, and sleep. He lived through a period of political upheaval and often found his studies interrupted by the need to stay on the move.
See also: Louis Pasteur 156–59

AMBROISE PARÉ
c1510–1590

Ambroise Paré spent 30 years working as a military surgeon in the French army, during which time he developed many new techniques, including the use of ligatures to tie arteries after amputation of a limb. He studied anatomy, developed artificial limbs, and produced one of the first medical descriptions of the condition known as "phantom limb," in which the patient feels sensation in a limb after it has been amputated. He also made artificial eyes from gold, silver, porcelain, and glass. Paré examined the internal organs of people who had died violent deaths and wrote the first legal medical reports, marking the beginning of modern forensic pathology. Paré's work raised the previously low social status of surgeons, and he acted as personal surgeon to four French kings. *Les Oeuvres* (*The Works*), a book detailing his techniques, was published in 1575.
See also: Robert Hooke 54

WILLIAM HARVEY
1578–1657

English physician William Harvey produced the first accurate description of the circulation of blood, showing that it flows rapidly through the body in one system pumped by the heart. Previously, there were thought to be two blood systems: the veins carried purple blood full of nutrients from the liver, while the arteries carried scarlet "life-giving" blood from the lungs. Harvey demonstrated blood flow in numerous experiments, and studied the heartbeats of various animals. However, he was opposed to the mechanical philosophy of Descartes, and believed that blood had its own life force. Initially resisted, by the time of his death, Harvey's theory of circulation was widely accepted. Smaller capillaries linking the arteries and veins were discovered under new microscopes in the late 17th century.
See also: Robert Hooke 54 ▪ Antonie van Leeuwenhoek 56–57

MARIN MERSENNE
1588–1648

The French monk Marin Mersenne is best remembered today for his work on prime numbers, showing that if the number 2^n-1 is prime, then n must also be prime. He also conducted extensive studies in many scientific fields, including harmonics, in which he figured out the laws that govern the frequency of vibrations of a stretched string. Mersenne lived in Paris, where he collaborated with René Descartes, and corresponded extensively with Galileo, whose works he translated into French. He strongly advocated experiment as the key to scientific understanding, stressing the need for accurate data and criticizing many of his contemporaries for their lack of rigor. In 1635, he founded the Académie Parisienne, a private scientific association with more than 100 members across Europe, which would later become the French Academy of Sciences.
See also: Galileo Galilei 42–43

RENÉ DESCARTES
1596–1650

The French philosopher René Descartes was a key figure in the Scientific Revolution of the 17th century, traveling widely across Europe and working with many of the prominent figures of his day. He helped European scientists to finally overcome Aristotle's nonempirical approach by applying a thorough scepticism to assumed knowledge. Descartes produced a four-pronged method of scientific inquiry, based on mathematics: accept nothing as true unless it is self-evident; divide problems into their simplest parts; solve the problems by moving from the simple to the complex; and, lastly, check your results. He also developed the Cartesian system of coordinates—with x, y, and z axes—to represent points in space using numbers. This allowed shapes to be expressed as numbers and numbers to be expressed as shapes, founding the mathematical field of analytical geometry.
See also: Galileo Galilei 42–43 ▪ Francis Bacon 45

HENNIG BRAND
c.1630–c.1710

Little is known about the early life of German chemist Hennig Brand. We do know that he fought in the Thirty Years' War and dedicated himself to alchemy on leaving the army, searching for the elusive philosopher's stone that would turn base metal into gold. In 1669, Brand produced a waxy, white material by heating the residue of boiled-down urine. He called this material "phosphorus" ("light-carrier") because it glowed in the dark. Phosphorus is highly reactive and never found as a free element on Earth, and this marked the first time that such an element had been isolated. Brand kept his method secret, but phosphorus was discovered independently by Robert Boyle in 1680.
See also: Robert Boyle 46–49

GOTTFRIED LEIBNIZ
1646–1716

The German Gottfrield Leibniz studied law at the University of Leipzig. During his studies, he became increasingly interested in science as he discovered the ideas of Descartes, Bacon, and Galileo, which marked the start of a lifelong quest to collate all human knowledge. He later studied mathematics in Paris under Christiaan Huygens, and it was here that he began to develop calculus—a mathematical means of calculating rates of change that was to prove crucial to the development of science. He developed calculus at the same time as Isaac Newton, with whom he corresponded and then fell out. Leibniz actively promoted the study of science, corresponding with more than 600 scientists across Europe and setting up academies in Berlin, Dresden, Vienna, and St. Petersburg.
See also: Christiaan Huygens 50–51 ▪ Isaac Newton 62–69

DENIS PAPIN
1647–1712

As a young man, French-born English physicist and inventor Denis Papin assisted both Christiaan Huygens and Robert Boyle in their experiments on air and pressure, and in 1679, he invented the pressure cooker. Observing how the steam in the cooker tended to raise the lid, Papin then came up with the idea of using steam to drive a piston in a cylinder, and produced the first design for a steam engine. Papin never built a steam engine himself, but in 1709, he constructed a paddle wheel that demonstrated the practicability of using paddles instead of oars in steam-powered ships.
See also: Robert Boyle 46–49 ▪ Christiaan Huygens 50–51 ▪ Joseph Black 76–77

STEPHEN HALES
1677–1761

English clergyman Stephen Hales conducted a series of pioneering experiments on plant physiology. He measured the water vapor emitted by the leaves of plants in a process called transpiration, and this led him to the discovery that transpiration drives a continuous upward flow of fluid from the roots that carries dissolved nutrients around the whole plant. Sap moves from an area of high pressure in the roots to areas of lower pressure where water vapor is transpiring. Hales published his results in 1727 in the book *Vegetable Staticks*. In addition, he conducted extensive experiments with animals, particularly dogs, measuring blood pressure for the first time. Hales also invented the pneumatic trough, an apparatus used to collect the gases emitted during chemical reactions.
See also: Joseph Priestley 82–83 ▪ Jan Ingenhousz 85

DANIEL BERNOULLI
1700–1782

Daniel Bernoulli was perhaps the most gifted in a remarkable family of Swiss mathematicians—his uncle Jakob and father Johann both did important work in developing calculus. In 1738, he published *Hydrodynamica*, in which he examined the properties of fluids. He formulated Bernoulli's principle, that a fluid's pressure decreases as its velocity increases. This principle is key to understanding how the wings of an airplane produce lift. He realized that a moving fluid must exchange some of its pressure for kinetic energy in order not to violate the principle of the conservation of energy. In addition to mathematics and physics, Bernoulli studied astronomy, biology, and oceanography.
See also: Joseph Black 76–77 ▪ Henry Cavendish 78–79 ▪ Joseph Priestley 82–83 ▪ James Joule 138 ▪ Ludwig Boltzmann 139

GEORGES-LOUIS LECLERC, COMTE DE BUFFON
1707–1788

From 1749 to the end of his life, French aristocrat and naturalist the Comte de Buffon worked tirelessly on his monumental work *Histoire Naturelle* (*Natural History*). His goal was to collate all knowledge in the fields of natural history and geology. The encyclopedia spanned 44 volumes when it was finally completed by his assistants 16 years after his death. Buffon constructed a geological history of Earth, suggesting that it was much older than previously assumed. He charted the extinction of species and suggested a common ancestor of humans and apes, predating Charles Darwin by a century.
See also: Carl Linnaeus 74–75 ▪ James Hutton 96–101 ▪ Charles Darwin 142–49

GILBERT WHITE
1720–1793

British parson Gilbert White was an unmarried curate who lived a quiet life in the small Hampshire village of Selborne. His 1789 book, *The Natural History and Antiquities of Selborne*, was a compilation of letters written to his friends. In his letters, White laid out a record of his systematic observations of nature and developed his ideas about the interrelationships of living things. He was, in effect, the first ecologist. White recognized that all living things have a role to play in what we would now call the ecosystem, noting of earthworms that they "seem to be the great promoters of vegetation, which would proceed but lamely without them." White's methods, including taking recordings in the same places over many years, were highly influential on subsequent biologists.
See also: Alexander von Humboldt 130–35 ▪ James Lovelock 315

NICÉPHORE NIEPCE
1765–1833

The oldest surviving photograph was taken in 1825 by French inventor Nicéphore Niepce of the buildings around his country estate in Saint-Loup-de-Varennes. Niepce had been experimenting for several years to find a technique to fix the image projected onto the back of a camera obscura. In 1816, he produced a negative image using paper coated with silver chloride, but the image disappeared when exposed to daylight. Then around 1822, he came up with a process he called heliography, which used a plate of glass or metal coated with bitumen. The bitumen hardened when it was exposed to light, and when the plate was washed with lavender oil, only the hardened areas remained. It took eight hours of exposure to fix the images. Near the end of his life, Niepce collaborated with Louis Daguerre on ways to improve the process.
See also: Alhazen 28–29

ANDRÉ-MARIE AMPÈRE
1775–1836

Upon hearing of Hans Christian Ørsted's accidental discovery of the link between electricity and magnetism in 1820, French physicist André-Marie Ampère started formulating a mathematical and physical theory that explained their relationship. In the process, he formulated Ampère's law, which states the mathematical relation of a magnetic field to the electric current that produces it. Ampère published his results in 1827, and his book, *Memoir on the Mathematical Theory of Electrodynamic Phenomena*, uniquely deduced from experience, gave a name to this new scientific field—electrodynamics. The standard unit of electric current, the ampere (or amp), is named after him.
See also: Hans Christian Ørsted 120 ▪ Michael Faraday 121

LOUIS DAGUERRE
1787–1851

The first practical photographic process was invented by the French painter and physicist Louis Daguerre. From 1826, Daguerre collaborated with Nicéphore Niepce on his heliographic process, but this needed at least eight hours of exposure. Following Niepce's death in 1833, Daguerre developed a process in which an image on an iodized silver plate was developed by exposure to mercury fumes and fixed using saline. This reduced the exposure time required to 20 minutes, making it practical to take photographs of people for the first time. Daguerre wrote a full description of his process, called the daguerreotype, in 1839, and it made him a fortune.
See also: Alhazen 28–29

AUGUSTIN FRESNEL
1788–1827

French engineer and physicist Augustin Fresnel is best known as the inventor of the Fresnel lens, which allows the light from a lighthouse to be seen over greater distances. He studied the behavior of light, building on the double-slit experiments of Thomas Young, with whom he corresponded. Fresnel conducted a great deal of important theoretical work on optics, producing a set of equations describing how light is refracted or reflected as it passes from one medium to another. The importance of much of his work was only recognized after his death.
See also: Alhazen 28–29 ▪ Christiaan Huygens 50–51 ▪ Thomas Young 110–11

CHARLES BABBAGE
1791–1871

British mathematician Charles Babbage conceived the first digital computer. Appalled by the number of errors in printed mathematical tables, Babbage designed a machine to calculate the tables automatically, and in 1823 hired engineer Joseph Clement to build it. His "Difference Engine" was to be an elegant contraption of brass cogwheels, but Babbage got only as far as a prototype before running out of money and energy. In 1991, scientists at London's Science Museum built a Difference Engine to Babbage's specification, using only technology that would have been available at the time, and it worked, though it tended to jam after a minute or two. Babbage also dreamed of a steam-powered "Analytical Engine," which would take instructions on punched cards, hold data in a "store," carry out calculations in the "mill," and print out the results. This might have been a real computer in the modern sense. His protégée Ada Lovelace (the daughter of poet Lord Byron) wrote programs for it, and has been called the world's first computer programmer. However, the Analytical Engine project never got off the ground.
See also: Alan Turing 252–53

SADI CARNOT
1796–1832

Nicolas-Léonard-Sadi Carnot was an officer in the French army who semiretired on half-pay to Paris in 1819 to devote himself to science. Hoping to see France catch up with Britain in the Industrial Revolution, Carnot began designing and building steam engines. His research led to his only publication, in 1824, *Reflections on the Motive Power of Fire*, in which he noted that the efficiency of a steam engine depends principally on the temperature difference between the hottest and coldest parts of the engine. This pioneering work on thermodynamics was later developed by Rudolf Clausius in Germany and William Thomson, Lord Kelvin in Britain, but was largely ignored in Carnot's lifetime. He died in relative obscurity during a cholera epidemic, at just 36.
See also: Joseph Fourier 122 ▪ James Joule 138

JEAN-DANIEL COLLADON
1802–1893

Swiss physicist Jean-Daniel Colladon demonstrated that light could be trapped by total internal reflection inside a tube, allowing it to travel along a curved path—a core principle behind modern-day optical fibers. In experiments conducted on Lake Geneva, Colladon demonstrated that sound travels four times more quickly through water than through air. He transmitted sound through water over a distance of 30 miles (50km), and proposed using this method as a means of communicating across the English Channel. He also conducted important work in the field of hydraulics, studying the compressibility of water.
See also: Léon Foucault 136–37

JUSTUS VON LIEBIG
1803–1873

The son of a chemical manufacturer in Darmstadt, Germany, Justus von Liebig conducted his first chemistry experiments as a child in his father's laboratory. He grew up to become a charismatic professor of chemistry whose laboratory-based teaching methods were hugely influential. Von Liebig discovered the importance of nitrates to plant growth and developed the first industrial fertilizers. He was also interested in the chemistry of food and developed a manufacturing process to produce beef extracts. The company he founded, the Liebig Extract of Meat Company, would later produce the trademarked Oxo stock cubes.
See also: Friedrich Wöhler 124–25

CLAUDE BERNARD
1813–1878

French physiologist Claude Bernard was a pioneer in experimental medicine. He was the first scientist to study the internal regulation of the body, and his work was to lead to the modern concept of homeostasis—the mechanism by which the body maintains a stable internal environment while the external environment changes. Bernard studied the roles of the pancreas and liver in digestion, and described how chemicals are broken down into simpler substances only to be built up again into the complex molecules needed to make body tissues. His major work, *An Introduction to the Study of Experimental Medicine*, was published in 1865.
See also: Louis Pasteur 156–59

WILLIAM THOMSON
1824–1907

Born in Belfast, physicist William Thomson became professor of natural philosophy at Glasgow University at 22 years old. In 1892, he was ennobled, and became Baron Kelvin, after the river that runs through Glasgow University. Kelvin viewed physical change as fundamentally a change in energy, and his work produced a synthesis of many areas of physics. He developed the second law of thermodynamics and established the correct value for "absolute zero," the temperature at which all molecular movement ceases, at –459.6°F (–273.15°C). The Kelvin scale, which starts at 0 at absolute zero, is named after him. He invented the mirror galvanometer to receive faint telegraph signals, and presided over the laying of the transatlantic cable in 1866. He also invented an improved mariner's compass and a tide-predicting machine. Lord Kelvin often courted controversy, rejecting Darwin's theory of evolution and making many bold statements—including the prediction that "no aeroplane will ever be practically successful," made one year before the Wright brothers' first flight in 1903. However, a quote widely attributed to Lord Kelvin stating that "there is nothing new to be discovered in physics now" is almost certainly apocryphal.
See also: James Joule 138 ▪ Ludwig Boltzmann 139 ▪ Ernest Rutherford 206–213

JOHANNES VAN DER WAALS
1837–1923

Dutch physicist Johannes van der Waals made a significant contribution to the field of thermodynamics with his 1873 doctoral thesis, in which he showed that there is a continuity between a liquid and gaseous state at a molecular level. Van der Waals showed not only that these two states of matter merge into one another, but also that they should be considered as essentially of the same nature. He postulated the existence of forces between molecules, which are now called the van der Waals forces, and which explain properties of chemicals such as their solubility.
See also: James Joule 138 ▪ Ludwig Boltzmann 139 ▪ August Kekulé 160–65 ▪ Linus Pauling 254–59

ÉDOUARD BRANLY
1844–1940

A physics professor at the Paris Catholic Institute, Édouard Branly was a pioneer in wireless telegraphy. In 1890, he invented a radio receiver known as the Branly coherer. The receiver was a tube with two electrodes inside it spaced a little apart, and metal filings in the space between the electrodes. When a radio signal was applied to the receiver, the resistance of the filings was reduced, allowing an electric current to flow between the electrodes. Branly's invention was used in later experiments on radio communication by Italian Guglielmo Marconi, and widely used in telegraphy up to 1910, when more sensitive detectors were developed.
See also: Alessandro Volta 90–95 ▪ Michael Faraday 121

IVAN PAVLOV
1849–1936

The son of a priest, Russian Ivan Pavlov abandoned plans to follow in his father's footsteps in order to study chemistry and physiology at the University of St. Petersburg. In the 1890s, Pavlov was studying salivation in dogs when he noticed that his dogs would salivate whenever he entered the room, even if he had no food with him. Pavlov realized that this must be a learned behavior, and started 30 years of experiments into what he called "conditioned responses." In one experiment, he would ring a bell every time he fed the dogs. He found that after a period of learning (conditioning), the dogs would salivate just when hearing the bell. In this work, Pavlov laid the groundwork for the scientific study of behavior, although physiologists today consider his explanations to be oversimplified.
See also: Konrad Lorenz 249

HENRI MOISSAN
1852–1907

French chemist Henri Moissan received the 1906 Nobel Prize in Chemistry for his work isolating the element fluorine, which he produced by electrolysing a solution of potassium hydrogen difluoride. When Moissan cooled the solution to −58°F (−50°C), pure hydrogen appeared at the negative electrode, and pure fluorine at the positive one. Moissan also developed an electric-arc furnace that could reach a temperature of 6,300°F (3,500°C), which he used in his attempts to synthesize artificial diamonds. He did not succeed, but his theory that diamonds could be made by putting carbon under high pressure at high temperatures was subsequently proved correct.
See also: Humphry Davy 114 ▪ Leo Baekeland 140–41

FRITZ HABER
1868–1934

The scientific legacy of German chemist Fritz Haber is mixed. On the positive side, Haber and his colleague Carl Bosch developed a process for synthesizing ammonia (NH_3) from hydrogen and atmospheric nitrogen. Ammonia is an essential ingredient of fertilizers, and the Haber–Bosch process allowed the industrial production of artificial fertilizers, greatly increasing food production. On the negative side, Haber developed chlorine and other deadly gases for use in trench warfare, and personally oversaw their use on battlefields during World War I. His wife Clara, also a chemist, killed herself in 1915 in opposition to her husband's involvement in the use of chlorine gas at Ypres.
See also: Friedrich Wöhler 124–25 ▪ August Kekulé 160–65

C. T. R. WILSON
1869–1959

Charles Thomson Rees Wilson was a Scottish meteorologist with a particular interest in the study of clouds. To help his studies, he developed a method of expanding moist air inside a closed chamber to produce the state of supersaturation needed for cloud formation. Wilson found that clouds formed in the chamber much more easily in the presence of dust particles. In the absence of dust, clouds only formed when the saturation of the air passed a critical high point. Wilson believed that clouds were forming on ions (charged molecules) in the air. To test this theory, he passed radiation through the chamber to see whether the resultant ion formation would cause clouds to form. He found that the radiation left a trail of condensed water vapor in its wake. Wilson's cloud chamber proved crucial for studies in nuclear physics, and won him the Nobel Prize in Physics in 1927. In 1932, the positron was first detected using a cloud chamber.
See also: Paul Dirac 246–47 ▪ Charles Keeling 294–95

EUGÈNE BLOCH
1878–1944

French physicist Eugène Bloch conducted studies in spectroscopy, and produced evidence in support of Albert Einstein's interpretation of the photoelectric effect using the idea of quantized light. During World War I, Bloch worked on military communications, developing the first electronic amplifiers for radio receivers. In 1940, he fell victim to the anti-Jewish laws of the Vichy government and was dismissed from his post as a professor of physics at the University of Paris. He fled to unoccupied southern France, but was captured by the Gestapo in 1944 and deported to Auschwitz, where he was killed.
See also: Albert Einstein 214–21

MAX BORN
1882–1970

In the 1920s, German physicist Max Born, while professor of experimental physics at the University of Göttingen, collaborated with Werner Heisenberg and Pascual Jordan to formulate matrix mechanics, a mathematical means of dealing with quantum mechanics. When Erwin Schrödinger formulated his wave function equation to describe the same thing, Born was the first to suggest the real-world meaning of Schrödinger's mathematics—it described the probability of finding a particle at a specific point on the space-time continuum. In 1933, Born and his family left Germany when the Nazis dismissed Jews from academic posts. He settled in Britain, becoming a British citizen in 1939. He was awarded a Nobel Prize in Physics for his work on quantum mechanics in 1954.
See also: Erwin Schrödinger 226–33 ▪ Werner Heisenberg 234–35 ▪ Paul Dirac 246–47 ▪ J. Robert Oppenheimer 260–65

NIELS BOHR
1885–1962

One of the leading early theorists of quantum physics, Dane Niels Bohr's first major contribution to the quantum revolution was to refine Ernest Rutherford's model of the atom. In 1913, Bohr added the idea that electrons occupy specific quantized orbits around the nucleus. In 1927, Bohr collaborated with Werner Heisenberg to formulate an explanation of quantum phenomena that came to be known as the Copenhagen interpretation. A concept central to this interpretation was Bohr's complementarity principle, which states that a physical phenomenon, such as the behavior of a photon or an electron, may express itself differently depending on the experimental set-up used to observe it.
See also: Ernest Rutherford 206–13 ▪ Erwin Schrödinger 226–33 ▪ Werner Heisenberg 234–35 ▪ Paul Dirac 246–47

GEORGE EMIL PALADE
1912–2008

Romanian cell biologist George Emil Palade graduated in medicine from the University of Bucharest in 1940. He emigrated to the US at the end of World War II, and did his most important work at the Rockefeller Institute in New York. Palade developed new techniques for tissue preparation that allowed him to examine the structure of cells under an electron microscope, and this work greatly advanced the understanding of cellular organization. His most important achievement was the discovery in the 1950s of ribosomes—bodies inside cells that were previously thought to be fragments of mitochondria, but are in fact the primary sites of protein synthesis, linking together amino acids in a specific sequence.
See also: James Watson and Francis Crick 276–83 ▪ Lynn Margulis 300–01

DAVID BOHM
1917–1992

American theoretical physicist David Bohm advanced an unconventional interpretation of quantum mechanics. He postulated the existence of an "implicate order" to the universe that is a more fundamental order of reality than the phenomena we experience as time, space, and consciousness. He wrote: "an entirely different sort of basic connection of elements is possible, from which our ordinary notions of space and time, along with those of separately existent material particles, are abstracted as forms derived from the deeper order." Bohm worked with Albert Einstein at Princeton University until the early 1950s, when his Marxist political views led him to leave the US—first for Brazil and later London, where he was a professor of physics at Birkbeck College from 1961.
See also: Erwin Schrödinger 226–33 ▪ Hugh Everett III 284–85 ▪ Gabriele Veneziano 308–13

FREDERICK SANGER
1918–2013

British biochemist Frederick Sanger is one of four scientists to have won two Nobel prizes, both in Chemistry. He won his first prize in 1958 for determining the sequence of amino acids that make up the protein insulin. Sanger's work on insulin provided a key to understanding the way that DNA codes for making proteins, by showing that each protein has its own unique sequence of amino acids. Sanger's second prize was awarded in 1980 for his later work sequencing DNA. Sanger's team sequenced human mitochondrial DNA—a set of 37 genes found on mitochondria that is inherited only from the mother. The Sanger Institute, now one of the world's leading centers of genomic research, was established in his honor near his home in Cambridgeshire, Britain.
See also: James Watson and Francis Crick 276–83 ▪ Craig Venter 324–25

MARVIN MINSKY
1927–

American mathematician and cognitive scientist Marvin Minsky was an early pioneer in artificial intelligence, co-founding in 1959 the AI laboratory at the Massachusetts Institute of Technology (MIT), where he spent the rest of his career. His work focused on the generation of neural networks—artificial "brains" that can develop and learn from experience. In the 1970s, Minsky and his colleague Seymour Papert developed the "Society of Mind" theory of intelligence, investigating the way in which intelligence can emerge from a system made solely of nonintelligent parts. Minsky defines AI as "the science of making machines do things that would require intelligence if done by men." He was an advisor on the film *2001: A Space Odyssey*, and has speculated as to the possibility of extraterrestrial intelligence.
See also: Alan Turing 252–53 ▪ Donald Michie 286–91

MARTIN KARPLUS
1930–

Increasingly, modern science is conducted using computers to model results. In 1974, American-Austrian theoretical chemist Martin Karplus and his colleague, American-Israeli Arieh Warshel, produced a computer model of the complex molecule retinal, which changes shape when exposed to light and is crucial to the working of the eye. Karplus and Warshel used both classical physics and quantum mechanics to model the behavior of electrons in the retinal molecule. Their model greatly improved the sophistication and accuracy of computer modeling for complex chemical systems. Karplus and Warshel shared the 2013 Nobel Prize in Chemistry with British chemist Michael Levitt for their achievement in this field.
See also: Augus Kekulé 160–65 ▪ Linus Pauling 254–59

ROGER PENROSE
1931–

In 1969, British mathematician Roger Penrose collaborated with physicist Stephen Hawking to show how matter in a black hole collapses into a singularity. Penrose subsequently worked out the mathematics to describe the effects of gravity on the space-time surrounding a black hole. Penrose has turned his attention to a wide range of topics, proposing a theory of consciousness based on quantum mechanical effects operating at a subatomic level in the brain, and more recently a theory of a cyclic cosmology, in which the heat death (end state) of one universe becomes the Big Bang of another, in an endless cycle.
See also: Georges Lemaître 242–45 ▪ Subrahmanyan Chandrasekhar 248 ▪ Stephen Hawking 314

FRANÇOIS ENGLERT
1932–

In 2013, Belgian physicist François Englert shared the Nobel Prize in Physics with Peter Higgs for independently proposing what is now known as the Higgs field, which gives fundamental particles their mass. Working with fellow Belgian Robert Brout, Englert first suggested in 1964 that "empty" space might contain a field that confers mass to matter. The Nobel Prize was awarded as a result of the detection in 2012 at CERN of the Higgs boson—the particle associated with the Higgs field—which confirmed Englert, Brout, and Higgs' predictions. Brout had died in 2011, and so missed out on the Nobel Prize, which is not awarded posthumously.
See also: Sheldon Glashow 292–93 ▪ Peter Higgs 298–99 ▪ Murray Gell-Mann 302–07

STEPHEN JAY GOULD
1941–2002

American paleontologist Stephen Jay Gould's specialized area of research concerned the evolution of land snails in the West Indies, but he wrote widely about many aspects of evolution and science. In 1972, Gould and colleague Niles Eldredge proposed the theory of "punctuated equilibrium," which proposed that, rather than being a constant, gradual process as Darwin had imagined, the evolution of new species took place in rapid bursts over periods as short as a few thousand years, which were followed by long periods of stability. To back up their claim, they cited evidence from the fossil record, in which patterns of evolution in various organisms support their theory. In 1982, Gould coined the term "exaptation" to describe the way in which a particular trait may be passed on for one reason, and then later come to be coopted for a very different function. His work widened understanding of the mechanisms by which natural selection takes place.
See also: Charles Darwin 142–49 ▪ Lynn Margulis 300–01 ▪ Michael Syvanen 318–19

RICHARD DAWKINS
1941–

British zoologist Richard Dawkins is best known for his popular science books, including *The Selfish Gene* (1976). His most significant contribution to his field is his concept of the "extended phenotype." An organism's genotype is the sum of the instructions contained in its genetic code. Its phenotype is that which results from the expression of that code. While individual genes may simply code for the synthesis of different substances in an organism's body, the phenotype should be considered to be everything that results from that synthesis. For example, a termite mound may be considered to be part of a termite's extended phenotype. Dawkins views the extended phenotype as the means by which genes maximize their chances of survival to the next generation.
See also: Charles Darwin 142–49 ▪ Lynn Margulis 300–01 ▪ Michael Syvanen 318–19

JOCELYN BELL BURNELL
1943–

In 1967, while working as a research assistant at Cambridge University, British astronomer Jocelyn Bell was monitoring quasars (distant galactic nuclei) when she discovered a strange series of regular radio pulses coming from space. The team she was working with jokingly called the pulses LGM (Little Green Men), referring to the remote chance that they were an attempt at extraterrestrial communication. They later determined that the sources of the pulses were rapidly spinning neutron stars, which were dubbed pulsars. Two of Bell's senior colleagues were awarded the 1974 Nobel Prize in Physics for the discovery of pulsars, but Bell missed out because she was only a student at the time. Many leading astronomers, including Fred Hoyle, objected publicly to her omission.
See also: Edwin Hubble 236–41 ▪ Fred Hoyle 270

MICHAEL TURNER
1949–

American cosmologist Michael Turner's research focuses on understanding what happened directly following the Big Bang. Turner believes that the structure of the universe today, including the existence of galaxies and the asymmetry between matter and antimatter, can be explained by quantum-mechanical fluctuations that took place during the rapid burst of expansion called cosmic inflation, which occurred moments after the Big Bang. In 1998, Turner coined the term "dark energy" to describe the hypothetical energy that permeates the whole of space and explains the observation that the universe is expanding in all directions at an accelerating rate.
See also: Edwin Hubble 236–41 ▪ Georges Lemaître 242–45 ▪ Fritz Zwicky 250–51

TIM BERNERS-LEE
1955–

Few living scientists have had as much impact on everyday life as British computer scientist Tim Berners-Lee, who invented the World Wide Web. In 1989, Berners-Lee was working at CERN, the European Organization for Nuclear Research, when he had the idea of establishing a network of documents that could be shared across the world via the Internet. A year later, he wrote the first web client and server, and in 1991, CERN built the first website. Today, Berners-Lee campaigns for open access to the Internet, free from government control.
See also: Alan Turing 252–53

GLOSSARY

Absolute zero The lowest possible temperature: 0K or –459.67°F (–273.15°C).

Acceleration The rate of change of velocity. Acceleration is caused by a force that results in a change in an object's direction and/or speed.

Acid A chemical that, when dissolved in water, liberates hydrogen ions and turns litmus red.

Algorithm In mathematics and computer-programming, a logical procedure for making a calculation.

Alkali A base that dissolves in water and neutralizes acids.

Alpha particle A particle made of two neutrons and two protons, which is emitted during a form of radioactive decay called alpha decay. An alpha particle is identical to the nucleus of a helium atom.

Amino acids Organic chemicals with molecules that contain amino groups (NH_2) and carboxyl groups (COOH). Proteins are made from amino acids. Each different protein contains a specific sequence of amino acids.

Angular momentum A measure of the rotation of an object, which takes into account its mass, shape, and spin speed.

Antiparticle A particle that is the same as a normal particle except that it has an opposite electrical charge. Every particle has an equivalent antiparticle.

Atom The smallest part of an element that has the chemical properties of that element. An atom was thought to be the smallest part of matter, but many subatomic particles are now known.

Atomic number The number of protons in an atom's nucleus. Each element has a different atomic number.

ATP Adenosine triphosphate. A chemical that stores and transports energy across cells.

Base A chemical that reacts with an acid to make water and a salt.

Beta decay A form of radioactive decay in which an atomic nucleus gives off beta particles (electrons or positrons).

Big Bang The theory that the universe began from an explosion of a singularity.

Black body A theoretical object that absorbs all radiation that falls on it. A black body radiates energy according to its temperature, so may not in fact appear black.

Black hole An object in space that is so dense that light cannot escape its gravitational field.

Bosons Subatomic particles that carry forces between other particles.

Brane In string theory, an object that has between zero and nine dimensions.

Cell The smallest unit of an organism that can survive on its own. Organisms such as bacteria and protists are single cells.

Chaotic system A system whose behavior over time changes radically in response to small changes to its initial condition.

Chromosome A structure made of DNA and protein that contains a cell's genetic information.

Cladistics A system for classifying life that groups species according to their closest common ancestors.

Classical mechanics Also known as Newtonian mechanics. A set of laws describing the motion of bodies under the action of forces. Classical mechanics gives accurate results for macroscopic objects that are not traveling close to the speed of light.

Color charge A property of quarks by which they are affected by the strong nuclear force.

Continental drift The slow movement of continents around the globe over millions of years.

Covalent bond A bond between two atoms in which they share electrons.

Dark energy A poorly understood force that acts in the opposite direction to gravity, causing the universe to expand. About three quarters of the mass-energy of the universe is dark energy.

Dark matter Invisible matter that can only be detected by its gravitational effect on visible matter. Dark matter holds galaxies together.

Diffraction The bending of waves around obstacles and spreading out of waves past small openings.

DNA Deoxyribonucleic acid. A large molecule in the shape of a double helix that carries genetic information in a chromosome.

Doppler effect The change in frequency of a wave experienced by an observer in relative motion to the wave's source.

Ecology The scientific study of the relationships between living organisms and their environment.

Electric charge A property of subatomic particles that causes them to attract or repel one another.

Electric current A flow of electrons or ions.

Electromagnetic force One of the four fundamental forces of nature. It involves the transfer of photons between particles.

Electromagnetic radiation A form of energy that moves through space. It has both an electrical and a magnetic field, which oscillate at right-angles to each other. Light is a form of electromagnetic radiation.

Electroweak theory A theory that explains the electromagnetic and weak nuclear force as one "electroweak" force.

Electron A subatomic particle with a negative electric charge.

Electrolysis A chemical change in a substance caused by passing an electric current through it.

Element A substance that cannot be broken down into other substances by chemical reactions.

Endosymbiosis A relationship between organisms in which one organism lives inside the body or cells of another organism to their mutual benefit.

Energy The capacity of an object or system to do work. Energy can exist in many forms, such as kinetic energy (movement) and potential energy (for example, the energy stored in a spring). It can change from one form to another, but never be created or destroyed.

Entanglement In quantum physics, the linking between particles such that a change in one affects the other no matter how far apart in space they may be.

Entropy A measure of the disorder of a system. Entropy is the number of specific ways a particular system may be arranged.

Ethology The scientific study of animal behavior.

Event horizon A boundary surrounding a black hole within which the gravitational pull of the black hole is so strong that light cannot escape. No information about the black hole can cross its event horizon.

Evolution The process by which species change over time.

Exoplanet A planet that orbits a star that is not our Sun.

Fermion A subatomic particle, such as an electron or a quark, that is associated with mass.

Field The distribution of a force across space-time, in which each point can be given a value for that force. A gravitational field is an example of a field in which the force felt at a particular point is inversely proportional to the square of the distance from the source of gravity.

Force A push or a pull, which moves or changes the shape of an object.

Fractal A geometric pattern in which similar shapes can be seen at different scales.

Gamma decay A form of radioactive decay in which an atomic nucleus gives off high-energy, short-wavelength gamma radiation.

Gene The basic unit of heredity of living organisms, which contains coded instructions for the formation of chemicals such as proteins.

General relativity A theoretical description of space-time in which Einstein considers accelerating frames of reference. General relativity provides a description of gravity as the warping of space-time by mass. Many of its predictions have been demonstrated empirically.

Geocentrism A model of the universe with Earth at its center.

Gravity A force of attraction between objects with mass. Massless photons are also affected by gravity, which general relativity describes as a warping of space-time.

Greenhouse gases Gases such as carbon dioxide and methane that absorb energy reflected by Earth's surface, stopping it from escaping into space.

Heat death A possible end state for the universe in which there are no temperature differences across space, and no work can be done.

Heliocentrism A model of the universe with the Sun at its center.

Higgs boson A subatomic particle associated with the Higgs field, whose interaction with matter gives matter its mass.

Hydrocarbon A chemical whose molecules contain one of many possible combinations of hydrogen and carbon atoms.

Ion An atom, or group of atoms, that has lost or gained one or more of its electrons to become electrically charged.

Ionic bond A bond between two atoms in which they exchange an electron to become ions. The ions' opposite electric charge attracts them to each other.

Leptons Fermions that are affected by all of the four fundamental forces except the strong nuclear force.

Magnetism A force of attraction or repulsion exerted by magnets. Magnetism is produced by magnetic fields or by the property of magnetic moment of particles.

Mass A property of an object that is a measure of the force required to accelerate it.

Mitochondria Structures within a cell that supply energy to the cell.

Molecule The smallest unit of a compound that has its chemical properties, made of two or more atoms.

Momentum A measure of the force required to stop a moving object. It is equal to the product of the object's mass and its velocity.

Multiverse A hypothetical set of universes in which every possible event happens.

Natural selection The process by which characteristics that increase an organism's chances of reproducing are passed on.

Neutrino An electrically neutral subatomic particle that has a very small mass. Neutrinos can pass right through matter undetected.

Neutron An electrically neutral subatomic particle that forms part of an atom's nucleus. A neutron is made of one up-quark and two down-quarks.

Nucleus The central part of an atom, comprising protons and neutrons. The nucleus contains almost all of an atom's mass.

Optics The study of vision and the behavior of light.

Organic chemistry The chemistry of compounds containing carbon.

Parallax The apparent movement of objects at different distances relative to each other when an observer moves.

Particle A tiny speck of matter that can have velocity, position, mass, and charge.

Pauli exclusion principle In quantum physics, the principle that two fermions (particles with mass) cannot have the same quantum state in the same point in space-time.

Periodic table A table containing all the elements arranged according to their atomic number.

Photoelectric effect The emission of electrons from the surfaces of certain substances when light hits them.

Photon The particle of light that transfers the electromagnetic force from one place to another.

Photosynthesis The process by which plants use the energy of the Sun to make food from water and carbon dioxide.

Pi (π) The ratio between the circumference of a circle and its diameter. It is roughly equal to 22/7, or 3.14159.

Pi bond A covalent bond in which the lobes of the orbitals of two or more electrons overlap sideways, rather than directly, between the atoms involved.

Plate tectonics The study of continental drift and the way in which the ocean floor spreads.

Polarized light Light in which the waves all oscillate in just one plane.

Polymer A substance whose molecules are in the shape of long chains of subunits called monomers.

Positron The antiparticle counterpart of an electron, with the same mass but a positive electric charge.

Pressure A continual force pushing against an object. The pressure of gases is caused by the movement of their molecules.

Proton A particle in the nucleus of an atom that has positive charge. A proton contains two up-quarks and one down-quark.

Quantum electrodynamics (QED) A theory that explains the interaction of subatomic particles in terms of an exchange of photons.

Quantum mechanics The branch of physics that deals with the interactions of subatomic particles in terms of discrete packets, or quanta, of energy.

Quark A subatomic particle that protons and neutrons are made from.

Radiation Either an electromagnetic wave or a stream of particles emitted by a radioactive source.

Radioactive decay The process in which unstable atomic nuclei emit particles or electromagnetic radiation.

Redshift The stretching of light emitted by galaxies moving away from Earth, due to the Doppler effect. This causes visible light to move toward the red end of the spectrum.

Refraction The bending of electromagnetic waves as they move from one medium to another.

Respiration The process by which organisms take in oxygen and use it to break down food into energy and carbon dioxide.

Salt A compound formed from the reaction of an acid with a base.

Sigma bond A covalent bond formed when the orbitals of electrons meet head-on between atoms. It is a relatively strong bond.

Singularity A point in space-time with zero length.

Space-time The three dimensions of space combined with one dimension of time to form a single continuum.

Special relativity The result of considering that both the speed of light and the laws of physics are the same for all observers. Special relativity removes the possibility of an absolute time or absolute space.

Species A group of similar organisms that can breed with one another to produce fertile offspring.

Spin A quality of subatomic particles that is analogous to angular momentum.

Standard model The theoretical framework of particle physics in which there are 12 basic fermions —six quarks and six leptons.

String theory A theoretical framework of physics in which pointlike particles are replaced by one-dimensional strings.

Strong nuclear force One of the four fundamental forces, which binds quarks together to form neutrons and protons.

Superposition In quantum physics, the principle that, until it is measured, a particle such as an electron exists in all its possible states at the same time.

Thermodynamics The branch of physics that deals with heat and its relation to energy and work.

Transpiration The process by which plants emit water vapor from the surface of their leaves.

Uncertainty principle A property of quantum mechanics that means that the more accurately certain qualities, such as momentum, are measured, the less is known of other qualities such as position, and vice versa.

Uniformitarianism The assumption that the same laws of physics operate at all times in all places across the universe.

Valency The number of chemical bonds that an atom can make with other atoms.

Velocity A measure of an object's speed and direction.

Vitalism The doctrine that living matter is fundamentally different from nonliving matter. Vitalism posits that life depends on a special "vital energy." It is now rejected by mainstream science.

Wave An oscillation that travels through space, transferring energy from one place to another.

Weak nuclear force One of the four fundamental forces, which acts inside an atomic nucleus and is responsible for beta decay.

INDEX

Numbers in **bold** indicate main entries.

A

H

IJ

K

L

M

N

O

P

T

UV

W

XYZ

ACKNOWLEDGMENTS

Dorling Kindersley and Tall Tree Ltd. would like to thank Peter Frances, Marty Jopson, Janet Mohun, Stuart Neilson, and Rupa Rao for editorial assistance; Helen Peters for the index; and Priyanka Singh and Tanvi Sahu for assistance with illustrations. Directory written by Rob Colson. Additional artworks by Ben Ruocco.

PICTURE CREDITS

The publisher would like to thank the following for their kind permission to reproduce their photographs:

(Key: a-above; b-below/bottom; c-center; f-far; l-left; r-right; t-top)

25 Wikipedia: Courant Institute of Mathematical Sciences, New York University (bl). **27 J D Maddy:** (bl). **Science Photo Library:** (tr). **38 Getty Images:** Time & Life Pictures (t). **39 Dreamstime.com:** Nicku (tr). **41 Dreamstime.com:** Nicku (tr). **43 Dreamstime.com:** Nicku (bl). **47 Dreamstime.com:** Georgios Kollidas (bc). **48 Chemical Heritage Foundation:** (bl). **Getty Images:** (cr). **51 Dreamstime.com:** Nicku (bl); Theo Gottwald (tc). **54 Wikipedia:** (crb). **55 Dreamstime.com:** Matauw (cr). **56 Science Photo Library:** R.W. Horne / Biophoto Associates (bc). **57 U.S. National Library of Medicine, History of Medicine Division:** Images from the History of Medicine (tr). **61 Dreamstime.com:** Georgios Kollidas (bl); Igor3451 (tr). **65 NASA:** (br). **68 Wikipedia:** Myrabella (tl). **69 Dreamstime.com:** Georgios Kollidas (tr). **NASA:** Johns Hopkins University Applied Physics Laboratory / Carnegie Institution of Washington (bl). **74 Dreamstime.com:** Isselee (crb). **75 Dreamstime.com:** Georgios Kollidas (bl). **77 Dreamstime.com:** Georgios Kollidas (tr). **Getty Images:** (bl). **79 Getty Images**: (bl). **Library of Congress, Washington, D.C.:** (cr). **81 NOAA:** NOAA Photo Library (cr). **83 Dreamstime.com:** Georgios Kollidas (tr). **Wikipedia:** (bl). **85 Getty Images:** Colin Milkins / Oxford Scientific (cr). **87 Dreamstime.com:** Georgios Kollidas (bl). **Wikipedia:** (tl). **89 Science Photo Library:** European Space Agency (tl). **92 Getty Images:** UIG via Getty Images (bl). **94 Getty Images:** UIG via Getty Images. **95 Dreamstime.com:** Nicku (tr). **99 Dreamstime.com:** Adischordantrhyme (b). **100 Dreamstime.com:** Nicku (bl). **101 Getty Images:** National Galleries Of Scotland (tr). **103 Dreamstime.com:** Deborah Hewitt (tl). **The Royal Astronomical Society of Canada:** Image courtesy of Specula Astronomica Minima (bl). **104 Dreamstime.com:** Es75 (crb). **111 Wikipedia:** (tr). **113 Dreamstime.com:** Georgios Kollidas (bl). **Getty Images:** SSPL via Getty Images (cr). **114 Getty Images:** SSPL via Getty Images (cr). **117 Getty Images:** After Henry Thomas De La Beche / The Bridgeman Art Library (bl); English School / The Bridgeman Art Library (tr). **121 Getty Images:** Universal Images Group (cr). **123 Wikipedia:** (bl). **124 iStockphoto.com:** BrianBrownImages (cb). **125 Dreamstime.com:** Georgios Kollidas (tr). **129 Dreamstime.com:** Whiskybottle (bl). **Library of Congress, Washington, D.C.:** James W. Black (tr). **132 Corbis:** Stapleton Collection (bl). **135 Science Photo Library:** US Fish And Wildlife Service (tr). **Wikipedia:** (bl). **137 Wikipedia:** (bl). **138 Getty Images:** SSPL via Getty Images (crb). **141 Corbis:** Bettmann (tr). **Dreamstime.com:** Paul Koomen (cb). **145 Dreamstime.com:** Georgios Kollidas (bc). **146 Dreamstime.com:** Gary Hartz (bl). **147 Image courtesy of Biodiversity Heritage Library. http://www.biodiversitylibrary.org:** MBLWHOI Library, Woods Hole (tc). **148 Getty Images:** De Agostini (bl). **Wikipedia:** (tr). **149 Getty Images:** UIG via Getty Images. **152 Getty Images:** (bl). **153 NASA:** Jacques Descloitres, MODIS Rapid Response Team, NASA / GSFC (tl). **154 Getty Images:** SSPL via Getty Images (tl). **155 Dreamstime.com:** Proxorov (br). **157 Science Photo Library:** King's College London (cr). **159 Dreamstime.com:** Georgios Kollidas (tr). **164 Science Photo Library:** IBM Research (tr). **165 Dreamstime.com:** Nicku (bl). **Wikipedia:** (tr). **168 Getty Images:** (bl). **iStockphoto.com:** RichardUpshur (tr). **171 Getty Images:** Roger Viollet (tl). **172 Dreamstime.com:** Skripko Ievgen (cr). **176 Alamy Images:** Interfoto (cr). **178 iStockphoto.com:** Cerae (tr). **179 iStockphoto.com:** Popovaphoto (tr). **183 123RF.com:** Sastyphotos (br). **185 Dreamstime.com:** Nicku (tr). **187 Getty Images:** SSPL via Getty Images (clb); Time & Life Pictures (tr). **192 Getty Images:** (bl). **193 Wikipedia:** (br). **195 Dreamstime.com:** Sophie Mcaulay (b). **197 U.S. Department of Agriculture: Electron and Confocal Microscope Unit, BARC:** (cr). **Wikipedia:** Delft University of Technology (bl). **204 NASA:** NASA / SDO (tr). **205 Getty Images:** (bl). **209 Getty Images**. **210 Getty Images:** SSPL via Getty Images (bl). **218 Bernisches Historisches Museum:** (bl). **221 Wikipedia:** (tr). **223 Wikipedia:** Bildarchiv Foto Marburg (bl). **225 U.S. National Library of Medicine, History of Medicine Division:** Images from the History of Medicine (tr). **228 Wikipedia:** Princeton Plasma Physics Laboratory (b). **231 Getty Images:** SSPL via Getty Images (tr). **232 Getty Images:** Gamma-Keystone via Getty Images (tr). **235 Getty Images:** (tr). **238 Getty Images:** (bl). **239 Alamy Images:** WorldPhotos (tl). **240 NASA:** JPL-Caltech (br). **243 Corbis:** Bettmann (tr). **244 NASA:** WMAP Science Team (br). **245 NASA:** (tr). **247 Getty Images:** Roger Viollet (bl). **249 Wikipedia:** © Superbass/CC-BY-SA-3.0 - https://creativecommons.org/licenses/by-sa/3.0/deed (via Wikimedia Commons) - http://commons.wikimedia.org/wiki/File:Christian_Moullec_5.jpg (cr). **251 Corbis:** Bettmann (tr). **253 Alamy Images:** Pictorial Press Ltd (tr). **259 National Library of Medicine:** (tr). **264 Wikipedia:** U.S. Department of Energy (bl). **265 USAAF. 271 123RF.com:** Kheng Ho Toh (cr). **273 Corbis:** Bettmann (tr). **278 Wikipedia:** © 2005 Lederberg and Gotschlich / Photo: Marjorie McCarty (bl). **280 Science Photo Library. 281 Alamy Images:** Pictorial Press Ltd (tc). **282 Wikipedia:** National Human Genome Research Institute. **285 Getty Images:** The Washington Post (tl). **289 Getty Images:** SSPL via Getty Images (cb). **290 University of Edinburgh:** Image from Donald Michie Web Site, used with permission of AIAI, University of Edinburgh, http://www.aiai.ed.ac.uk/~dm/donald-michie-2003.jpg (bl). **291 Corbis:** Najlah Feanny / CORBIS SABA. **292 SOHO (ESA & NASA):** (br). **293 Getty Images:** (tr). **295 National Science Foundation, USA:** (bl). **297 NASA:** NASA Langley Research Center (cb). **Science Photo Library:** Emilio Segre Visual Archives / American Institute Of Physics (tr). **299 © CERN:** (tc). **Wikipedia:** Bengt Nyman. File licensed under the Creative Commons Attribution 2.0 Generic licence: http://creativecommons.org/licenses/by/2.0/deed.en (bl). **300 Science Photo Library:** Don W. Fawcett (cb). **301 Javier Pedreira:** (tr). **305 Getty Images:** Peter Ginter / Science Faction. **307 Getty Images:** Time & Life Pictures (tr). **313 Wikipedia:** Jbourjai. **316 Dreamstime.com:** Cflorinc (cr). **319 Science Photo Library:** Torunn Berge (cr). **321 Wikipedia:** Trevor J. Simmons (tr). **325 Getty Images:** (bl); UIG via Getty Images (cr).